# Magnetic Domain Walls
# in Bubble Materials

# Applied Solid State Science

ADVANCES IN

MATERIALS AND DEVICE RESEARCH

*Editor Raymond Wolfe*

BELL LABORATORIES
MURRAY HILL, NEW JERSEY

**Supplement 1**
Magnetic Domain Walls in Bubble Materials
*A. P. Malozemoff and J. C. Slonczewski*

# Magnetic Domain Walls in Bubble Materials

*A. P. Malozemoff*

IBM THOMAS J. WATSON RESEARCH CENTER
YORKTOWN HEIGHTS, NEW YORK


*J. C. Slonczewski*

IBM THOMAS J. WATSON RESEARCH CENTER
YORKTOWN HEIGHTS, NEW YORK

ACADEMIC PRESS   New York   London   Toronto   Sydney   San Francisco   1979

*A Subsidiary of Harcourt Brace Jovanovich, Publishers*

ACADEMIC PRESS, INC.
111 Fifth Avenue, New York, New York 10003

*United Kingdom Edition published by*
ACADEMIC PRESS, INC. (LONDON) LTD.
24/28 Oval Road, London NW1 7DX

ISSN 0194-2891

ISBN 0-12-002951-0

PRINTED IN THE UNITED STATES OF AMERICA

79 80 81 82    9 8 7 6 5 4 3 2 1

# Contents

## VI.   Wall Dynamics in Three Dimensions

## VII.   Low-Velocity Dynamics with Vertical Bloch Lines

## VIII.   Nonlinear Wall Motion in Two Dimensions

## IX.   Nonlinear Bubble Translation

## X.   Wall Waves and Microwave Effects

# Preface

There is a relatively small community of scientists who have been active in the physics of domain walls in bubble domain materials, but the attention to this specialty has grown far broader with time. On the one hand, the elegance of the physical phenomena has attracted interest in the fundamental physics community. On the other hand, the impact on devices has attracted interest in the large community of engineers and materials researchers involved in development of practical bubble devices.

At the time of this writing, bubble devices look very promising. We are at the start of a new phase with the first products already on the market and with many laboratories pursuing more advanced product ideas. Thus we believe that a comprehensive review of the physics will have relevance and significance at this time.

In beginning the list of our acknowledgments, we must first of all mention the whole bubble physics community whom we have come to know over the years at conferences and laboratory visits throughout the world. We have appreciated the stimulating and relatively open environment that has been maintained in spite of the pressures of competition and that has contributed immensely to the rapid development of the field. We will not attempt to list all the scientists whose feedback has aided us both in our work and, directly or indirectly, in the preparation of this book. Here we limit our explicit mention to several collaborators who have contributed by extensive readings and criticisms of sections of the manuscript, and whom we heartily thank: B. E. Argyle, M. Cohen, and J. C. DeLuca of IBM Research, Yorktown, M. H. Kryder of Carnegie-Mellon University, Pittsburgh, S. Maekawa of Tohoku University, Sendai, R. Wolfe of Bell Laboratories (our editor), and F. H. deLeeuw of Philips Laboratories, Eindhoven. We also reserve special thanks to Laurie Vassari for her patience and expertise in typing and retyping the many drafts of the manuscript.

A. P. Malozemoff

J. C. Slonczewski

# I

# Introduction

Magnetic domain walls are regions in a magnetic medium where the magnetic moment or spin vector rotates rapidly as a function of position. The walls form boundaries between domains, that is, between regions of gradual or zero spin rotation. Theoretical study of magnetic domain walls began with the work of Bloch,[1] Landau and Lifshitz,[2] and Néel[3] who first derived the basic static internal spin structures. Landau and Lifshitz[2] also proposed the basic equation, named after them, for spin dynamics in a ferromagnetic continuum. This equation forms the foundation for the theory of domain wall dynamics. Early experimental work, reviewed by Kittel and Galt[4] and Dillon,[5] was largely confined to bulk single crystal or polycrystalline materials in which precise domain structures were difficult to determine or control. A fresh impetus came with the development in the 1950s of thin metallic films out of such materials as permalloy, composed of iron and nickel, in which individual domain walls could be studied more easily. Experimental results of this period are reviewed, for example, by Middelhoek,[6] Prutton,[7] Soohoo,[8] Cohen,[9] and Craik and Tebble,[10] and the theory has been reviewed by Hubert.[11]

Isolated bubble domains were studied as early as 1960 by Kooy and Enz,[12] but their potential for memory and other device applications was brought out first in 1967 by Bobeck.[13] General descriptions of bubble domains and their device applications have been given by Nalecza,[14] O'Dell,[15] Smith,[16] Bobeck and Della Torre,[17] Chang,[18] and Lachowicz et al.[19] The device potential of bubbles has led to a new surge of interest in domain-wall physics.

Because of the high quality of the materials and certain simplifications afforded by the particular geometry of bubble domain materials, it has been possible to achieve a rewardingly detailed theoretical and experimental understanding of the domain-wall structure and dynamics. Brief reviews of bubble domain-wall dynamics in particular have been given by Hage-dorn,[20] Konishi,[21] Oteru,[22] de Leeuw,[23] and Slonczewski and Maloze-moff.[24] In this present work we give a more comprehensive review of both domain-wall statics and dynamics, including theory and experiment. Our emphasis is on the micromagnetic internal structure and dynamics of the walls themselves. The larger scale properties of individual domains and multi-domain patterns are discussed only as background.

Enough introductory material is given, particularly in Chapter II, to permit a technical reader, unacquainted with bubbles, to follow the rest of the text. While this review is intended primarily for the specialist working theoretically or experimentally in this field, an effort is made to summarize the implications of the field for people working on the periphery, in materials development (Chapter II) or in bubble devices (Section 21). An effort is also made to stand-ardize notation and sign conventions in what has been up to now a frus-tratingly diverse literature.

A "bubble-domain material" is a platelet or thin film of magnetic material characterized by a thickness $h$, spontaneous magnetization $M$, exchange stiffness $A$, and uniaxial anisotropy $K$ with easy axis perpendicular to the plane of the film. In addition to these static parameters, bubble materials are also characterized by the following dynamic parameters: a viscous damp-ing constant $\alpha$, gyromagnetic ratio $\gamma$, and coercive field $H_c$ for domain-wall motion. We use cgs units throughout. Further description of these param-eters will be given in Chapter II. In terms of these parameters, bubble mate-rials may be more specifically defined in terms of three requirements:

(1) $h \sim 4l$, where $l$ is the "material length parameter" given by $l = \sigma_0 / 4\pi M^2$, where $\sigma_0$ is the wall energy $4(AK)^{1/2}$ The quantity $8l$ is a measure of the characteristic minimum domain size that can be achieved at optimum thick-ness in the material. Therefore this requirement means that the material thickness must be comparable to domain dimensions. Technological interest has focused on ever smaller bubbles, from $\sim 50$ $\mu$m in the late 1960s to $\sim 2$ $\mu$m today, to $\sim 0.5$ $\mu$m for the future. This implies that particularly the small bubble materials must be preparable in thin film form, e.g., by epitaxial growth or sputtering onto a magnetically inert substrate.

(2) $Q = K/2\pi M^2 \gtrsim 1$, where $Q$ is called the "quality factor." The sig-nificance of this requirement, coupled with the thin film geometry, is that the domain magnetization generally points perpendicular to the film plane. If $Q < 1$, demagnetizing energy tends to cause the magnetization to lie in the plane of the film and entirely different kinds of domain and wall structures can occur as exemplified by permalloy films. In a bubble material, however, the domains are allowed only two magnetic orientations—either "up" or "down" along the easy axis of anisotropy perpendicular to the film plane.

For example the bubble is a cylinder of down magnetization surrounded by a region of up magnetization. This and other domain configurations characteristic of bubble materials will be briefly reviewed in Section 2. The requirement $Q > 1$ is also of the greatest significance for domain-wall theory. In the opposite case of $Q < 1$, the structure of the domain wall is often dominated by the magnetostatic energy, which, being a nonlocal energy arising from the long-range interaction of magnetic poles, is difficult to calculate. In the bubble material case, however, the wall energy is dominated by the local anisotropy and exchange energies, and the magnetostatic energy can be approximated in a simple way. This simplification has considerably facilitated theoretical progress. In this review we usually limit our presentation of the theory to the high-$Q$ limit, referring to the literature in those cases where corrections of order $Q^{-1}$ or better have been attempted.

(3) $H_c/4\pi M \lesssim 0.05$. This requirement of low coercivity is of particular importance for the theory of domain-wall dynamics. There are at least two types of energy loss mechanisms in bubble materials: coercive loss arising from wall "pinning" at defects or other nonuniformities, and viscous loss arising from the damping of spin precession. If coercive loss is dominant, domain-wall motion tends to become irregular and difficult to account for in a theoretical framework. Low coercivity is also essential for most bubble devices, which rely on bubble motion over long distances within the material rather than external access to fixed bubble locations.

In this review we consider primarily three classes of materials which are known to have the requisite properties of bubble materials: the uniaxial garnets, the orthoferrites, and amorphous films of mixed rare-earths and transition metals. Of these, the garnets have been studied most so far. Hexaferrites and magnetoplumbites have also been studied but not used for bubble applications. We give only brief consideration to materials such as cobalt because of its low $Q$ and MnBi because of its relatively high coercivity.

Bubble devices are based on the fact that bubbles can easily be moved in any direction within the bubble material. The basic concept is that the bubbles represent bits of information which can be moved into and out of storage areas on a chip and up to and away from reading and writing stations. The main appeal of these devices so far is that by virtue of the small size of bubbles and the consequent high information storage densities that can be achieved, bubble devices offer a low-cost nonmechanical solid-state storage that may compete with standard magnetic recording media for certain applications. In general they do not compete with semiconductor memory in a computer main frame because of the relatively greater access time connected with low domain wall velocities, which are a topic of this review.

Bubble devices may be classed in terms of the mode of information storage.

The most common is to use the presence or absence of a bubble to represent a bit of information. This method demands that bubble locations be sufficiently far apart, usually four bubble diameters, to prevent bubble interactions. Such devices may be termed "isolated bubble devices" and have been developed to the point of useful commercial application. Work on these devices has provided considerable incentive to domain-wall physics, particularly in the area of nonlinear domain-wall motion, so as to determine the material limitation on device operating frequency. These nonlinearities are discussed in detail in Chapters VIII and IX. An alternative mode of information storage is to code information by distinguishing more than one kind of bubble, which permits bubbles to be packed more closely together, thus offering ultimately greater storage density and lower cost.[25-27a] Such devices may be termed "bubble lattice devices" and are in an exploratory stage of development. The principal coding method has been by means of the internal domain-wall structure of the bubble. This application has also provided an incentive to domain-wall physics, particularly in the area of static bubble wall structures and the dynamic properties whereby they can be distinguished. These topics are discussed in detail in Chapters IV, VII, and IX. Bubble devices can also be classed in terms of the method of applying forces to the bubbles, e.g., by electric current lines or by permalloy structures polarized by an external field or by "charged walls." Some of the physics involved in these methods of bubble propagation is reviewed briefly in Section 6b.

This review develops theory alongside relevant experiments, chapter by chapter. It is organized as follows: Chapter II summarizes, by way of introduction, the microscopic origins and characteristics of the material parameters and reviews briefly the principles of domain statics. It also introduces the Landau–Lifshitz equation, which is the basic equation of magnetization dynamics, and describes its physical significance. Chapter III surveys the experimental techniques, both static and dynamic, used in studying domain walls. Chapter IV describes the static internal structure of bubble-domain walls. including the complications due to stray fields, Bloch lines, Bloch points, and surface layers. These effects give rise to phenomena such as hard bubbles, dumbbells, integral and non-integral winding states, and controlled state-switching.

The remainder of the review is devoted to domain-wall dynamics. Chapter V discusses Bloch-wall dynamics based on one-dimensional solutions of the Landau–Lifshitz equation. The theory here is largely classical, involving the well-known concepts of linear mobility, Döring mass, and Walker limiting velocity. However, more than one dimension must be considered to correctly account for many observed phenomena. Chapter VI extends wall-motion theory to three dimensions and gives a physical picture of why

Bloch lines become so important in the dynamics. Chapter VII applies the theory to low velocity phenomena in domain walls containing vertical Bloch lines. Thus are explained the remarkable phenomena of bubble deflection, dumbbell rotation, Bloch-line wind-up, and wall mobility reduction by Bloch lines. Chapter VIII discusses high-velocity radial and quasi-planar wall motions in terms of a simple Bloch-line model and also more general models. Theory is compared to experimental observations of velocity peaks, velocity saturation, and domain-wall mass. Chapter IX covers the particularly important case of "nonlinear" bubble translation including dynamic conversion, ballistic effects, and gradientless bubble propulsion, and it also describes the implications of the theory for bubble motion in devices. Chapter X surveys special phenomena involving vibrations and wave motions of walls and effects of microwave-frequency fields on walls.

# II

# Resume of Classical Magnetism and Bubble Domain Statics

## 1. Static Material Parameters

Throughout this review we treat domain walls in a micromagnetic framework,[28] that is, we assume the material is a magnetic continuum characterized by a spontaneous magnetization $M$, exchange stiffness $A$, and uniaxial anisotropy $K$. Given a magnetization distribution $\mathbf{M}(\mathbf{x})$ as a function of position vector $\mathbf{x}$ in the medium, there exist static energies that arise from each of these parameters: field energy (demagnetizing and external), exchange energy, and anisotropy energy, respectively. These energies determine the static equilibrium structure of domain walls and, if the structure is a nonequilibrium one, the motive forces on them.

In this section we review by way of introduction the microscopic origin and typical values of the parameters $M$, $A$, and $K$ in bubble materials, and we also give formulas for the static energies that they give rise to. More information can be found in general magnetism texts like those by Chikazumi,[29] Morrish,[30] and Tebble and Craik,[10] or in the reviews mentioned in the introduction.

### A. Magnetization and Demagnetizing Energy

The most fundamental magnetic parameter is the magnetization $M$, the magnetic-dipole strength per unit volume, which arises from the mutual alignment of atomic magnetic dipoles. The latter dipoles arise primarily

7

from electron spins. Although the orbital motion of electrons usually contributes less to the dipole strength, it plays a significant role in magnetic anisotropy and in the gyromagnetic ratio (see Sections 1,C and 3,B).

Most bubble materials are ferrimagnetic, that is, they contain different atomic subnetworks antiferromagnetically oriented with more magnetic spins in one subnetwork than the other. For example the garnets,[31] which are cubic crystals with a chemical formula $R_3M_2N_3O_{12}$, have per formula unit: two octahedral sites denoted by M, three tetrahedral sites denoted by N, and three dodecahedral sites denoted by R. In the case of YIG (yttrium iron garnet), for example, the R sublattice is populated by the nonmagnetic Y ions, whereas the octahedral and tetrahedral sites are all filled with trivalent iron, which has a half-filled magnetic shell. The tetrahedral and octahedral sites are antiferromagnetically coupled, and there is a net magnetization corresponding to one iron ion per formula unit at 0 K. The system orders at 560 K and the magnetization rises to $4\pi M = 1800$ G at room temperature and 2400 G at 0 K.

The magnetization of YIG may be decreased by doping with nonmagnetic ions such as trivalent gallium or tetravalent germanium, which preferentially occupy the tetrahedral sites. Magnetic compensation, that is $M = 0$, is thus reached by doping with approximately one nonmagnetic atom per formula unit. Actually only $\sim 90\%$ of the gallium goes to tetrahedral sites while the remaining 10% goes to octahedral sites. By contrast almost all of the germanium goes to tetrahedral sites. Therefore one finds that $\sim 25\%$ more gallium than germanium is required for magnetic compensation; this fact is of importance for the Curie temperature and exchange stiffness (see Section 1,B). The tetravalent germanium is usually charge-compensated by an equal amount of divalent calcium, which goes onto the dodecahedral sites. Thus one speaks of the "CaGe system," which has become preferred in technological applications over the "Ga system" because of the larger Curie temperature and exchange stiffness that can be obtained.[32] The magnetization may also be affected by doping magnetic rare earth ions into the dodecahedral sites. These ions magnetize in the exchange field of the iron sublattices, in most cases in a direction opposite to the magnetization of the tetrahedral sublattice. The trivalent gadolinium ion in GdIG (gadolinium iron garnet) has a particularly large temperature-dependent magnetization and causes a compensation to occur just below room temperature. Typically, a combination of such dodecahedral and tetrahedral doping schemes are used to tailor $4\pi M$ from about 100 to 1000 G for bubbles ranging from 10 to 1 $\mu$m diameter at room temperature. Since almost all garnet bubble compositions are based on YIG, we adopt the convention of referring to any given composition as $D_i$YIG, where $D_i$ are the chemical symbols of the various dopants. Examples are EuGaYIG, SmLuCaGeYIG, etc. The films are usually grown by liquid phase epitaxy on gadolinium gallium garnet (GGG) substrates oriented in the cubic [111]-direction.

The amorphous rare-earth transition-metal films [33,34] are also ferrimagnetic. For example $GdCo_x$ is usually composed of a roughly 4:1 formula ratio of cobalt to gadolinium, and the cobalt and gadolinium moments are antiparallel. The Curie temperature, as extrapolated from data taken below the amorphous-crystalline transition temperature, is in the range of 600–700 K. The magnetization curve behaves like that of GdIG, showing an increasing magnetization due to cobalt ordering as temperature is reduced below the Curie temperature, then a decrease as the gadolinium ions gradually magnetize with decreasing temperature in the cobalt exchange field. Magnetic compensation can be obtained near room temperature by proper adjustment of the Co:Gd ratio. Furthermore as in the garnet case, the magnetization can be tailored by doping; atoms such as Au simply dilute the magnetic structure but atoms such as Mo can affect the electron population of the cobalt $d$-bands and thus more rapidly reduce the cobalt magnetization. The magnetization can typically be tailored from 500 to 5000 G for bubbles ranging from 2 to 0.1 $\mu$m in diameter at room temperature. The films are usually prepared by diode sputtering in an inert gas such as argon.

In contrast to these ferrimagnetic systems, the orthoferrites[35,36] are basically antiferromagnetic, but the two equal sublattice moments are found to be canted at a small angle $\beta$ from the principal antiferromagnetic axis, as shown in Fig. 1.1a. Thus if the magnetization of each sublattice alone is

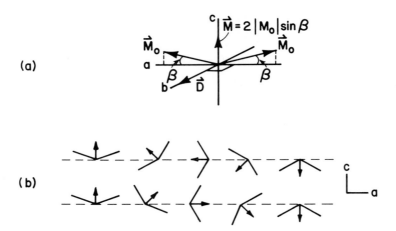

**Fig. 1.1.** (a) Schematic magnetic structure of orthoferrite sublattices above the reorientation temperature. (b) Schematic magnetic structure of the two possible Néel walls of an orthoferrite (see Section 7) with $a$ axis perpendicular to the wall plane.

$M_0$, the net magnetization pointing perpendicular to the principal magnetic axis is $2M_0 \sin \beta$. The orthoferrites are orthorhombic crystals with the chemical formula $RMO_3$, where M is usually trivalent iron and R is either trivalent yttrium or a trivalent rare-earth ion. Magnetic ordering occurs at 600–700 K and the canting angle is of order $0.5°$, so that the net $4\pi M$ at room temperature is typically only 100 G. Since bubble size depends roughly inversely on $M$, the orthoferrites typically have large bubbles of diameter 25 $\mu$m or more. The samples used for exploratory bubble applications have generally been polished platelets cut from bulk single crystals.

Having reviewed the materials aspects of the magnetization, we next describe the two kinds of magnetostatic energies it gives rise to: demagnetizing energy and external-field energy. The magnetostatic energy term is

$$W_m = (8\pi)^{-1} \int H^2 \, dV, \tag{1.1}$$

where $H$ is the magnetic field as defined in electrodynamics, and where the volume integral is carried over all space. Since domain-wall velocities are typically less than $10^{-5}$ times the velocity of light, $\mathbf{H(x)}$ is governed by the static limit of Maxwell's equations:

$$\nabla \times \mathbf{H} = \mathbf{j}, \qquad \nabla \cdot (\mathbf{H} + 4\pi\mathbf{M}) = 0, \tag{1.2}$$

where $\mathbf{j}$ is the current density, assuming $\mathbf{j(x}, t)$ and $\mathbf{M(x}, t)$ are known.

It is often convenient to distinguish the "applied-field" $\mathbf{H_a}$, whose source is $\mathbf{j}$, from the "demagnetizing" or "stray" field $\mathbf{H_d}$, whose source is $\mathbf{M}$. Thus we write

$$\mathbf{H} = \mathbf{H_a} + \mathbf{H_d}, \tag{1.3}$$

$$\nabla \times \mathbf{H_a} = \mathbf{j}, \qquad \nabla \cdot \mathbf{H_a} = 0, \tag{1.4}$$

$$\nabla \times \mathbf{H_d} = 0, \qquad \nabla \cdot (\mathbf{H_d} + 4\pi\mathbf{M}) = 0, \tag{1.5}$$

with the boundary conditions $\mathbf{H_d} \to 0$ at $|\mathbf{x}| \to \infty$. By the standard methods of electrodynamic theory[28] one finds that Eq. (1.1) becomes

$$W_m = (8\pi)^{-1} \int H_a^2 \, dV - \int \mathbf{M} \cdot \mathbf{H_a} \, dV + W_d, \tag{1.6}$$

where $-\mathbf{M} \cdot \mathbf{H_a}$ is the "Zeeman energy" density, and where the demagnetizing energy $W_d$ may be written in any one of three equivalent forms:

$$W_d = \tfrac{1}{2} \iint dV_1 \, dV_2 [\nabla \cdot \mathbf{M(x_1)}][\nabla \cdot \mathbf{M(x_2)}](|\mathbf{x_1} - \mathbf{x_2}|)^{-1}, \tag{1.7}$$

$$W_d = (8\pi)^{-1} \int H_d^2 \, dV = -\tfrac{1}{2} \int \mathbf{M} \cdot \mathbf{H_d} \, dV. \tag{1.8}$$

The first form, Eq. (1.7), is conceptually useful in exhibiting explicitly the fact that the total magnetostatic energy of the system [Eq. (1.6)] depends only on the distribution $\mathbf{H_a(x}, t)$, which is given, and $\mathbf{M(x}, t)$ which is to be calculated, while $\mathbf{H_d(x}, t)$ is in principle eliminated. The second and third forms, Eq. (1.8) taken in combination with Eq. (1.5), are more practical for the explicit determination of $W_d$.

The double-integral character of Eq. (1.7) means that the demagnetizing field is a long range force; minimization of the system energy including this term leads to integral, rather than differential, equations for $M(x, t)$. Since integral equations are notoriously intractable, demagnetizing energy generally causes much difficulty in micromagnetic theory. Fortunately, in domain-wall theory, $M$ usually depends strongly on the coordinate $n$, normal to the wall, and only weakly on coordinates orthogonal to $n$. Assuming that $H_d$ and $M$ depend only on $n$ and that $M_n$ vanishes at $n = \pm\infty$, Eqs. (1.5) integrate trivially and (1.8) reduces to

$$H_{dn} = -4\pi M_n, \qquad W_d = 2\pi \int M_n^2 \, dV, \qquad (1.9)$$

with the components of $H_d$ orthogonal to $n$ vanishing. The fact that $W_d$ is now a single integral means that demagnetization is effectively local, leading to much simplification in this special case.


## B. EXCHANGE STIFFNESS

The basic interaction which causes cooperative magnetic ordering is the exchange interaction $-2J\,S_1 \cdot S_2$, where the exchange coefficient $J$ can be positive or negative and $S_1$ and $S_2$ are interacting spins or magnetic moments of two electrons occupying specific atomic-orbital wave functions. The exchange interaction is a manifestation of the Coulomb interaction between electron charges and the quantum-mechanical antisymmetry principle. The magnitude of $J$ determines the magnetic ordering temperature $T_c$ according to the formula (assuming one dominant magnetic interaction only)

$$T_c = fz|J|S(S + 1)/k, \qquad (1.10)$$

where $S$ is the spin quantum number, $z$ is the number of nearest neighbors (assuming a localized interaction model), $k$ is the Boltzmann constant, and $f$ is a factor of order unity depending on the atomic arrangement of the magnetic atoms. The magnitude of $J$ can thus be deduced from the fact that $T_c$ of most bubble materials lies in the range 400–700 K.

In addition to causing magnetic ordering, the exchange interaction gives rise to the micromagnetic "exchange energy density"

$$w_A = A(\nabla m)^2 \equiv A \sum_{ij} (\partial m_i / \partial x_j)^2, \qquad (1.11)$$

where $m$ is the unit vector $M|M|^{-1}$, and $A$ is the "exchange stiffness" coefficient. Equation (1.11) is valid in the limit of a magnetic continuum, i.e., provided the actual angle difference between spins on neighboring sites is small. As we shall see in Chapter IV, the exchange energy is a key ingredient in domain-wall structure, its effect being to make more gradual any transition between different magnetization directions in different domains or different parts of domain walls.

The magnitude of $A$ is of order $|J|\langle S\rangle^2/a$, where $a$ is the lattice parameter, and $\langle S\rangle$ may be interpreted crudely as a thermal average over the subnetwork spin. Therefore $A$ drops to zero at the Curie temperature and can be considerably smaller at room temperature than at absolute zero, even though $J$ itself is usually to very good approximation independent of temperature. It is important to note that $A$ does not go to zero at a magnetic compensation temperature, even though the net $M$ is zero, because $A$ depends on the degree of subnetwork spin alignment, which, for subnetwork $k$, is proportional to $M_k$, not to the algebraic net $M = \sum M_k$ of the ferrimagnetic system. $A$ also depends on the doping into—or dilution of—the transition metal sublattices, which usually causes a reduction in Curie temperature $T_c$, as can be seen by considering the factor $z$ in Eq. (1.10).

Detailed calculations of $A$ for mixed garnets at $T = 0$ K, based on quantum statistics, are available.[37] A simpler molecular-field discussion is useful at arbitrary values of $T$.[37a] An approximate formula for the room-temperature $(T_r)$ exchange stiffness $A$ of YIG-based garnet with Curie temperature $T_c$ is[38]

$$A = A_{YIG}(T_c - T_r)/(T_{c,YIG} - T_r), \qquad (1.12)$$

where $A_{YIG}$, the exchange stiffness of pure YIG at room temperature, is $4.15 \times 10^{-7}$ erg/cm, and where $T_{c,YIG}$, the Curie temperature of YIG, is 560 K. According to spin-wave measurements, $A(T_r)$ of LaGaYIG[38a] and CaGeYIG[39] fall a bit below this equation. Typically the $A$ values for garnets in practical applications range from 1 to $4 \times 10^{-7}$ erg/cm.[37a,39a]

The exchange effect arising from pure Coulomb interactions, as discussed above, is fundamentally isotropic. In other words, the energy does not change if all spin vectors are rotated simultaneously through the same angle about a common axis in three-dimensional space. However, anisotropic corrections to the Coulomb exchange interaction arise from the atomic spin–orbit effect of relativistic origin. One form of anisotropic exchange, known as antisymmetric, Dzialoshinski, or Moriya exchange, is important in orthoferrites because without it the vector sum $\mathbf{M} = \mathbf{M}_1 + \mathbf{M}_2$ would vanish (see Fig. 1.1). In this case the exchange energy density for uniform magnetization of the two sublattices in the unit vector directions $\mathbf{m}_1$ and $\mathbf{m}_2$ can be written

$$w_{\lambda D} = \lambda \mathbf{m}_1 \cdot \mathbf{m}_2 + \mathbf{D} \cdot (\mathbf{m}_1 \times \mathbf{m}_2), \qquad (1.13)$$

where $\lambda$ is the isotropic exchange constant, and $\mathbf{D}$ is the Dzialoshinski exchange vector, which is found to point along the $b$ axis of the orthorhombic structure (see Fig. 1.1). Minimizing this energy, one finds the equilibrium canting angle $\beta = \frac{1}{2} \tan^{-1}(D/\lambda) \approx D/2\lambda$ (defined in Fig. 1.1a), which gives rise to the small net magnetization discussed earlier. Typically $D$ is $4 \times 10^7$ ergs/cm$^3$ and $\lambda$ is $2.5 \times 10^9$ ergs/cm$^3$; so $\beta$ is only $0.5°$. In orthoferrites and other canted antiferromagnets, the magnitude of $\mathbf{M}$ cannot be regarded as independent of its orientation. Therefore micromagnetic theory based on the assumption of constant $M$, including most of the theory in this review, may not apply to orthoferrites.

Equation (1.12) shows that at room temperature $A$ in garnets is very sensitive to Curie temperature. For example, if $T_c$ increases by 15% from 350 to 400 K, $A$ will increase by $\sim 100\%$. $T_c$ in turn decreases roughly proportionally with doping, as can be seen from Eq. (1.10) by regarding $z$ as the *average* number of *magnetic* neighbors. As mentioned earlier, the Ga system requires more doping than the CaGe system to achieve the same magnetization, and thus one can understand the ratios of up to two between $A$ values in the two systems. Similarly, smaller bubble compositions with higher $M$ have less doping and consequently have higher $A$ values. These differences are of great practical importance because a high Curie temperature facilitates design of materials with good temperature stability in the typical device operating range of 0–100°C and because raising the exchange stiffness increases the domain-wall saturation velocity (see Section 19,A). Values of $A$ for orthoferrites and amorphous materials are similar to those of garnets and depend strongly on the doping.

## C. ANISOTROPY

As discussed in the Introduction, a key requirement of bubble domain materials is that they possess a uniaxial anisotropy with its easy axis perpendicular to the plane and with a magnitude $K$ larger than $2\pi M^2$. The effective uniaxial anisotropy energy density can be written

$$w_K = K \sin^2 \theta, \tag{1.14}$$

where $\theta$ is the polar angle of the net magnetization from the easy axis. Because the garnets have a cubic crystal structure and because amorphous materials have no inherent preferred directions, uniaxial anisotropy would appear to be ruled out by symmetry in these cases. In the garnets, for example, one would only expect a cubic anisotropy. The cubic anisotropy is magnetocrystalline in nature, that is, it arises from the interaction of the atomic magnetic moments with the intrinsic symmetry of their crystalline environment via a spin–orbit interaction. For the common (111) orientation of epitaxial garnet bubble films, the cubic anisotropy has the form

$$w_{K_1} = K_1\left(\frac{1}{4}\sin^4\theta + \frac{1}{3}\cos^4\theta + \frac{\sqrt{3}}{2}\sin^3\theta\cos\theta\cos 3\phi\right), \qquad (1.15)$$

where $\phi$ is the angle from the $[11\bar{2}]$-direction in the plane of the film. Higher order cubic anisotropy terms are usually negligible. For typical garnet films $K_1$ ranges from 500 to 10,000 ergs/cm³. Cubic anisotropy is by itself insufficient to sustain bubble domains because it leads to the formation of 109° walls between {111}-axes, rather than 180° walls. However, it can make a small contribution to the net energy of 180° walls in those cases where uniaxial anisotropy dominates.[40]

The two principal mechanisms for uniaxial anisotropy in the garnets are the "stress" and "growth" mechanisms, in which cases the cubic symmetry is lifted either by a noncubic stress or by the surface symmetry of a growing crystal face.[41] In the first mechanism a biaxial stress $\tau$ parallel to the film plane gives rise to uniaxial anisotropy through magnetostriction $\lambda$. For example for a (111) oriented film, one finds[29]

$$K = -\tfrac{3}{2}\lambda_{111}\tau. \qquad (1.16)$$

The stress arises from a lattice mismatch between the gadolinium gallium garnet (GGG) substrate and the film. In the second mechanism,[42] one must consider that for each of the three magnetic sites of the garnet crystal structure, tetrahedral, octahedral, and dodecahedral, there are in fact multiple sites of lower symmetry all related to each other by cubic symmetry operations. Different groups of these sites become preferentially populated during growth by defects or different magnetic species. The microscopic mechanism whereby this grown-in asymmetry gives rise to uniaxial anisotropy has been difficult to ascertain. It could arise from single-ion magnetocrystalline anisotropy, dipolar energy, or anisotropic exchange. When the films are annealed, the populations of symmetry-related sites can become homogenized and then growth anisotropy disappears. In typical device quality garnet materials, growth anisotropy is dominant over stress anisotropy. The anisotropy is found to range in magnitude up to $2 \times 10^5$ ergs/cm³ for room temperature garnets and to change with temperature roughly as the square or cube of the sublattice magnetization. No anomaly in $K$ is expected at magnetic compensation ($M = 0$), although the "anisotropy field" $2K/M$ diverges there and so $K$ becomes difficult to determine by standard techniques. The anisotropy of GdCo amorphous films attains values up to $6 \times 10^5$ ergs/cm³, and the mechanisms are even less well understood than in the garnets. Possible sources of anisotropy are columnar inhomogeneities, stress, and pair ordering, which, like garnet "growth" anisotropy, may act through dipolar energies or anisotropy of exchange.

Of course if the film plane is intrinsically asymmetric, as may occur in garnets grown as a (110)-film or with the [111]-axis misoriented at some angle to the surface normal, one may also find "in-plane" or "orthorhombic" anisotropy of the form

$$w_{K_p} = K_p \sin^2 \theta \sin^2(\phi - \phi_p). \tag{1.17}$$

Here $\phi_p$ represents the azimuthal direction of the easy axis of in-plane anisotropy. Values of $K_p/2\pi M^2$ as large as 50 have been achieved in (110) garnets even while maintaining adequate values of $Q = K/2\pi M^2$.[43–47] For the case of slightly misoriented (111) films, surprisingly large $K_p$'s have been found.[48] For example, in EuGa YIG epitaxial films, $K_p \approx 0.1$ K per degree of misorientation has been reported. Such in-plane anisotropy can have a large effect on domain-wall dynamics (see Sections 11,C and 16,C). Another effect in misoriented films is that the uniaxial anisotropy axis can be tilted away from the surface normal, sometimes by an angle significantly larger than that of the original crystallographic misorientation.[48,49] Such "tilt anisotropy" can cause tilted walls and effective in-plane fields.

The orthoferrites have orthorhombic symmetry; so one may expect an orthorhombic anisotropy of magnetocrystalline origin. In $YFeO_3$ at room temperature, the perpendicular anisotropy is $1.5 \times 10^5$ erg/cm$^3$, favoring the net magnetization pointing along the c axis. In some orthoferrites with magnetic rare earths, the easy axis is found to switch to the a axis at a "reorientation" temperature; for example, in $SmFeO_3$ this occurs at $\sim 500$ K. In the vicinity of the reorientation temperature, the net anisotropy favoring the easy axis can become very small. The Dzialoshinski exchange also creates an effective in-plane anisotropy, assuming the b axis lies in the plane of a platelet. As can be seen from Eq. (1.13), if one attempts to turn the net magnetization parallel to the Dzialoshinski exchange vector along the b axis, the canting must vanish and stabilization energy of order $D\beta$ is lost. Thus one may consider the in-plane anisotropy to be $K_p = D\beta$, although this result is only valid when $\phi - \phi_p$ is small in Eq. (1.17)

In closing this section, we comment that magnetostriction, except for its effect in generating uniaxial anisotropy through the stress mechanism [Eq. (1.16)], is not expected to contribute significantly to domain wall structure in bubble materials. The magnetostrictive energy is of order $0.5Ce^2$, where C is the elastic modulus and e the strain that would result from magnetostriction in an unclamped medium. Considering characteristic values for garnets, one finds an energy of order 1000 ergs/cm$^3$, which is generally small compared to the uniaxial anisotropy. Furthermore while the magnetization in domains points parallel to the easy axis, it points perpendicular to the easy axis in the midplane of the domain wall. Magnetostrictive contraction in this latter direction is prevented by the clamping effect of the surrounding domains. Moreover, if the moment of a single-domain film is rotated away from the easy direction, as by means of a large external field, the state of strain is held constant by the clamping effect of the attached substrate.

## 2.   Domain Statics

### A.   INTRODUCTION

Here we review briefly some basic aspects of static domain structures in bubble materials which are necessary background for subsequent chapters. For more complete discussions we refer the reader to an extensive literature.[12,50-59] As mentioned already in Chapter I, the high $Q$ of bubble materials, coupled with their thin film geometry, gives rise to a domain structure with only two kinds of domains. One kind has magnetization pointing "up" and the other "down" along the easy anisotropy direction perpendicular to the plane. Furthermore the walls separating the domains stand upright, spanning the distance between film surfaces. They lie nearly parallel to the domain magnetization, to minimize the wall surface and to avoid surface poles on the wall that would otherwise give rise to extra magnetostatic energy. The domains have a characteristic width determined by a balance of wall energy $\sigma$ tending to minimize the number of walls per unit length along the film and magnetostatic energy tending to maximize the

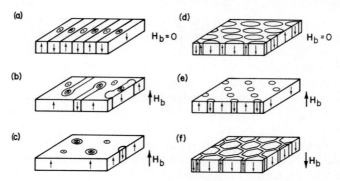

**Fig. 2.1.**   Schematic domain configurations in a bubble material with bias field $H_b$. (a) Regular stripe array. (b) Contracting stripes. (c) Isolated bubbles. (d) Hexagonal bubble lattice. (e) Bubble lattice contracted by field. (f) Honeycomb lattice.

number of walls. One can see these conflicting tendencies in the expression for the length parameter $l = \sigma_o/4\pi M^2$, which is a measure of the characteristic domain width whenever the film thickness is comparable in magnitude.

A great variety of different domain configurations is consistent with these simple requirements. In zero field, two simple configurations are the stripe array or hexagonal bubble array, illustrated schematically in Figs. 2.1a and 2.1d. Actual photomicrographs of such structures, observed in a garnet film by means of a polarizing microscope (see Section 4), are shown in Fig. 2.2 If an upward "bias" field is applied, the upward-oriented domains grow at the expense of the downward ones. In practice the stripes are curved and finite in length. An unfavored stripe can contract not just in width but also in length (Fig. 2.1b), and above a critical field the ends run in to form practically isolated bubbles as shown in Fig. 2.1c. The bubble is stable over a narrow bias field range above which it collapses and below which it distorts elliptically and runs back out into a stripe. In the case of the bubble array in

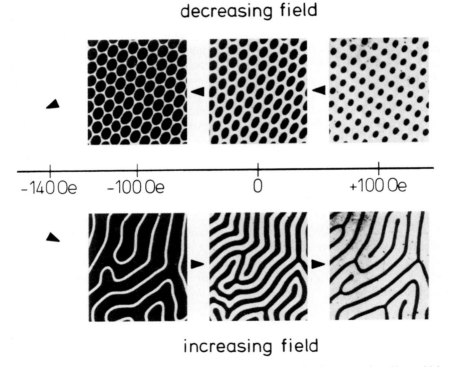

**Fig. 2.2.** Faraday contrast micrographs of bubble lattice and stripe arrays in a 40 $\mu$m thick $Gd_{2.32}Tb_{0.59}Eu_{0.09}Fe_5O_{12}$ garnet film as a function of bias field (after de Jonge and Druyvesteyn[54]). Bubbles in zero field are $\sim 30$ $\mu$m in diameter.

the figures, an upward field shrinks the bubbles, but a downward field creates a "honeycomb" array as the sides of the expanding bubbles press against each other (Fig. 2.1e,f).

The Faraday-contrast micrographs in Fig. 2.2 show the sequence of domain configurations that occurs during one cycle of external field applied to a 40 $\mu$m film of composition $Gd_{2.32}Tb_{0.59}Eu_{0.09}Fe_5O_{12}$. To form the initial lattice from the virgin stripe array with zero bias, a sequence of field pulses with gradually decreasing amplitude normal to the film plane is first applied in the absence of static bias. The bias is then increased to 100 Oe, providing the initial structure at the upper right of Fig. 2.2. Subsequent variation of the bias from $+100$ to $-140$ Oe and then back to $+100$ Oe again produces the sequence of domain patterns (lattice $\rightarrow$ stripes) shown by the counterclockwise sequence indicated in Fig. 2.2.

The static equilibrium dimensions of each of these configurations are determined by a balance of effective fields or pressures acting on the walls.[13,56,57] A general magnetic field $H$ parallel to the easy anisotropy axis exerts a force per unit area or "pressure" $2MH$ on a domain wall, as can be seen from the fact that if the wall is displaced by $\delta q$, magnetization flips from antiparallel to parallel to the field, giving an energy reduction of $-2MH\delta q$ per unit area. The wall is also subject to pressures arising from demagnetizing fields and from the surface tension of curved walls, which can also be interpreted in terms of effective fields. Let us consider a static external bias field $H = H_b$. The total of these three fields may be termed the "restoring force field" $H_k(q)$ because it is the effective field that restores a perturbed configuration back to its static equilibrium. Equilibrium is determined by $H_k = 0$. A linear or Hooke's law restoring force constant $k$ may be defined as

$$k = -2M\, dH_k/dq, \tag{2.1}$$

where $q$ is the coordinate normal to the wall, measured from the equilibrium position defined by $H_k = 0$. Obviously, this origin depends on $H_b$, as does $k$ itself. The sign is defined by taking positive $H_k$ to point up along the $z$ (up) direction and positive $q$ to point from the $+z$ (up) to $-z$ (down) domain. For example, if in addition to the static bias, a pulse field $H_a(t)$ is applied along $+z$ to a wall initially at equilibrium, the pressure $P_A$ on the wall is represented by

$$P_A = 2MH_a - kq, \tag{2.2}$$

so that as the wall moves in response to the pressure, $q$ increases and the pressure drops linearly. This linear approximation is valid, of course, only for $q$ small relative to domain dimensions. The restoring force constant is of great importance in domain-wall dynamics and depends strongly on the particular domain configuration. As we shall see in Chapter III, the experimentalist can choose a particular domain configuration to obtain widely varying restoring forces, depending on the information he seeks.

The restoring force is also related to the susceptibility $\chi$ defined by

$$\chi = M_{av}/H_a, \tag{2.3}$$

where $M_{av}$ is magnetization normal to the plane, averaged over the sample, and where $H_a$ is the applied field normal to the plane. In the case of a static array of stripes of width $w$ at $H_b = 0$, as shown in Fig. 2.1a,

$$\chi = 2Mq/H_a w, \tag{2.4}$$

where $q$ is the average domain wall displacement induced by the field $H_a$. Since $k = 2MH_a/q$ under static conditions, the dc susceptibility $\chi_0$, is

$$\chi_0 = 4M^2/kw. \tag{2.5}$$

Thus the susceptibility is inversely related to the restoring force constant and the stripe-width.

## B. ISOLATED BUBBLE IN A UNIFORM BIAS FIELD

For example, consider one cylindrical domain with radius $r$ and with magnetization pointing down $(-z)$ inside, as indicated in Fig. 2.1c. We note that a positive change in $q$ corresponds to a negative change in $r$. A positive applied bias field $H_b$ along $+z$ creates an inward pressure $2MH_b$. Similarly if the bubble radius decreases by $dq$, wall energy decreases by $2\pi\sigma h\,dq$, where $\sigma$ is the wall energy per unit area and $h$ is the film thickness. Therefore the wall curvature pressure is $\sigma/r$ and its effective field is $\sigma/2Mr$. Opposing these pressures is the pressure $2MH_d$ from the demagnetizing field $H_d$ given by[51]

$$H_d = -4\pi Mhd^{-1}F(d/h), \tag{2.6}$$

where $F(d/h)$ is the "Thiele force function," and $d$ is the bubble diameter. $F(d/h)$ can be expressed in terms of elliptic integrals, but for many purposes it can be simply approximated by[58]

$$F(d/h) = (d/h)/[1 + (3d/4h)] \tag{2.7}$$

for the range $0 < d/h < 10$. Thus the net effective field on the wall may be written

$$H_k = H_b + 4\pi M\{(l/2r) - [1 + (3r/2h)]^{-1}\}, \tag{2.8}$$

using the approximation of Eq. (2.7). This function is plotted in Fig. 2.3a. Provided $H_b$ is small enough, there are two solutions corresponding to

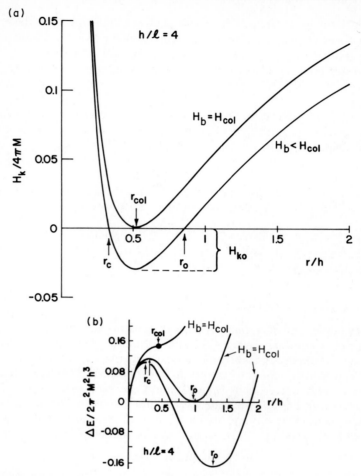

**Fig. 2.3.** (a) Isolated-bubble restoring-force field, Eq. (2.8). (b) Normalized bubble energy $\Delta E$ versus radius for three different bias fields $H_b$ (after Cape and Lehman[52]).

$H_k = 0$. This condition may also be written in a more familiar normalized form[51]

$$(l/h) + (d/h)(H_b/4\pi M) - F(d/h) = 0. \qquad (2.9)$$

The solution in Fig. 2.3a with larger radius $r_0$ is the static equilibrium solution, while the solution with smaller radius $r_c$ is an unstable solution as can be seen from the fact that if the radius is slightly smaller than $r_c$, $H_k$

becomes positive, tending to contract the bubble still further. As $H_b$ is increased the curve in Fig. 2.3a is raised and eventually a critical field $H_{col}$ is is reached at which there is only one solution. At higher fields there is no solution, and $H_k$ is always positive, tending to contract the bubble to zero radius. Thus $H_{col}$ is the "collapse" field because above this field the bubble collapses. Further insight into this effect may be obtained from plots of total energy relative to the saturated state, as shown in Fig. 2.3b. The equilibrium solution corresponds to the bottom of the well in the figure. The unstable solution corresponds to the top of the energy barrier. Above a critical collapse field the barrier disappears and the bubble spontaneously contracts to zero radius, where it is assumed to annihilate. The product $H_k r$ is proportional to the derivative of the energy curve of this figure. Plots of the dependence of stable bubble diameter on bias field are shown in Fig. 2.4 for different material parameters.[54]

The restoring force constant for a bubble can be determined from Eqs. (2.1) and (2.8):

$$k = k_{max} \cdot 16(l/h)\{3[1 + (3r/2h)]^{-2} - (lh/r^2)\}, \qquad (2.10)$$

$$k_{max} = \pi M^2/4l.$$

$k_{max}$ is the maximum attainable restoring force for given material parameters, and it occurs at a critical thickness $h_{crit} = 6l$ and a critical radius $r_{crit} = 2h_{crit}/3$. The reason for the maximum in restoring force as a function of $r$ may be seen from the energy plots of Fig. 2.3b. As the bias field is increased from a low value, the bubble size decreases and the curvature at the bottom of the well increases. The restoring force is proportional to this curvature. However at sufficiently large fields ($H_b = H_{col}$), the well disappears altogether and the restoring force must clearly drop back to zero.

For larger excursions from equilibrium, the full "restoring force field" $H_k$ of Eq. (2.8) must be considered. If a bubble is biased at a field $H_b = H_{col} - H_{ko}$ below its static collapse field $H_{col}$, then the negative minimum of the restoring field $H_k$ is just $H_{ko}$, as is apparent from Fig. 2.3a. When $H_{ko}$ is small, a parabolic approximation to $H_k$ becomes appropriate:

$$H_k = -4H_{ko}(r_o - r)(r - r_c)/(r_o - r_c)^2, \qquad (2.11)$$

with the minimum value at $r_{min} = (r_o + r_c)/2$. These formulas are useful in connection with the dynamic bubble collapse experiment discussed in Section 5,D.

An exact analysis has been given of the restoring force for radial distortions as well as for other distortion modes of an isolated bubble—elliptical,

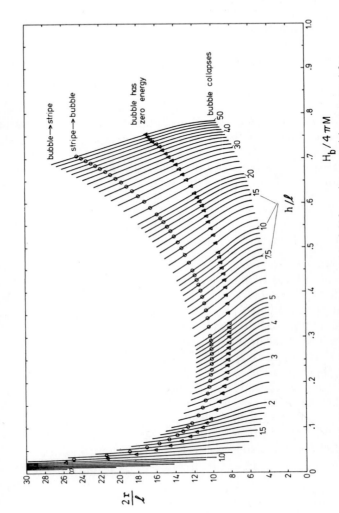

**Fig. 2.4.** Normalized bubble diameter versus bias field for various material parameters (after de Jonge and Druyvesteyn[54]).

triangular, and so forth.[51] Let the bubble domain radius be expanded in the form

$$r(\beta) = r_0 + \delta r_0 + \sum_{n=1}^{\infty} \delta r_n \cos[n(\beta - \beta_n)], \qquad (2.12)$$

where $\beta$ is the azimuthal angle describing positions on the wall circumference, and where $\delta r_n$ are small variations. Then the energy variation is

$$\delta W / (8\pi^2 M^2 h^3) = [(l/h) + (d/h)(H_b/4\pi M) - F(d/h)](\delta r_0/h)$$
$$- (h/d)[(l/h) - S_0(d/h)](\delta r_0/h)^2$$
$$+ \tfrac{1}{2} \sum_{n=2,\infty} (n^2 - 1)(h/d)[(l/h) - S_n(d/h)](\delta r_n/h)^2, \qquad (2.13)$$

where the functions $F(d/h)$ and $S_n(d/h)$ are the demagnetizing "force functions" given by Thiele.[51] We have already considered the terms in $\delta r_0$, using the approximation of Eq. (2.7) to replace $F(d/h)$ and $S_0(d/h)$. Considering the second-order term in $\delta r_2$, describing elliptical distortions, we note that the condition $l/h = S_2(d/h)$ determines the diameter beyond which the bubble spontaneously becomes elliptical. This is the "runout diameter" which determines in turn a "runout field" as indicated by the bubble-to-stripe transition in Fig. 2.4. At smaller diameters, the coefficient of $(\delta r_2)^2$ in Eq. (2.13) determines the restoring force for elliptical perturbations, which are of importance in dynamic experiments (see Section 19,B).

The idealized nature of the bubble description in terms of a reverse domain having a boundary with wall surface density should be constantly borne in mind. A truly valid representation of a bubble domain requires solution of the nonlinear partial differential equations of magnetism obtained by extremalizing the total system energy $W = W_m + W_A + W_K$, where the magnetostatic, exchange, and anisotropy terms are given by Eqs. (1.6), (1.11), and (1.14), respectively.[28] Numerical solutions of this problem reveal features such as the thickening of the wall near the film surfaces and the slight "barrel" character of the shape.[59a]

## C. Motive Forces on Bubbles and Stripes

We consider next two field configurations of particular importance for domain wall dynamics because they exert motive forces on bubbles and stripes, respectively. Consider a bubble subjected to an external field whose film-normal component $H_z(x, y)$ depends so regularly on the in-plane coordinates $x$ and $y$ that the in-plane gradient vector $\nabla H_z$ is substantially constant. We suppose that the mean of $H_z(x, y)$ averaged over the cylindrical

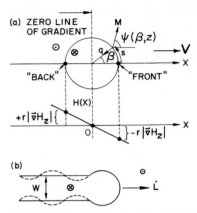

**Fig. 2.5.**   (a) Schematic bubble domain in a gradient. (b) Schematic stripe running out, with dotted line showing buckling of side walls.

wall which is equal to $H_z = H_b$ evaluated at the domain center, is balanced by the other effective fields (Section 2,B) which determine the domain radius $r$. Letting $H_g = r|\nabla H_z|$ be a measure of the strength of the gradient, we see that for a negative gradient along the positive $x$ axis in Fig. 2.5a, the unbalanced field $H_z - H_b$ on the front $(+x)$ edge of the bubble is $-H_g$, the field on the back is $+H_g$ and the field on the flanks $(x = 0$ in the figure) is zero. Such a field distribution clearly puts a net pressure on the bubble to move forward in the $+x$ direction (assuming the magnetization inside the bubble is down and outside is up). Consider a displacement of the bubble by $d\mathbf{X}$. The average bias field change on the bubble is $dH_z = \nabla H_z \cdot d\mathbf{X}$, leading to a change in field energy $W$ of $2M \, dH_z$ times the bubble volume $\pi r^2 h$. Thus the force on the bubble is[51]

$$\mathbf{F} = -dW/d\mathbf{X} = -2\pi r^2 h M \nabla H_z. \qquad (2.14)$$

There is no restoring force in this case, as can be seen from the fact that even though the bubble radius may change linearly with $d\mathbf{X}$, the bubble energy arising from this radius change will go quadratically. This obvious but important conclusion means that a bubble is free to move under the influence of a bias field gradient.

Consider the motive force tending to expand or contract a single long straight stripe domain of length $L$, illustrated in Fig. 2.5b, in the presence of a uniform external field $H_z = H_b$. Let $e$ be the internal (i.e., wall plus demagnetization) energy per unit length of the stripe at points far from its ends. Also let the stripe width $w$ always have the equilibrium value that minimizes the total energy per unit length, including the external-field energy. Then both $e$ and $w$ depend on $H_b$. The change in total energy $dW$ corresponding to the change in length $dL$ is[52]

$$dW = [e(H_b) - 2hMH_bw(H_b)]\, dL, \qquad (2.15)$$

since one can imagine that a new segment of length $dL$ is inserted at some point far from the ends. The critical value $H_b = H_{ri}$, at which $dW$ vanishes and $L$ is in stable equilibrium, satisfies the condition

$$e(H_{ri}) = 2hMH_{ri}w(H_{ri}). \qquad (2.16)$$

For $H_b$ near $H_{ri}$ we can approximate, in Eq. (2.15), $w(H_b)$ with $w(H_{ri})$ and $e(H_b)$ with $e(H_{ri})$ as expressed in Eq. (2.16). Thus the force on an end is approximately

$$F_{st} = -dW/dL = -2whM(H_b - H_{ri}), \qquad (2.17)$$

which is in a direction such as to increase $L$ when $H_b$ is less than $H_{ri}$. There is no restoring force. Above the so-called run-in field $H_{ri}$, the total stripe energy is positive relative to the saturated state, and the stripe will contract along its length. Below this field its energy is negative and it will extend itself, theoretically without limit. In practice, the stripe bends, folding back on itself many times until it "fills" the available space. As indicated in Fig. 2.4, run-in (stripe → bubble) always occurs at a higher bias than run-out (bubble → stripe), which is determined by the onset of elliptical distortion in a bubble. Therefore, this degree of "hysteresis" in cycling between bubble and stripe states is inevitable.

## D. STRAIGHT WALLS

The simplest possible domain wall configuration for experiments on bubble materials is that of a single straight isolated wall, which can be obtained in a sufficiently strong applied $z$-field gradient.[60] The domain wall rests at the zero-field point of the gradient, which can be created by a pair of opposed permanent c-shaped magnets. Gradients $\nabla H_z$ of up to $3 \times 10^4$ Oe/cm have been conveniently achieved. The restoring force constant (Eq. 2.1) is given simply by $k = 2M\nabla H_z$. In addition to its simplicity, another advantage of this configuration for experiments is that small in-plane fields can be applied in an arbitrary orientation relative to the wall without destabilizing it, in contrast to the case with stripe domains (see Section 5,C).

Stability of the single wall is limited by the fact that a certain minimum gradient is needed to prevent the wall from buckling along its length.[60] The value of that critical gradient can be calculated from magnetostatic theory[60,61] and increases rapidly as $l$ decreases. Thus orthoferrites, which have large bubbles and accordingly large $l$ values, require gradients in the range 500 Oe/cm, conveniently within experimental capability, but most garnets and amorphous films, which have much smaller $l$ values, require gradients of $10^5$ Oe/cm or more. The problem of determining the gradient at which a wall buckles is just one aspect of the more general problem of determining the stiffness or restoring force of a wall to sinusoidal perturbation of arbitrary wavelength (see Section 22).

Next we consider stripe arrays as shown in Fig. 2.1a. In the limit of film thickness $h$ large compared to stripe width $w$, it has been found that[55]

$$w = 2.72(lh)^{1/2}. \tag{2.18}$$

In this limit $\chi_0$ must approach $(4\pi)^{-1}$, for the degmanetizing field is just $4\pi M_{av}$ and there is no effective field from wall energy, so for $H_k = 0$, $4\pi M_{av}$ must equal the applied perpendicular field $H$. Combining Eqs. (2.5) and (2.18) with $\chi_0 = (4\pi)^{-1}$, one finds

$$k = 5.88\pi M^2(lh)^{-1/2}. \tag{2.19}$$

As the thickness decreases, $\chi_0$ increases while $w$ goes through a minimum, as described in more detail in the literature.[52,54,55] These effects imply that for given material parameters there is a maximum in $k$ as a function of $h$, much as in the bubble case (Eq. 2.10), and it is of order $M^2/l$.

As mentioned earlier, if a bias field is applied to a stripe domain pattern, the unfavorably oriented stripes will first shrink in width and move apart. Eventually, above a critical "run-in" field they will contract along their length into bubbles. This field is indicated as the stripe-to-bubble transition in Fig. 2.4. However, if for some reason stripes cannot contract, as when their ends are pinned at defects, or as in the honeycomb pattern[54] of Fig. 2.1f in which they have no ends, an interesting situation occurs with regard to the restoring force constant.[62,63] Assuming an infinite isolated stripe, one can show that the width $w$ depends on the field $H$ according to the relation[13]

$$H/4\pi M = \pi^{-1}\{2\arctan(h/w) - (w/h)\ln[1 + (h/w)^2]\}, \tag{2.20}$$

which predicts that $w$ shrinks to zero monotonically at $H$ approaches $4\pi M$. The corresponding restoring force is given by substituting the above expression into Eq. (2.1), with $q$ replaced by $w/2$:

$$k = (16M^2/h)\ln[1 + (h/w)^2]. \tag{2.21}$$

Note that both the stripe array, Eq. (2.19), and the single stripe, Eq. (2.21), provide markedly larger values of $k$ than the isolated bubble, Eq. (2.10), for ordinary values of $h(\sim 4l)$ and $w(\sim 4l)$. As will be seen in Section 5,C, large restoring forces are advantageous in wall-resonance studies.

### E.  MULTILAYER DOMAIN STRUCTURES

In addition to the static domain properties of single uniform bubble films, there has been considerable interest in multilayer structures. In the case of a bilayer, a " capping layer" is somehow created on top of the primary bubble layer. Such structures fall into two groups, depending on whether the capping layer has its magnetization pointing perpendicular to the film or lying in-plane.[64-67] In the first case, there is an additional subdivision of two possibilities,[64] illustrated in Fig. 2.6. If the compensation temperatures of both layers are on the same side of the ambient temperature, the film is called type 1. If they are on opposite sides, the film is called type 2. It is obvious from Fig. 2.6 that in a type 1 film, a domain wall must exist "capping" the bubble and connecting with the wall around the circumference of the bubble. On the other hand, in a type 2 film, if the net magnetization of both layers points in the same direction, the subnetwork magnetizations oppose each other so that some rotation must occur between the two regions. This magnetic structure at the interface or "compensation plane" is called a "compensation wall."[68] It carries surface energy per unit area much as a normal magnetic domain wall. The main difference from the normal domain wall is that it cannot move but is restricted to the compensation plane and carries no net magnetization. If the net magnetizations of the two layers are oppo-

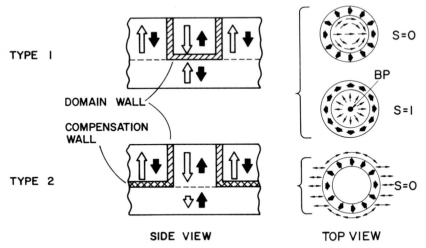

**Fig. 2.6.**  Schematic sublattice and domain-wall structure in double-layer films in which the upper layer contains a bubble. In Type 1, the two layers are on the same side of compensation, in Type 2 they are on opposite sides (after Bobeck *et al.*[64]). In the top view, fat arrows represent vertical-wall magnetization near the interface and small arrows represent horizontal-wall magnetization in the interface. BP, Bloch point; S, bubble winding number.

sitely directed, then there is no wall at the compensation plane. These facts imply that the domain wall of a bubble in a type 2 multilayer structure is connected with a compensation wall spreading outside it as shown in Fig. 2.6. The spin structures of the capping and compensation walls are also shown in the figure and will be discussed in Section 9,C. We refer to the literature for calculations on the static stability of bubbles in one or both layers.[69–73]

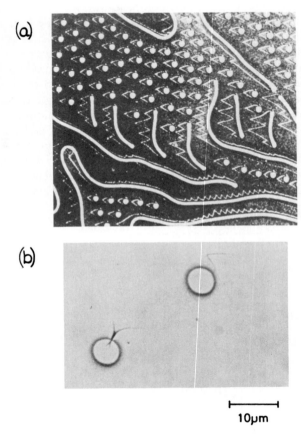

**Fig. 2.7.** (a) Ferrofluid Bitter patterns of domains in an ion-implanted (111) film of GdTm-GaYIG in a bias field of 85 Oe. The white circles and strips are associated with bubbles and stripe domains in the underlying garnet. The fine features show the walls of planar closure domains in the 0.6 $\mu$m thick implanted layer. Photograph was made with transmitted unpolarized light (not Faraday effect). The circles are ~8 $\mu$m in diameter (after Wolfe and North[74]). (b) Ferrofluid Bitter pattern of domains in a GdTmGaYIG film overcoated with 80 Å of permalloy (after Suzuki *et al.*[76]).

A rather different kind of domain wall structure exists if the second layer has in-plane magnetization.[74–80] Here again there are different subdivisions corresponding to the case of permalloy capping layers which have high $4\pi M$ ($\sim 10,000\ G$) and ion-implanted capping layers with lower $4\pi M$ ($<1000\ G$). In the case of ion-implanted layers in [111]-garnets, the characteristic in-plane domain structure in the absence of in-plane field is illustrated in Fig. 2.7a where the white spots show bubbles and the lighter V's indicate closure domains of the in-plane layer.[74] The presumed magnetization distribution is indicated schematically in Fig. 2.8a. The closure domain allows the magnetization to follow the bubble stray field in the vicinity of the bubble wall. The in-plane domain walls carry magnetic charge and hence are termed "charged walls."[75] Their negative charge is balanced by a concentration of positive charge above the center of the bubble. In the presence of an in-plane field, the closure domain points approximately in the direction of in-plane field, for the polarity of Fig. 2.8, although it may be deflected as indicated, depending on the orientation of the cubic crystallographic axes.[79,80] As the in-plane field is increased, the closure domain shrinks in size and finally disappears as the in-plane layer saturates (Fig. 2.8b). This process is called

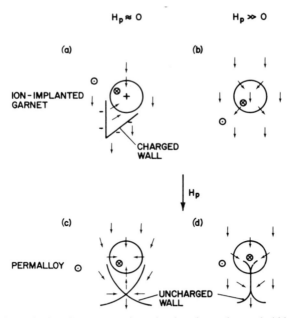

**Fig. 2.8.** Schematic domain structures in an in-plane layer above a bubble layer. In (a) the bias field is assumed to point in the [111] direction of a garnet film and the in-plane field along [11$\bar{2}$]. The closure domain magnetization is assumed to follow the [$\bar{2}$11] direction, which, along with [11$\bar{2}$] and [1$\bar{2}$1], is favored by cubic anisotropy.

a "cap switch" and it occurs in the range of 100 Oe for typical 5 $\mu$m garnet ion-implanted layers.[67,75a] The reverse process of switching from a saturated state back to the closure domain state occurs at a lower in-plane field, indicating a kind of hysteresis in the process as a function of in-plane field. The cap switch process is of great significance for bubble wall states in ion-implanted films, as will be discussed further in Section 9,D.

Since charged walls create a local nonuniform field distribution, they can be used to propagate bubbles (see Section 6,B). This concept is the basis for a class of devices called "contiguous disk" or "ion-implanted propagation pattern" (IIPP) devices. The statics and dynamics of charged walls are just beginning to be studied but are beyond the scope of this review.[74-75e]

The in-plane domain-wall structure for the case of a permalloy overlayer several hundred angstroms in thickness is illustrated in Fig. 2.7b. The reason for the difference with the ion implanted case is presumably the large magnetization of permalloy, which makes the characteristic magnetostatic energy $2\pi M^2$ two orders of magnitude larger than in ion-implanted layers. Therefore charged walls are strongly disfavored. In low in-plane field the X-structure of Fig. 2.7b is favored and the presumed magnetization distribution, which reduces magnetic charges on the in-plane domain walls, is shown in Fig. 2.8c.[76] At larger in-plane fields, the Y-structure becomes preferred, and this eventually is transformed into the saturated state with increasing in-plane field. These structures are sensitive to permalloy thickness and anisotropy; we refer the reader to the literature for further details on this complex subject.[74-80]

### 3. Landau–Lifshitz Equation and Dynamic Material Parameters

In this section we introduce the Landau–Lifshitz equation,[2] which underlies the domain-wall dynamics theory, and we discuss its physical significance. We also review by way of introduction the microscopic origin and typical values of the dynamic material parameters, namely the gyromagnetic ratio $\gamma$ and the viscous damping constant $\alpha$, which enter into the Landau–Lifshitz equation.

### A. LANDAU–LIFSHITZ EQUATION

Magnetization dynamics is based on the fundamental mechanical law stating that the time-rate of angular momentum change is equal to the torque **T**. For a unit of volume located at any point **x** in a magnetic material, the angular momentum is that of electron spin, with corrections for orbital motion. It differs from **M** only by a constant of proportionality $-\gamma$ called the gyromagnetic ratio ($\gamma$ is usually positive). Thus the basic equation of motion may be written

$$-\dot{\mathbf{M}}/\gamma = \mathbf{T},\tag{3.1}$$

where the superscript dot here and henceforth indicates a time derivative. Any torque $\mathbf{T}$ on the magnetic moment $\mathbf{M}$ can be written in terms of an effective field $\mathbf{H}_e$ considered to provide the torque:

$$\mathbf{T} = \mathbf{M} \times \mathbf{H}_e.\tag{3.2}$$

The effective field can be conveniently expressed as the sum of two terms thusly:

$$\mathbf{H}_e = -\frac{\delta w}{\delta \mathbf{M}} - \frac{\alpha \dot{\mathbf{M}}}{\gamma M},\tag{3.3}$$

where $M$ is the magnitude of $\mathbf{M}$. The first term is derived fundamentally by functional differentiation from the total static stored energy, whose volume density is $w$. It includes the magnetic field $\mathbf{H}$ of ordinary electrodynamics, and effects of magnetic anisotropy, exchange stiffness, etc., as described in Section 1. The second term $\alpha\dot{\mathbf{M}}/\gamma M$ is a special and approximate phenomenological representation of dissipative effects, which inherently cannot be derived from a stored energy. By analogy to Ohm's law one considers an effective field tending to oppose the rate of change of $\mathbf{M}$. It is thus proportional to $\dot{\mathbf{M}}$ and has a negative sign. The Gilbert[81] or viscous damping parameter $\alpha$, which is dimensionless, measures the strength of this effect, with the factor $\gamma M$ present in Eq. (3.3) only for dimensional reasons.

Combining Eqs. (3.1)–(3.3), one finds the Landau–Lifshitz equation[2] (although the damping is written in a different form from their original paper, see Section 3,C):

$$\dot{\mathbf{M}} = \gamma\mathbf{M} \times \frac{\delta w}{\delta \mathbf{M}} + \frac{\alpha\dot{\mathbf{M}} \times \mathbf{M}}{M}.\tag{3.4}$$

This equation is the starting point for all discussions of bubble dynamics. Next we consider in more detail the physical significance of the various terms in Eq. (3.4).

The term $\dot{\mathbf{M}}$ on the left and $\alpha\mathbf{M} \times \dot{\mathbf{M}}/M$ on the right may be considered "dynamic" terms because they contain a time derivative of $\mathbf{M}$. By contrast the term $\gamma\mathbf{M} \times \delta w/\delta \mathbf{M}$ is essentially a static term. If this term is zero, then Eq. (3.4) can be satisfied with $\dot{\mathbf{M}} = 0$, so that no spin motion or "precession" occurs. The static torque term is nonzero whenever the effective field $\mathbf{H}_e$ has a component normal to the spin direction. The special case in which

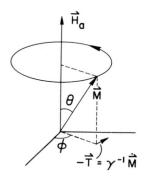

**Fig. 3.1.** Schematic precession of magnetization **M** around a static effective field **H**$_a$ in the direction predicted by the right-hand rule.

there is a constant applied field $\mathbf{H}_a = -\delta w/\delta \mathbf{M}$ is illustrated in Fig. 3.1, with $\theta$ defined as the polar angle between **M** and **H**$_a$. Let us first ignore the viscous damping term ($\mathbf{H}_e = \mathbf{H}_a$). Then Eq. (3.4) predicts that $\dot{\mathbf{M}}$ is always normal to the plane of **M** and **H**$_a$. This means that the spin precesses around the field direction in a cone, with $\theta$ remaining constant as shown in the figure. This kind of precession is called "Larmor precession" or "gyrotropic precession" because it is analogous to the motion of a gyroscope. From Eq. (3.4) the precession occurs at an angular frequency

$$\omega = \gamma H_a. \tag{3.5}$$

Another feature of this behavior is that the static energy $-\mathbf{M} \cdot \mathbf{H}_a$ does not change, and hence the motion is called "conservative." In the general case with $\mathbf{H}_e$ not constant, the total energy represented by the volume integral of $w$ is conserved when $\alpha = 0$. A simple example of Larmor precession occurs in ferromagnetic resonance where a maximum absorption of microwave energy occurs at the Larmor frequency. We shall see many other more surprising gyrotropic effects in bubble dynamics.

The viscous damping term of Eq. (3.4) permits a precessing moment to lose energy and approach its static equilibrium direction along **H**$_a$. This can be seen from the fact that $\dot{\mathbf{M}}$ points normal to the plane of **M** and **H**$_e$ [Eqs. (3.1) and (3.2)]; so $\alpha \mathbf{M} \times \dot{\mathbf{M}}$ must lie in that plane, causing **M** to approach **H**$_e$, and as $|\dot{\mathbf{M}}|$ decreases, **H**$_e$ approaches **H**$_a$ [Eq. (3.3)]. Obviously the viscous damping term is "nonconservative" since $-\mathbf{M} \cdot \mathbf{H}_a$ diminishes. The energy dissipation density can be written:

$$\dot{w} = (\delta w/\delta \theta)\dot{\theta} + (\delta w/\delta \phi)\dot{\phi}, \tag{3.6}$$

where $\theta$ and $\phi$ are the polar and azimuthal angles (see Fig. 3.1), and $\delta/\delta\theta$

and $\delta/\delta\phi$ are functional derivatives [see Eq. (7.3)]. Now the Landau–Lifshitz equation in component form is

$$\dot{\theta} = -(\gamma/M)(\delta w/\delta\phi) \sin \theta - \alpha\dot{\phi} \sin \theta, \tag{3.7}$$

$$\dot{\phi} \sin \theta = (\gamma/M)(\delta w/\delta\theta) + \alpha\dot{\theta}. \tag{3.8}$$

Solving Eqs. (3.7) and (3.8) for $\delta w/\delta\phi$ and $\delta w/\delta\theta$, and substituting in Eq. (3.6), one finds

$$\dot{w} = -(\alpha/\gamma M)\dot{\mathbf{M}}^2,$$
$$= -(M\alpha/\gamma)(\dot{\theta}^2 + \dot{\phi}^2 \sin^2 \theta). \tag{3.9}$$

This expression is termed the "dissipation function." Thus the viscous damping implies a pseudo-Ohmic energy loss proportional to the square of the rate of change of $\mathbf{M}$.

In summary Eq. (3.4) may be viewed as requiring a balance of a static torque term $\gamma\mathbf{M} \times \delta w/\delta\mathbf{M}$ by dynamic terms $\dot{\mathbf{M}}$ and $-\alpha\mathbf{M} \times \dot{\mathbf{M}}/M$, which are called, respectively, the gyrotropic and viscous-damping terms. These dynamic terms are also called "dynamic reaction forces" because they react to the static force or torque on the spin. The same terminology will apply when we treat Bloch walls and Bloch lines as units in domain-wall dynamics theory, for these units experience effective static forces which must be balanced by dynamic reaction forces arising from the corresponding terms of the Landau–Lifshitz equation.

## B.  GYROMAGNETIC RATIO

The crystal structure of a magnetic garnet, orthoferrite, or other magnetic insulator generally includes more than one sublattice, each comprised of identical atoms. Let Eq. (3.1) refer to some one of these sublattices. Then the gyromagnetic ratio $-\gamma$ appearing in this equation is the ratio of the atomic magnetic moment $-g\beta J$ to the atomic angular momentum $\hbar J$ on this sublattice. The constant $\gamma$ is related to an effective dimensionless $g$-factor by the formula

$$\gamma = g\beta/\hbar, \tag{3.10}$$

where $\beta = 0.927 \times 10^{-20}$ emu is the Bohr magneton, and $\hbar = 1.05 \times 10^{-27}$ erg sec is Planck's constant divided by $2\pi$. For a free electron with spin but no orbital motion, the $g$-factor is 2.0023 and $\gamma$ is $1.76 \times 10^7$ Oe$^{-1}$ sec$^{-1}$. For a free multielectron atom, there are both spin and orbital angular momenta that are coupled relativistically. The atom is then characterized by the total spin angular momentum $\hbar S$, total orbital angular momentum $\hbar L$, and total angular momentum $\hbar J$, where $S$, $L$, and $J$ are the well-known dimensionless quantum numbers. In this case the $g$-factor is given by the Landé formula:

$$g_J = [3J(J + 1) + S(S + 1) - L(L + 1)]/[2J(J + 1)]. \qquad (3.11)$$

$S$, $L$, and $J$ are given in standard texts for the ions of interest to us.[29,30] The ground states of the ions $Fe^{3+}$ and $Gd^{3+}$ are essentially pure spin states ($L = 0$, $S = J$) with $g = 2$, like free electrons. On the other hand the orbital contribution is of importance for other trivalent magnetic rare-earth ions including the "light rare earths" from $Ce^{3+}$ to $Eu^{3+}$ and the "heavy rare earths" from $Tb^{3+}$ to $Yb^{3+}$. The $g$-factors of the heavy rare earths lie between 2 and 1, while those of the light rare earths lie between 1 and zero. $Eu^{3+}$ is an important exception. Its $g$-factor is undefined in Eq. (3.11) because $J = 0$ and $S = L$. In a solid, $Eu^{3+}$ is known to behave as though it has a magnetic moment and as though it contributes no angular momentum,[83] so one may take $g = \infty$. Other trivalent ions like $Y^{3+}$, $La^{3+}$, and $Lu^{3+}$ are nonmagnetic. In an oxide host lattice these "free-ion" results are still approximately valid for all rare-earth ions and $Fe^{3+}$, because the perturbing energy of the electric fields of neighboring ions is generally small compared to the energy of the next highest $SLJ$ level.

The sense of the Larmor precession is determined by the sign of $\gamma$. For the magnetic moment of electron spins ($\gamma > 0$) the sense is given by the right-hand rule as shown in Fig. 3.1. The fact that this case is the most common one accounts for the negative sign convention in Eq. (3.1), although in multisublattice systems, the apparent net $\gamma$ can change sign as will be shown in the following.

Equation (3.1) can also be applied to a multi-sublattice system provided the magnetic sublattices are strongly coupled to each other by exchange interactions. The net gyromagnetic ratio $\gamma_T$ is then the ratio of the total magnetization $M_T = \sum_i M_i$ (where $M_i$ carries a sign) to the total angular momentum per unit volume $\hbar J_T N = \sum_i \hbar J_i N_i$, where $i$ refers to the $i$th sublattice and where $N_i$ is the number of $i$ atoms per unit volume. Since $\hbar J_i N_i = M_i/\gamma_i$, we find the "Wangsness formula"[82]:

$$\gamma_T = (\sum_i M_i)/\sum_i (M_i/\gamma_i). \qquad (3.12)$$

Consider the case of EuGaYIG in which just enough Ga has been doped into the iron sublattices for the tetrahedral and octahedral site sublattices of $Fe^{3+}$ ions to be at magnetic compensation. That is, defining $M_1$ and $M_2$ to be the net magnetization of the iron and rare earth sublattices respectively, we have $M_1 = 0$. But because the exchange fields from the octahedral and tetrahedral iron sites are different in magnitude, it is possible, and is in fact observed,[83] that the $Eu^{3+}$ ions remain magnetized even when the octahedral

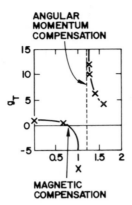

**Fig. 3.2.** Total $g$-factor of $Eu_3Fe_{5-x}Ga_xO_{12}$ at 4.2 K versus amount $x$ of gallium (after Le Craw *et al.*[83]). A similar curve applies at room temperature, but with the compensation points shifted to higher gallium values.

and tetrahedral sites are at compensation; thus $M_2 \neq 0$. However since $\gamma_2$ of the $Eu^{3+}$ ions is effectively infinite as mentioned before, the total angular momentum is zero, and so $\gamma_T$ must be infinite. Values of $\gamma_T$ as high as 30 have been observed experimentally in such systems.[83] The dependence of $g_T = \hbar\gamma_T/\beta$ on composition is illustrated in Fig. 3.2. In a certain range, $\gamma_T$ becomes negative and all precession directions should change sign. In addition we see that $\gamma_T$ falls to zero at a point of magnetic compensation; but this result is of no interest for bubble materials, since no bubbles can exist without a net magnetization. Angular-momentum compensated ($\gamma_T \to \infty$) garnet bubble films of composition EuCaSiGeYIG have been prepared for device use.[84-86] Similar effects are seen in amorphous GdCo films near their compensation point,[88,89] and in garnet films containing high damping rare-earth ions.[87] Angular-momentum compensated films show remarkably high bubble velocities, as will be discussed in Section 11,C, but their practical use has been limited by a high temperature dependence of the magnetization. Another problem is that among these systems only the $Eu^{3+}$-based garnet systems have high mobility, but in this case the magnetization at angular-momentum compensation is solely due to $Eu^{3+}$. At room temperature and full concentration, the magnetization is therefore at most 400 G, which is insufficient for bubbles below a few microns in diameter.

The gyromagnetic ratio $\gamma$ is most simply obtained from a ferromagnetic resonance (FMR) experiment. Microwave fields applied transversally to the moment of a single-domained sample excite uniform spin precession around a net constant field composed of the applied field and the internal anisotropy and demagnetizing fields.[30] For example, at a microwave frequency $\omega$ one expects resonant absorption at a constant $z$-field $H_z$ of[90]

$$H_z = \frac{\omega}{\gamma} - \frac{2K}{M} - \frac{4|K_1|}{3M} + 4\pi M, \qquad (3.13)$$

where $2K/M$ and $4|K_1|/3M$ are the relevant uniaxial and cubic anisotropy fields, respectively, for a [111] oriented garnet film. Knowing the anisotropies and magnetization, one can thus determine the gyromagnetic ratio $\gamma$. Since the anisotropies are often not known beforehand, measurements can be done with fields along different directions to obtain simultaneous equations to solve for $\gamma$, $K$, and $K_1$. Some FMR experiments on high-g garnets reveal two resonance peaks.[87] Presence of two peaks is connected with the fact that the weakness of the exchange interaction between the rare-earth and iron sublattice moments brings an excited resonance mode into the range of the spectrometer. Discussion of microwave effects in the presence of domain walls is given in Chapter X.

## C.   VISCOUS DAMPING PARAMETER

The "viscous" or "Gilbert damping parameter" $\alpha$ has been defined as the phenomenological coefficient of the nonconservative term of Eq. (3.4). Landau and Lifshitz[2] originally introduced a seemingly different nonconservative term with the Landau–Lifshitz damping parameter $\lambda$. However, their equation, taken as a whole, is exactly equivalent to Eq. (3.4), though in general $\gamma$ has different values in the two equations. However, the $\gamma$'s are the same in the limit $\alpha \ll 1$, in which case $\alpha$ is related to $\lambda$ by

$$\alpha = \lambda/\gamma M. \qquad (3.14)$$

The physical significance of $\alpha$ or $\lambda$ is most readily understood in terms of the dissipation function (3.9) whose coefficient $M\alpha\lambda^{-1}$ ($\equiv \lambda\gamma^{-2}$) is proportional to the rate of energy loss. In insulating materials like the garnets and orthoferrites, the loss mechanisms are often dominated by localized losses arising from spin–lattice coupling at the local magnetic ion sites. If there are different kinds of ions contributing to the loss, then the $i$th magnetic sublattice contributes an additional power loss of the form (3.9). If one assumes that all the spins are coupled tightly by exchange, then all the $\theta_i$'s and $\phi_i$'s are equal and one has the sum rule for the total damping[20] $\alpha_T$ or $\lambda_T \gamma_T^{-2}$

$$\lambda_T \gamma_T^{-2} = \sum_i \lambda_i \gamma_i^{-2}, \qquad (3.15)$$

$$\alpha_T = \left(\sum_i \alpha_i M_i \gamma_i^{-1}\right) / \left(\sum_i M_i \gamma_i^{-1}\right). \qquad (3.16)$$

These formulas are analogous to the Wangsness formula (3.12) for the total gyromagnetic ratio in a multisublattice system.

Equation (3.16) shows that the contributions of each sublattice to the total $\alpha$ are weighted by the fraction of angular momentum $\hbar J_i N_i = M_i/\gamma_i$ of each sublattice. Clearly at angular momentum compensation, $\alpha_T$ must diverge, just as $\gamma_T$ does.[86] We can see the physical reason for this result from Eq. (3.9). The rate of energy loss is assumed to be a single-ion effect that behaves smoothly through any compensation point. Thus if $\gamma_T$ diverges, then $\alpha_T$ must also diverge to keep $M_T \alpha_T/\gamma_T$ finite.

The damping constant can most simply be extracted from microwave resonance experiments. If $\Delta H_{pp}$ is the peak-to-trough linewidth of the derivative signal of the uniform resonance, and if a Lorentzian line shape is assumed, $\alpha$ is given by

$$\alpha = \sqrt{3}\gamma \Delta H_{pp}/4\pi f, \qquad (3.17)$$

where $f$ is the microwave frequency. The damping can also be determined from domain-wall mobility experiments, as will be discussed in considerable detail in Chapter V. However as we shall see further on it is not certain that the microwave and domain-wall damping constants should be the same for low damping materials.

The atomistic origin of the local damping is generally attributed to a "slow spin relaxation" for those ions with significant spin–orbit coupling.[20,91] In this mechanism, as the moment of a magnetic ion rotates it interacts with its nonspherical environment in such a way as to shift its energy levels. The populations of these levels adjust toward their appropriate Boltzmann values by gaining energy from and losing energy to phonon and electron excitations; the irreversible net energy loss constitutes the damping. Crudely, the smaller the orbital angular momentum $\hbar L$, the more spherical an ion, the smaller will be the level shifts, and so the smaller the damping. Thus $Fe^{3+}$ and $Gd^{3+}$, having $L = 0$, and $Eu^{3+}$, with a singlet ground state, are expected to show low damping. On the other hand ions with large $L$ such as $Tb^{3+}$, $Ho^{3+}$, and $Dy^{3+}$ are highly nonspherical, and therefore they have high damping.

The expression "slow spin relaxation," which is commonly used, should not be misconstrued. It means that the relaxation frequency $\omega_{rel}$ is smaller than the $Fe^{3+}$ rare-earth exchange interaction ($\hbar\omega_{rel} \ll J_{RFe}$). However, it is usually *fast* compared to the rate of macroscopic moment precession $\omega$ in an FMR or wall-motion experiment ($\omega \ll \omega_{rel}$). In those cases where $\omega \ll \omega_{rel}$ is not satisfied, the Gilbert (or Landau–Lifshitz) term in Eq. (3.4) is not correct.

Table 3.1 shows values of $\lambda/\gamma^2$ obtained in a study of rare-earth iron garnets.[92,93] The domain-wall mobility value for damping from high-fre-

TABLE 3.1

ROOM TEMPERATURE DAMPING PARAMETERS $\lambda/\gamma^2$ OF UNDILUTED RARE-EARTH IRON GARNETS
OBTAINED FROM HIGH-FREQUENCY DOMAIN-WALL SUSCEPTIBILITY ("MOBILITY") AND
MEASUREMENTS OF THE FERROMAGNETIC RESONANCE LINE WIDTH [FMR (A)][a]

| | $\lambda/\gamma^2$ ($\times 10^{-7}$ Oe$^2$ sec rad$^{-1}$) | | | |
| Rare earth | Mobility | FMR (A) | FMR (B) | FMR (C) |
|:---:|:---:|:---:|:---:|:---:|
| Y | 0.52 | 0.006 | ... | ... |
| Gd | 0.52 | 0.19 | ... | ... |
| Tm | 1.2 | 1.3 | 2.3 | 1.8 |
| Eu | 2.1 | 2.2 | 1.9 | 1.3 |
| Yb | 4.2 | 2.2 | 2.5 | 2.2 |
| Er | 7.0 | 8.5 | 7.8 | 5.5 |
| Pr | 12 | ... | 24 | ... |
| Dy | 26 | ... | 52 | 11 |
| Ho | 42 | ... | 58 | 29 |
| Tb | 48 | ... | 142 | 16 |

[a] Columns FMR (B) and FMR (C) represent extrapolations from FMR linewidths of YIG doped lightly with the different rare earth ions (after Vella-Coleiro et al.[92]).

quency wall-susceptibility measurements (see Section 5,B) is compared to the value from FMR measurements using fully doped crystals (A) or extrapolating from lightly doped crystals (B and C). In general the trends suggested by the slow relaxation mechanism are followed. The corresponding $\alpha$ values range from near 1 for compositions with lossy ions, to below 0.01 for compositions like LaGaYIG with no lossy ions. The origins of damping are less well understood in the amorphous materials, but $\alpha$ values of 0.1–0.5 appear to be typical for materials with 1 $\mu$m bubble size. Eddy current damping can generally be ignored because of the high resistivity of the amorphous materials.

A problem arises in such results for the lowest damping samples.[92-98] For example, as can be seen for YIG in the table, there is a discrepancy of a factor of ten between wall-motion damping and FMR. Discrepancies of a factor of three and four have been reported in careful studies on LaGaYIG and LuGdAlIG, respectively. Such discrepancies are not surprising considering that for low damping samples, other loss mechanisms such as dipolar or pseudodipolar scattering by spin disorder, sample inhomogeneities, surface pits, or defects are expected to predominate. Under these conditions it is not clear that the local energy loss formulation of the Landau–Lifshitz equations should be the correct one and there might be a considerable difference between results at microwave frequencies (FMR) and radio frequencies (domain-wall motion). An alternative picture involving energy loss through flexural vibrations of the domain wall is described in Chapter X. What is surprising in fact is how well the usual Landau–Lifshitz formulation with constant $\alpha$ works to describe domain-wall dynamics results, even in very low damping films like LaGaYIG. In most of our text we use this formulation exclusively to see how far it can be pushed. Future work may find more compelling need to correct the Landau–Lifshitz equations for low damping materials.

# III

## Experimental Techniques

### 4. Techniques of Domain-Wall Observation

An early review of experimental techniques for domain wall studies in bubble materials has been given by Shaw *et al.*[99] However the variety of techniques has considerably increased since that report. Here we briefly summarize the various techniques which have been used to observe or detect walls in bubble materials,[100,101] deferring until Sections 5 and 6 the description of specific domain configurations used in dynamics experiments.

The most common observation technique uses a magnetooptic effect and a polarizing microscope. Linearly polarized light incident normal to a magnetic bubble film undergoes a rotation of its plane of polarization in a sense depending on the direction of the domain magnetization normal to the film. For transparent materials such as orthoferrites and garnets, transmitted light can be used and the effect is called Faraday rotation. Rotations of $0.5°$ at saturation are typical for $5 \ \mu m$ garnet device films and rotations an order of magnitude larger have been achieved in special garnet compositions containing bismuth. For opaque materials such as amorphous GdCo, one uses reflected light and the effect is called the polar Kerr effect. Rotations are generally in the range of 5 to 10 min.[102]

If the analyzer of the polarizing microscope is slightly uncrossed with respect to the polarizer, one domain appears darker than the other. This method, called the "domain-contrast" mode, is the most convenient way of observing domain configurations visually. The light may be incoherent or coherent, continuous or pulsed. An example of static domain structures observed using continuous incoherent light is shown in Fig. 2.2.[18a] Some dynamic states of a pulsed bubble array, observed using pulsed laser illumination, are shown in Fig. 4.1.[103] The domain-contrast mode can also be used

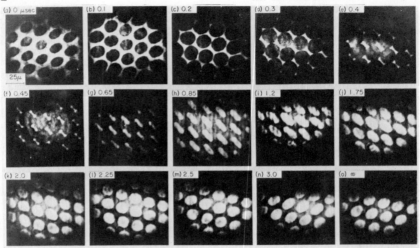

**Fig. 4.1.** High-speed laser photographs of a bubble array in a EuGaYIG film taken at the indicated time delays relative to the onset of a 200 Oe bias field pulse 0.42 $\mu$sec long. The reversal of bubble polarity between the first and last photograph is known as "topological switching" (after Malozemoff and Papworth[103], copyright Institute of Physics, 1975).

for photometric studies of dynamic wall displacements if the average intensity from an area of the sample can be made linearly proportional to the difference in area of up and down domains. This is achieved by making the polarizer–analyzer angle large compared to the Faraday rotation angle, but in this case there is a large dc background signal from both domains.

Another mode for domain observation is the "wall-contrast" mode in which the polarizer and analyzer are exactly crossed. In this case the domains are of equal intensity when viewed in a microscope but the domain walls appear dark. The width of this dark region is observed to be considerably larger than the theoretically expected domain-wall width.[104] Instead of indicating a basic flaw in the wall width theory discussed in Section 7, the dark region is most likely due to some kind of optical scattering. Interesting diffraction effects can occur in the wall-contrast mode.[105-107] For example, a parallel stripe array acts as an "antiphase" grating in which the intensities of even-order diffraction beams are proportional to the square of the difference between the widths of up and down domains. When applied to photometric measurements, this effect offers a method for reducing the large dc background signals that are characteristic of the domain-contrast mode.[108]

Let us next consider in more detail the photometric detection techniques for domain-wall motion experiments. The principle problem for high-speed experiments is photodetector shot noise. The signal-to-noise ratio can be improved by maximizing the intensity of illumination (limited by sample heating and by photomultiplier current capacity) and by maximizing the area swept out by the wall motion relative to the area illuminated. For example in pulsed bias-field experiments on isolated bubbles, good signal-to-noise has been achieved by focusing a cw laser into a small disc just surrounding the bubble; the laser spot also serves the purpose of holding the bubble in a thermal well.[109] With a parallel stripe geometry, one can obtain a signal-to-noise advantage even with normal illumination (xenon or mercury arc lamp) by observing an area including a large number of stripes.[110,111] In those samples where a single wall can be stabilized in a static bias-field gradient, the signals are equally good because larger displacements make up for the lack of multiple domain walls.[94]

The detection is accomplished with a fast-rise photomultiplier or photo-diode and wide-band amplifier, which have permitted signal rise times down to a few nanoseconds. For repetitive motions signal averaging can be used with either sinusoidal or repetitive pulsed bias-field excitation. For sinusoidal excitation, a phase detection scheme using a vector voltmeter, or a heterodyne detection scheme measuring differential amplitude response, have been used.[113,114] These methods have problems interpreting nonlinear signals for which a better approach is the use of time-domain detection of domain-wall response to a repetitive bias-field pulse. The optical signal, proportional to domain-wall displacement, is scanned repetitively using a sampling scope and the output is averaged with an RC circuit.[110,111] DC drift can be eliminated by triggering the sampling scope at double the repetition rate of the drive pulses and using lock-in detection on the sampled output signal.[112] Alternatively, a transient digitizer or multichannel analyzer has been used.[109] Typical results[109,113] using the latter method on a bubble in a EuGaYIG film subjected to a bias-field step are shown in Fig. 4.2.

While photometric detection is most effective for small-amplitude motion

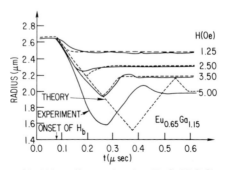

**Fig. 4.2.** Dependence of bubble radius on time in a EuGaYIG film in response to steps of bias field $H$, as determined by photometric measurements. Solid line is data and dashed line is theory according to Eqs. (17.8)–(17.11) (after Brown et al.[109] and Malozemoff[115]).

experiments, various visual or photographic means are preferable if the domain-wall excursion is larger because one can then monitor the domain-wall shape. In certain experiments such as dynamic bubble collapse (Section 5,D) or gradient bubble propagation (Section 6,C), one simply observes the change in domain configuration before and after the application of a field pulse. If the motion is repetitive, one may observe a blur indicating the time-averaged domain position. This simple technique has been used to measure the amplitude of bubble-radius response to sinusoidal excitation[59,116,117] or the time-averaged bubble position in devices.[118] A more sophisticated approach is a stroboscopic one, applicable equally to sinusoidal or pulsed excitation. Light from a continuous source may be chopped by an electrooptic modulator[119,120] or on detection by a gated image tube.[121] Alternatively one may use a pulsed light source such as a light-emitting diode driven by an avalanche transistor,[122] or a pulsed laser.[123–130] In the latter category, the two most effective sources have been a superradiant dye laser pumped by a pulsed nitrogen laser, giving ~7 nsec time resolution[124–127] and a mode-locked argon-ion laser giving 1 nsec resolution[128–130] An example of experimental results obtained with the latter technique is shown in Fig. 4.3, where the change in radius of a bubble in response to short bias pulses is plotted as a function of time from stroboscopic observations using a laser repetition rate of 3 kHz.

The stroboscopic technique can also be extended to device studies.[120,126,127,131–133a] If the device operating frequency is an integral multiple of the laser frequency, the domains on the device structure will appear "frozen" at a given point in the device operating cycle (see Section 6,B). An example is shown in Fig. 4.4 for a T and I bar device with a bubble nucleator on a garnet film operating at 180 kHz.[127] The diminishing contrast of the bubble domains away from the bubble nucleator indicates statistical failures of the device. The bubble shapes and positions through the operating cycle may be determined by scanning the phase lag of the device frequency with respect to the laser frequency. A problem in this technique is the elimination of diffraction fringes of the coherent laser light off the device structures. If sufficient intensity is available, a spinning ground-glass disk in the path of the laser beam can destroy spatial coherence.[127] Alternatively, a thin metal layer (e.g., 200 Å of chromium) deposited between a garnet film and the device structure acts like a half-silvered mirror which (1) reflects laser light passed through the garnet film to reveal the bubbles and (2) passes incoherent light from the other side to reveal the device structure.[126,133]

In addition to stroboscopic experiments, single-shot laser photography with light pulses shorter than 10 nsec is possible for nonrepetitive phenomena. For example, the 5–10 kW output of a nitrogen-laser-pumped dye laser of Rhodamine 6G can be focused in a microscope to a spot on a sample

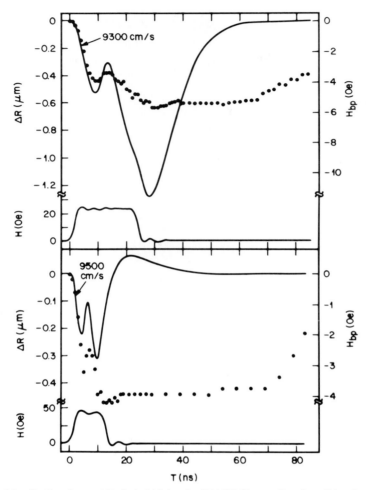

**Fig. 4.3.** Radius change $\Delta R$ of a bubble in a LuGdAlIG film as a function of time in response to a bias field pulse $H$, determined from high-speed stroboscopic observations. Dotted line is data and solid line is the prediction of the one-dimensional theory of Sections 10 and 11 (after Vella-Coleiro[130]).

100 $\mu$m across, giving sufficient light to expose high-speed Polaroid film at 500× magnification in a single shot.[124,125] When the experiment is performed several times and each time the laser flash is delayed a different amount relative to the onset of a drive-field pulse, the time sequence of domain shapes in a given experiment can be recorded, as illustrated for the topological-switching experiment in Fig. 4.1. These pictures record the effect

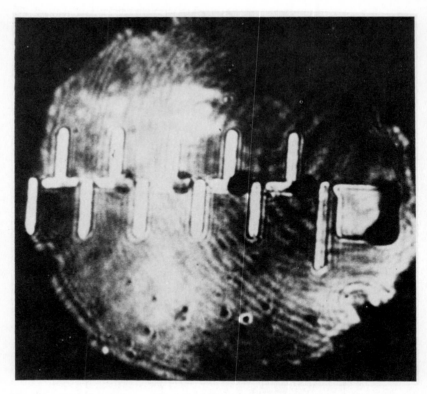

**Fig. 4.4.** Stroboscopic image of bubble device consisting of T and I bars and a bubble generator. Bubbles are also seen along the T and I bar propagation path, with diminishing contrast indicating statistical failures of the device (after Kryder and Deutsch[127]).

of a 200 Oe, 420 nsec bias field pulse on a bubble array in a EuGaYIG film, and they will be discussed further in Section 5,D. It should be emphasized that such a series of photos is not a "movie" because each photo is a time cut of a separate experiment. The continuity of the series relies on the reproducibility of the experimental results from pulse to pulse. Nevertheless such an experiment could not be recorded stroboscopically because slight shifts in the overall bubble array position from experiment to experiment would lead to a blurring of a stroboscopic picture.

One problem with this method is that heat from the absorbed laser light may be sufficient to noticeably affect the domain-wall motion after the light pulse, but it is generally considered that the heat does not affect the wall motion up to or during the light pulse. The heat pulse can be reduced and larger-area illumination obtained by photographing onto fine-grain film at lower magnification and then enlarging.[126] An image intensifier tube can also reduce the required light input level.[128-134] An image intensifier has also been used to make possible real-time observation of straight-wall motion or bubble collapse using a streak camera.[134,135] A high repetition-rate Q-switched NdYAG laser has been used to obtain a series of light pulses 150 nsec long at 7.5 μsec intervals.[401] This technique has been used to record in real time the phenomenon of stripe contraction in garnet films. Alternatively, a semiconductor injection laser has been used to obtain a series of 15 nsec pulses at 500 nsec intervals.[134a]

In addition to determining wall positions, it is of great interest to determine internal wall structure, particularly in the presence of vertical Bloch lines, which will be described at length in Chapter IV. So far direct optical detection of Bloch lines in bubbles by examination of local wall contrast has not been successful (although they can be seen in walls separating in-plane domains[135a,135b]). However indirect dynamic means can be used, based on the fact that wall regions containing Bloch lines move more slowly than normal wall regions. Bloch-line clumps have been revealed as lagging regions in high-speed photography of stripes and bubbles.[137-139] Usually when the walls relax after a drive pulse to a static configuration, wall surface tension straightens up the walls and masks the presence of Bloch lines, although in one study the coercivity of the Bloch-line regions was sufficient to leave static indentations in stripe walls.[140]

We turn now from the magnetooptic methods of domain observation to the Bitter-pattern method, in which magnetic particles are spread over the sample, for example by a ferrofluid suspension. The particles become oriented in the presence of a bias field. Let us assume the bias field points up and the suspension is on top of the sample. Then the particles will be repelled from any negative magnetic poles. Observed in transmitted unpolarized light, bubbles, stripes, or other downward oriented domains will appear bright as shown in Fig. 2.7a. The garnet sample of this figure was ion-implanted, and the ferrofluid also stains the "charged walls" of the surface layer, which appear like closure spikes around the domains and like zig-zags around the stripes (see Section 2,E). Another example is shown in Fig. 2.7b, where bubbles are observed in conjunction with the planar domain structure of a thin permalloy overlayer. The Bitter-pattern method suffers from the fact that the magnetic particles can measurably affect the properties of the bubbles that are being studied, particularly in dynamics experiments. However the method has been invaluable in observing static in-plane domain structures, which are less easily observed by magnetooptic techniques.

Electron microscopy offers another method for observing domain structures. In scanning electron microscopy,[141] a beam of electrons is scanned over a sample and the variations in the intensity of secondary electrons are monitored. The variations of interest arise from the electrons deflected by the Lorentz force

$$\mathbf{F} = q\mathbf{V} \times \mathbf{B},$$                                              (4.1)

where $q$ is the electronic charge, $\mathbf{V}$ is the electron velocity, and $\mathbf{B}$ is the magnetic induction $\mathbf{H} + 4\pi\mathbf{M}$. If the electrons are incident normally (along $z$) to the sample surface, there will be no interaction with the domain magnetization (along $z$) because of the cross product form of Eq. (4.1). The dominant deflection comes from the surface stray field, that is, from the demagnetizing fields at the surfaces near the domain boundaries (see Section 8,E). Domain contrast can be obtained if electrons are incident at an angle to the surface. Scanning electron microscopy has been used relatively little so far but may find use eventually in looking at bubbles of suboptical dimensions. Stroboscopic experiments are also possible with electron microscopy.[142]

All of the techniques discussed so far in this section, while permitting observation of domains and domain walls, are insensitive to the internal structure of the walls.[143-147] By contrast, transmission Lorentz microscopy has given evidence of internal wall structure. In this configuration one focuses on the transmitted electron intensity just off the sample surface. As discussed in the preceding, electrons incident normal to the sample are undeflected by the domain magnetization. The electrons are also relatively insensitive to the stray fields, for deflection according to Eq. (4.1) is along the domain wall since the stray field is perpendicular to it. Furthermore in transmission the deflection at one surface is partly canceled out by deflection at the other because the direction of the stray field on the two surfaces are opposite. Thus the principal deflection comes from the magnetization within the domain wall itself. As will be discussed in Chapter IV, the wall magnetization lies on the average in the planes of the wall and the film and has two possible directions. According to Eq. (4.1), electrons passing through the wall region are therefore deflected perpendicular to the wall by a small angle, leading to an accumulation of electron density on one side of the wall and to a depletion on the other, in a plane somewhat away from the sample.

This effect accounts for the black–white bands marking the domain walls in Figs. 4.5 and 4.6. The first figure shows bubbles in a $\sim 3000$ Å film of $PbFe_{12}O_{19}$, while the second shows an irregular domain array in a high-Q, high-coercivity 1000 Å film of amorphous GdCoAu near its compensation point.[145] Close examination of the black–white wall contrast reveals changes of magnetic polarity along the wall. Such observations provided the first evidence[143] of the existence of vertical Bloch lines and of different chiral bubble states, as will be discussed in more detail in the next chapter. Because of this unique sensitivity to wall structure, it would be of great interest to

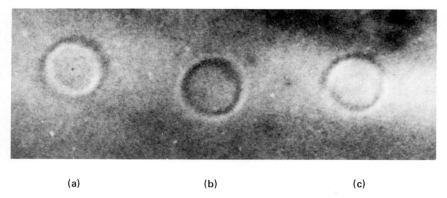

(a)     (b)     (c)

**Fig. 4.5.** Photograph by transmission Lorentz microscopy of bubble domains in a 3000 Å thick film of $PbFe_{12}O_{19}$ (after Grundy and Herd[144]). Cases (a) and (b) show the two $S = 1$ chiral bubbles, while (c) shows a bubble with two Bloch lines, which could be $S = (0, 2)$ or $(1, 2)^{\pm}$ (see Fig. 8.5).

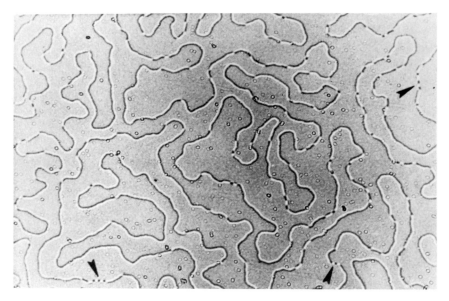

**Fig. 4.6.** Photograph by transmission Lorentz microscopy of domain structure in a 1000 Å thick film of amorphous GdCoAu (after Chaudhari and Herd [145]). The film is near magnetic compensation. The small circles are defects but the breaks in the domain-wall lines indicate Bloch lines. Arrows show groups of Bloch lines. (Copyright 1976 by International Business Machines Corporation, reprinted with permission.)

perform dynamic experiments by Lorentz microscopy. Unfortunately the technique has proven limited so far because samples commonly used in devices are too thick to transmit electrons while the smaller bubble films, which are thinner, are of poorer quality and less well characterized. For this reason, virtually all the dynamics experiments we describe in this review have been effected with magnetooptics on larger-bubble materials. However, considering the trend of bubble technology to smaller bubbles, it is likely that electron microscopy will be used much more in the future.

## 5. Dynamical Techniques with a Restoring Force

In the next two sections we discuss various experimental configurations for domain-wall dynamics experiments, breaking them down into two groups— with and without magnetostatic restoring forces. A summary of the relative advantages and disadvantages of the different techniques from the point of view of the materials researcher or device engineer is given in Section 6,E.

### A.  PHENOMENOLOGICAL EQUATION; COERCIVITY

The classical harmonic oscillator equation provides a useful phenomeno- logical framework for describing the results of many domain-wall dynamics experiments in which a magnetostatic restoring force is present.[4] As in Section 2,A, we define $H_a$ as the applied drive field along the $z$ direction parallel to the easy anisotropy direction, and $q$ as the wall normal displace- ment coordinate, which is positive for wall displacement towards the $-z$ domain. Then the oscillator equation is written

$$m\ddot{q} + b\dot{q} + kq = 2M(H_a - H_c \operatorname{sgn} \dot{q}), \tag{5.1}$$

where the sign function is defined by

$$\operatorname{sgn} \dot{q} = \begin{cases} +1, & \dot{q} > 0 \\ -1, & \dot{q} < 0. \end{cases} \tag{5.2}$$

Here $m$ is the effective wall mass per unit area, $b$ is the viscous drag per unit area, $k$ is the restoring force constant, $M$ is the magnetization, and $H_c$ is a phenomenological "coercive field" for domain-wall motion. Just as $2MH_a$ and $-kq$ are static pressures acting on the wall (see Section 2,A), the other terms of Eq. (5.1) may be termed "dynamic reaction pressures," arising from wall mass ($m\ddot{q}$), viscous damping ($b\dot{q}$), and coercivity ($H_c \operatorname{sgn} \dot{q}$), respectively.

While the force constant $k$ is determined by the domain configuration and is therefore under the experimentalist's control (see Section 2), the parameters $m$, $b$, and $H_c$ are material parameters determined by loss mech-

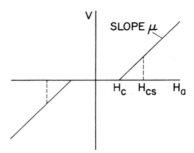

**Fig. 5.1.**   Schematic domain-wall velocity-versus-drive curve according to Eq. (5.3), show-ing dynamic and static coercive fields $H_c$ and $H_{cs}$.

anisms and spin precession effects. If $k = 0$ and dynamic equilibrium has been reached ($\ddot{q} = 0$) with constant $|H_a| > H_c$, Eq. (5.1) reduces to

$$\dot{q} = (2M/b)(H_a - H_c \operatorname{sgn} \dot{q}). \tag{5.3}$$

This relation is plotted in Fig. 5.1. The slope of the velocity-versus-drive curve in the high-drive region of this plot is the linear wall mobility, given by

$$\mu = 2M/b. \tag{5.4}$$

The mobility, and also the wall mass, are related to the static and dynamic material parameters discussed in Chapter II, and the derivation of such relationships will occupy much of Chapters V–IX.

By contrast the coercive field $H_c$, which is the back-extrapolated intercept of the velocity versus drive curve of Fig. 5.1, presumably arises from the interaction of the domain wall with defects or inhomogeneities and is analogous to the notion of dynamic friction in classical mechanics. If the wall is at rest, ($\dot{q} = 0$) the coercive pressure $H_c \operatorname{sgn} \dot{q}$ is undefined in Eqs. (5.1)–(5.2). Under these conditions the coercive field arises physically from the local restoring force of the defects which pin the wall and prevent it from moving. Thus, its magnitude is just such as to balance all other pressures from applied fields ($2MH_a$), wall mass ($-m\ddot{q}$), or longer range restoring forces ($-kq$), provided the combination of these pressures is less than some threshold. We define the threshold pressure to be $2MH_{cs}$, which is in general larger than $2MH_c$. This threshold is the counterpart of *static* friction in domain-wall motion. The static coercive field $H_{cs}$ is indicated schematically by the dotted line in Fig. 5.1. At drives below $H_{cs}$, there is no long-range motion, but above $H_{cs}$ the velocity-versus-drive curve rises rapidly toward Eq. (5.3). Typically $H_c$ is a few tenths of an oersted in device quality garnets and a few oersteds in amorphous GdCo device materials. $H_{cs}$ is often a factor of two or three greater.

Careful studies of coercivity in bubble materials have only recently begun. We digress in the next two paragraphs to mention briefly some of the remarkable effects that have been observed in $H_c$ and $H_{cs}$. In one study on a SmLuCaGeYIG film,[148] the minimum drive field for bubble motion (essentially the same as $H_{cs}$) was found to depend on the pulse length of the experiment, for example being 1 Oe at 10 $\mu$sec but 2 Oe at 100 nsec. The dependence was logarithmic in the inverse pulse length, suggesting a thermally activated process for the wall-motion threshold. The minimum drive field also depends on capping layers. In a study on a hydrogen-implanted

EuYbCaGeYIG film,[79] the minimum drive field exhibited a threefold symmetry relative to the direction of an in-plane field, varying from a minimum of 1.5 to a maximum of 3 Oe. This effect occurs in (111)-oriented samples because the bubbles must drag in-plane domain structures, which have a three-fold symmetry arising from cubic anisotropy (e.g., see Fig. 2.8a). Even more remarkable was a study on neon-implanted SmLuCaGeYIG films[149] in which the minimum drive field was reduced by up to a factor of two by the implantation. This occurred when the implantation energy was sufficient (e.g., 200 keV for neon) to cause the capping layer to be buried under the surface, so that it in some sense "shielded" the bubble from surface imperfections. The minimum drive field increased to its original value when an in-plane field saturated the capping layer. Further evidence for a surface contribution to coercivity comes from the fact that coercivity is usually observed to decrease with increasing film thickness.[150,150a] In one study, nonlinear dependence of wall velocity on field in the low-drive region was interpreted as arising from two coercive fields, one from the bulk of the film and one from an interface layer.[150b]

The problem becomes even more complicated if one goes beyond the simple coercivity model of Eqs. (5.1) and (5.2) and if one considers explicitly the local nonuniformity which causes the coercivity. In this case it can be shown that the apparent coercivity, determined as the back-extrapolation of a displacement vs. drive curve, decreases as the restoring force on the domain wall increases.[150c] For example the apparent coercivity of a wall in a maze pattern, where the restoring force is large, is generally significantly smaller than the coercivity of an isolated propagating bubble, where the restoring force is zero [see Section 6,A; $4/\pi$ correction of Eq. (6.2) taken into account]. Yet another effect is the increase of $H_c$ and $H_{cs}$ due to the presence of Bloch-lines in the domain wall; examples are given in Sections 13,A, 13,B, 14,B, and 16,B. These comments conclude our digression on coercivity. Henceforth we treat coercivity simply as the phenomenological parameter introduced in Eq. (5.1).

If Eq. (5.1) applies with fixed coefficients $m$ and $b$ or $\mu$, one speaks of the wall motion as "linear"; otherwise it is termed "nonlinear." Linear motion prevails when $q$ is small, ("small-signal motion"), which usually occurs when the drive field is small, though larger than the static coercive field. Sometimes the critical field for the onset of nonlinearity may be smaller than $H_{cs}$, in which case the linear regime may be masked altogether. The nonlinear regime is often characterized by a saturation velocity $V_s$ of the form

$$V_s = V_{so} + \mu_h H, \tag{5.5}$$

where $V_{so}$ is a saturation velocity back-extrapolated to zero drive field $H$, and $\mu_h$ is a "high drive mobility."

## B.   Bulk Techniques

Historically, the earliest domain-wall dynamics experiments were on bulk samples. In the Sixtus–Tonks experiment,[151] a single domain wall is nucleated at the end of a long rod-shaped sample and sensed in the course of its propagation by pickup coils located at appropriate intervals. In the picture-frame experiment,[5,152–155] what is effectively a toroid is milled from a single crystal and a pickup loop monitors the rate of change of flux in the sample as a single domain wall sweeps across radially to switch the magnetization direction of the toroid. In both methods, the force constant $k \approx 0$ and dynamic equilibrium is usually attained; so Eq. (5.3) is followed, permitting determination of $\mu$ and $H_c$. Such methods have been used for studying orthoferrites, where they prove convenient for measuring extremely high velocites and variations with temperatures over a wide range.[156-158] However, considering the difficulties of sample preparation and the uncertainties in the actual domain-wall configuration, there has been little incentive to apply these techniques to other bubble materials.

Another classical technique for bulk samples is the determination of the high-frequency magnetic susceptibility. An ac field is applied parallel to the easy axis and the real (in-phase) and imaginary (out-of-phase) parts of the susceptibility $\chi'$ and $\chi''$ are measured as a function of frequency. It turns out that domain-wall resonances and relaxations usually occur at megacycle frequencies; so standard rf techniques can be used. For example, $\chi'$ and $\chi''$ can be extracted from the response of a series tuned circuit containing a solenoid into which a sample can be inserted.[159,160] An example of resonant behavior observed in GdYIG platelets is shown in Fig. 5.2. Such a technique is useful for surveying sample compositions before preparing them in thin film form.[161] The results are also easy to interpret because surface effects, which complicate domain-wall behavior in thin films (see Chapter VIII), are usually negligible.

The results of such experiments may be interpreted in terms of the oscillator equation (5.1), which is well-known to give rise to resonant or relaxation behavior depending on the relative magnitudes of $m$, $b$, and $k$. Using $\chi = 2Mq/H_a w$ [see Eq. (2.4)], where $w$ is the average static domain width,

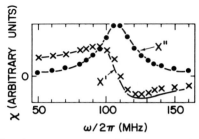

**Fig. 5.2.** Real and imaginary components of the ac magnetic susceptibility of GdGaYIG platelets as a function of frequency (after Vella-Coleiro et al.[160]). Solid lines represent a fit to Eqs. (5.6) and (5.7).

taking the oscillating applied field to be $H_a \exp(i\omega t)$, and ignoring coercivity, one finds from Eq. (5.1):

$$\chi' = \chi_0[1 - (\omega/\omega_r)^2]/\{[1 - (\omega^2/\omega_r^2)]^2 + (\omega/\omega_c)^2\}, \qquad (5.6)$$

$$\chi'' = \chi_0(\omega/\omega_c)/\{[1 - (\omega^2/\omega_r^2)]^2 + (\omega/\omega_c)^2\}, \qquad (5.7)$$

where

$$\chi_0 = 4M^2/kw, \qquad (5.8)$$

$$\omega_r = (k/m)^{1/2}, \qquad (5.9)$$

$$\omega_c = k/b. \qquad (5.10)$$

By fitting these equations to the data, one can determine the low frequency susceptibility $\chi_0$, the resonance frequency $\omega_r$, and the relaxation frequency $\omega_c$. Assuming $M$ is known, one still needs to determine the domain spacing $w$ before being able to solve for $k$, $m$, and $b$. In the case of Fig. 5.2, this was done by directly measuring the domain spacing of single-crystal platelets in a microscope.

In the susceptibility measurements, the question arises, what determines whether relaxation or resonant behavior is observed? From Eqs. (5.6) and (5.7), the criterion for resonance is that $\omega_r = (k/m)^{1/2}$ should be less than $\omega_c = k/b$. In other words, smaller $b$ and larger $m$ is favorable for resonant behavior. This also means that, given $b$ and $m$, resonant behavior can be obtained if $k$ is made large enough. As discussed in Section 2,D, there is a maximum, as a function of platelet thickness, in $k$ of a stripe array. Thus if resonant behavior is of interest for determining the domain wall mass, one must optimize sample thickness to maximize $k$, although if $m$, $b$, and $k_{max}$ are unfavorable, it may still be impossible to achieve resonant behavior. It is for these reasons that resonance has not been reported in orthoferrites in spite of the low damping of such compounds as $YFeO_3$ (see, however, reference 162), and that even in the garnets, domain-wall resonance has only been seen in compounds containing low damping ions like La, Y, Gd, and Eu. If one is only interested in the mobility, one can work with thick platelets having a low $k$, so that the relaxation will occur at low frequencies, masking out any resonance. Experimental techniques also become simpler at lower frequencies.

## C.   SMALL-SIGNAL EXPERIMENTS ON FILMS

Experiments on thin films permit more detailed study of dynamic domain-wall behavior than experiments on bulk samples because the domain configurations can be better controlled. Possible domain configurations include a single wall stabilized in an external field gradient, a stripe array, a bubble array, or an isolated bubble, and the restoring forces of each of these configurations have been described in Section 2. In dynamic experiments, the single-wall configuration has the advantage of a well-defined equilibrium location, but its use is limited in practice to bubble materials with large $l$-values ($l/h \gtrsim 0.5$).[94] Another well-defined configuration is the isolated bubble, but a problem is that, in the presence of the repetitive pulsing needed for signal averaging, the bubble may "wander" unless there is a potential well to hold it in place. Such potential wells can be supplied by defects,[163] by a magnetic pole from a permalloy bar, by a localized light beam heating the sample,[109,164] or by a localized current distribution created by a stripline configuration such as illustrated by the inner conductor pair of Fig. 5.3.[165,166] The effect of the potential well on the bubble restoring force must be taken into account. Alternatively, a hexagonal array of bubbles can be used, so that the well is provided by the neighboring bubbles. As shown in Section 2,D, the latter configuration can provide a particularly large restoring force.[62,63]

Another domain configuration of widespread use in dynamical experiments is a periodic array of stripes. Since the domain spacing is well-defined, the restoring force can be determined experimentally from the low-frequency susceptibility according to Eq. (5.8). The dynamic stability of such arrays however poses a problem. In the presence of repetitive drive pulses, the stripes often appear to "thrash about," that is, to shift in an uncontrolled manner.[112] An in-plane field lying along the walls tends to stabilize the configuration, while a field perpendicular to the walls tends to cause a static "zigzag" instability.[167] Observation of "thrashing" raises questions about the meaningfulness of experimental results because of possible interference from parametric excitation of wall-wave oscillations. Wall waves will be considered in more detail in Chapter X.

To study the small-signal response of domain walls in any of the aforementioned configurations, one typically uses either sinusoidal or pulsed bias-field excitation, corresponding to frequency- or time-domain signal processing. As described in Section 4, the response can be determined (a) visually from a blur in the time-averaged wall position, (b) photometrically, or (c) by stroboscopic or high-speed photography. Two examples of the response of an isolated bubble to a bias-field pulse were shown in Figs. 4.2 and 4.3. In principle, if Eq. (5.1) were followed for a low-damping sample, one would expect an underdamped harmonic oscillator response with oscillations dying away exponentially with a time constant $2m/b$. From the oscillation frequency one could determine $(k/m)^{1/2}$. Knowing $k$ from a low-frequency susceptibility measurement, one could then determine the wall mass $m$ and

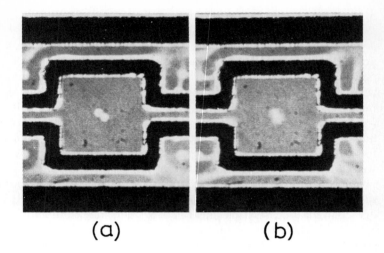

(a)            (b)

(c)

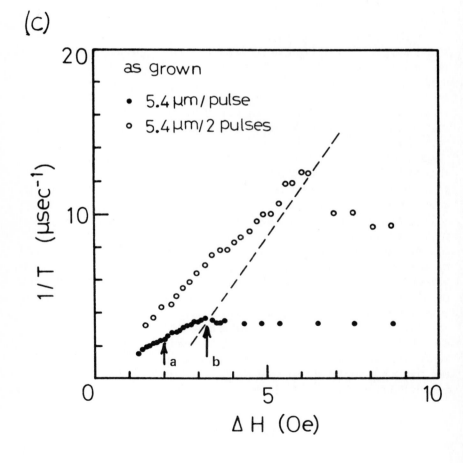

the linear mobility $\mu = 2M/b$. Fig. 4.2 shows oscillations, but ones rather different from those predicted by Eq. (5.1), indicating nonlinear wall motion characterized roughly by a saturation velocity of 380 cm/sec. Further discussion of such results is given in Section 17,B.

## D. DYNAMIC BUBBLE EXPANSION, COLLAPSE, AND TOPOLOGICAL SWITCHING

Next we consider the effect of stronger bias-field pulses that cause such large wall motions that Hooke's law is no longer sufficient to describe the restoring force. One experiment relying on high-speed photography is the domain expansion experiment in which a strong bias pulse is applied so as to expand the domain, which could be either a stripe[168] or a bubble.[169] An example of bubble-expansion results on a EuGaYIG film is shown in Fig. 5.4.[103] Typically the walls initially move at a constant (saturation) velocity, in spite of the rapidly changing restoring force. Eventually the bubble begins to distort, and multilobed shapes are observed, as illustrated by the four-lobed shape of the figure. Such shapes can be understood in terms of the higher order bubble distortion modes of Eqs. (2.12) and (2.13).[170] A problem with the bubble-expansion experiment is that because the restoring force is large the drive must be very large to expand the bubble, and therefore the motion is usually nonlinear in a range of drive field far beyond what is of interest for devices.

One of the most widely used techniques for studying wall motion in bubble materials is dynamic bubble collapse.[171] It takes advantage of the metastable nature of bubble domains, illustrated by the plot of bubble energy versus radius in Fig. 2.3b. At a given static field, the bubble sits in a potential well at its equilibrium radius $r_0$, stabilized from collapsing to the energy minimum at $r = 0$ by an energy barrier whose maximum is at $r_c$. The idea of the dynamic collapse experiment is to dynamically impel the bubble radius past the hump at $r_c$, by means of a bias field pulse; once past the hump, the bubble will collapse of its own accord. One can easily measure how long it takes the bubble to go from $r_0$ to $r_c$ by applying field pulses of steadily increasing length and simply observing in a microscope at what pulse length ($T$) the bubble collapses. Alternatively, one can gradually increase the pulse strength or the bias field, which has the effect of decreasing $(r_0 - r_c)$ until the bubble collapses.[172] The mean wall velocity $V_{ave}$ can then be evaluated roughly

---

**Fig. 5.3.** (a) and (b) Photographs of "rocking" experiment in a EuTmCaGeYIG film at the drives labeled $a$ and $b$ in (c), which shows a plot of reciprocal gradient pulse width versus drive $\Delta H = |2r\nabla H_z|$ for one and two pulses, respectively. (b) shows the "fuzziness" that indicates the onset of Bloch-line punch-through (after Konishi et al. [166]).

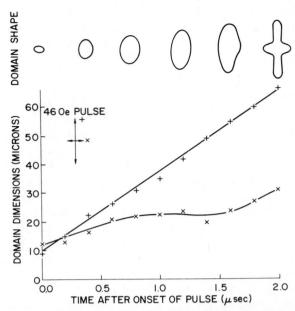

**Fig. 5.4.** Domain shape and dimensions of an isolated bubble in the EuGaYIG film of Fig. 4.1, as a function of time after the onset of a 46 Oe bias pulse 2 μsec long, determined by high-speed laser photography (after Malozemoff and Papworth,[103] copyright Institute of Physics, 1975).

from $V_{ave} = (r_0 - r_c)/T$ and the mobility $\mu = V_{ave}/H_a$ from the dependence of this velocity on pulsed field strength $H_a$. This calculation requires knowledge of $r_0$, which is simply the starting bubble size, and $r_c$, which cannot easily be observed experimentally since it is an unstable condition, but which can be calculated from the magnetostatic theory (see Section 2,B).[51,58,173] Drive fields of hundreds of oersteds can easily be achieved with hand-wound pancake coils or photolithographically configured hairpin loops. Pulse times as short as several nanoseconds can be obtained using commercial pulse generators.[174] Typical experimental results are shown in Fig. 5.5. The SmGdTbIG sample shows straightforward linear-mobility behavior, but the GaYIG sample shows nonlinearity.[163]

A more detailed interpretation of the dynamic bubble-collapse experiment is complicated by the fact that the restoring force is a nonlinear function of wall displacement rather than simply linear ($kq$) as in the small-signal experiments described earlier. The restoring force field $H_k$ is roughly parabolic in form, as sketched in Fig. 2.3a and discussed in Section 2,B, with a maximum magnitude $H_{ko} = H_{col} - H_o$, where $H_{col}$ is the static collapse field and $H_b$ is the dc bias field of the experiment. This maximum occurs roughly half-way between $r_0$ and $r_c$, but the bubble radius must shrink from $r_0$ all the way to $r_c$ before the restoring force changes sign so as to drive the bubble to collapse once the external drive pulse has turned off. Let us ignore wall

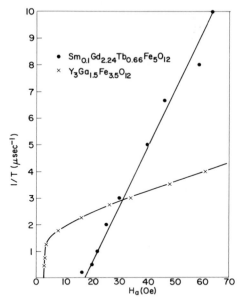

**Fig. 5.5.**   Reciprocal collapse time $T^{-1}$ versus pulsed bias field $H_a$ in a dynamic bubble-collapse experiment in platelets of SmGdTbIG and GaYIG (after Calhoun et al. [163]).

mass and coercivity and assume the domain-wall response is completely characterized by a linear drag $b$ or mobility $\mu = 2M/b$. The instantaneous velocity $v$ is then a linear function of the net drive $H$ on the wall, which is the difference between the pulse field $H_a$ and the restoring field $H_k$ of Fig. 2.3a. Clearly if the applied pulse field $H_a$ is less than $H_{ko}$, the net drive and velocity will drop to zero at some point in the required wall-displacement range and the bubble will not be able to collapse no matter how long the pulse. For pulse fields larger than $H_{ko}$ the net drive is positive throughout the wall-displacement range and the bubble can always collapse if the pulse is made long enough. For very large pulse fields, the percentage variation in the drive due to the restoring force becomes small; so to a first approximation the average velocity over the wall-displacement range is simply

$$V_{\mathrm{ave}} = \mu(H_a - H_i), \qquad (5.11)$$

where

$$H_i = (r_0 - r_c)^{-1} \int_{r_0}^{r_c} H_k(r)\, dr \qquad (5.12)$$

is the average of the restoring force acting to reduce the effective drive field seen by the bubble. If $H_{ko}$ is small, $H_k(r)$ becomes parabolic [Eq. (2.11)] and $H_i = 2H_{ko}/3$. The above analysis gives a simple prescription for analyzing bubble-collapse data[175]: At pulse fields $H_a$ large compared to $H_{ko}$, a plot of $(r_0 - r_c)/T$ versus $H_a$ approaches an asymptote whose slope is the mobility and whose intercept on the field axis is $H_i$. Thus the mobility can be extracted without any detailed fitting procedure. The precise theory including deviations from Eq. (5.11) when $H_a - H_{ko}$ is small, is given by Callen and Josephs.[58]

This simple analysis no longer pertains when the instantaneous velocity-field characteristic $v(H)$ is nonlinear. Let us assume that $v(H)$ is linear up to a critical field $H_n$, where nonlinearities abruptly set in. To avoid these nonlinearities and to measure the linear mobility region it is clearly necessary to apply pulse fields less than $H_n$. But this means $H_{ko}$ must also be made less than $H_n$; otherwise the bubble will not collapse. As a result one must often work at the very edge of the static collapse field where experimental scatter increases, collapse times become inconveniently short and the theoretical values of $r_0 - r_c$ become less reliable.[174,175] If the pulse field is allowed to become greater than $H_n$, then one can have the following complicated situation: Initially the motion is nonlinear, but as the wall moves, $H_k$ increases and the effective drive field drops; so the motion may fall back to the linear regime; finally as the bubble approaches collapse, the drive increases once again, and nonlinear behavior reoccurs. In such a case a computer fit may be required to extract dynamic parameters from the data.[175] Often even a general nonlinear $v(H)$ curve is insufficient to explain the details of the bubble collapse curves because inertial effects may also be involved. For example the effect of a wall mass will be to carry the bubble wall beyond the point it reached at the end of the bias pulse.[176,177] Such effects are further discussed in Section 17,B.[115,172] Another complication in these experiments is that different bubbles are observed to collapse with different collapse times.[175,178,179] Such effects have been attributed to differences in initial wall states as will be discussed more fully in Section 14,C. Most recently, evidence has been obtained by high-speed photography of nonuniform modes of collapse[180] in which one layer, presumably at a surface, switches completely and forms a domain wall capping the bubble. In some cases the cap appears to form even before the wall reaches $r_c$; then the theoretical analysis given above must be affected. It is not yet known how general this collapse process is.

One can envision analogs to the bubble-collapse experiment. For example it is often observed that with pulses of appropriate length and strength, it is possible to cut stripes into bubbles. A plot of inverse pulse length versus strength might be expected to give similar information to bubble collapse. A comparable experiment[103,181,182] has been conducted in the geometry of a regular elliptically distorted bubble array in which bubbles are expanded to cut the effective stripes that separate them, as shown by the high-speed photographs of the transient domain configurations during the drive pulse in Fig. 4.1. At a critical pulse length, the result of the pulse is to effectively switch the topology of the array from black bubbles on a white background to white bubbles on a black background. If the pulse ends too soon, the bubble walls do not make contact and the array relaxes to its original shape when the pulse turns off. If the pulse lasts too long, the dogbone-shaped nuclei in the figure can become cut, leading to a pattern of triangular domains of double the number required for regular topological switching, or the nuclei become annihilated altogether. The drawback of this method is that the restoring force is very large. According to Eq. (2.20) applied fields equal to $4\pi M$ are required to cut stripes. Furthermore, when two walls come together on the scale of a wall width, micromagnetic structures could be formed in which exchange energy dominates the restoring force. For these reasons, the stripe-cutting or topological-switching experiments require extremely large pulse fields—several hundred oersteds in typical 5 $\mu$m garnet materials. The advantages of the bubble-collapse geometry can now be fully appreciated. The instability point is of a purely magnetostatic nature and does not involve any micromagnetic structure on the scale of the wall width. One can envision the development of a "vortex" in the final stages of bubble collapse but the details of its disappearance are irrelevant to the bubble collapse threshold. Also the maximum restoring force is far smaller than in any of its alternatives; so lower drive-field effects can be explored.

## 6. Gradient Propagation of Bubbles: The Case of Zero Restoring Force

### A. PHENOMENOLOGICAL EQUATION

As discussed in Section 2,C, the restoring force on a bubble in a bias-field gradient is zero. This special condition leads to several particularly important experimental techniques in domain-wall dynamics. It is useful to obtain a phenomenological equation for this case, by integrating the oscillator equation (5.1) around the wall of the bubble, taking $k = 0$. We define $X$ as the coordinate of the bubble center along the direction of the gradient $-\nabla H_z$, and take $H_g = r|\nabla H_z|$. $\beta$ is the azimuthal angle of the bubble's cylindrical coordinate system, and is defined counterclockwise relative to the $x$ axis as shown in Fig. 2.5. Then the local field from the gradient at any point on the bubble circumference is $H = -H_g \cos \beta$. We assume as usual that the bubble has magnetization down inside and up outside. Since $q$ in Eq. (5.1) represents a displacement radially inward, $q = -X \cos \beta$. Furthermore each of the terms of Eq. (5.1) represents a pressure, whose component in the $x$ direction may be obtained by multiplying by $\cos \beta$. Integrating this $x$-component of pressure around the bubble circumference, one obtains directly[57,117]:

$$m\ddot{X} + b\dot{X} = 2M(H_g - H_{cb}).$$ (6.1)

This relation is similar to Eq. (5.1), having the identical values of $m$ and $b$. However no restoring force is present and the bubble coercive field is now

$$H_{cb} = 4\pi^{-1}H_c.$$ (6.2)

A common convention in the literature is to define the bubble drive field as the field difference $2r|\nabla H_z|$ between $H_z$ values across the diameter $d = 2r$. This is twice the value of $H_g = r|\nabla H_z|$ used in Eq. (6.1). We adopt the $H_g$ convention because $H_g$ is the actual drive field experienced locally at the front and back of the bubble.

The absence of a restoring force in Eq. (6.1) makes possible long-term steady-state motion at the velocity

$$\dot{X} = 2Mb^{-1}(H_g - H_{cb}).\tag{6.3}$$

The bubble mobility $2Mb^{-1}$ is the same as the wall mobility considered before [Eq. (5.4)]. A similar calculation for an elliptical bubble described by $r(\beta) = r_0 + r_2 \cos 2\beta$ gives the following result for steady-state motion[183,184] to first order in $r_2/r_0$:

$$\dot{X} = 2Mb^{-1}\{H_g[1 + (3r_2/2r_0)] - H_{cb}[1 + (r_2/2r_0)]\},\tag{6.4}$$

where $H_g = r_0|\nabla H_z|$. Elliptical bubbles can occur in materials with in-plane anisotropy or in the presence of in-plane fields.[103,185-187]

While the above results are plausible and have been widely used in the interpretation of experimental data, it is important to recognize that they ignore details of the spin structure of the bubble wall, which give rise to some of the most interesting effects in bubble physics like bubble skew deflection and mobility changes. Also, at high drives, saturation velocities of the form of Eq. (5.5) can occur. A more complete treatment of bubble translation including the spin structure and velocity saturation is given in Chapters VII and IX.

Another configuration in which a domain wall can move with no restoring force is that of an expanding stripe described in Section 2,C. If $H_{ri}$ is the run-in field of a stripe, then the drive field experienced by the stripe head is just $H_b - H_{ri}$, where $H_b$ is the bias field [Eq. (2.7)]. If we make the simple approximation that the stripe head is rectangular, then, in analogy to Eq. (6.3),

$$b\dot{L} = 2M(H_{ri} - H_b - H_c),\tag{6.5}$$

where $L$ is the length of the stripe expanding in one direction. Results for other stripe-head geometries are given in the literature.[188] One problem is that a free stripe subjected to such a drive will not only run out but also buckle and grow side arms as indicated schematically in Fig. 2.5b. Alternatively the stripe can be contracted, in which case buckling does not occur.[189,190] This distortion is also suppressed in a device structure such as a "chevron expander," in which the stripe is localized to a long narrow field trough, created by an array of permalloy bars in the shape of chevrons. Such a structure is widely used in devices to expand bubbles to stripes in preparation for magnetoresistive sensing. The velocity of growth of the stripe can be measured stroboscopically or electronically via the magnitude of the induced magnetoresistance signal in the permalloy detector.[121,191-193]

## B.  PROPAGATING FIELD WELL AND BUBBLE DEVICES

Most present-day bubble devices operate by applying a bias-field gradient to displace the bubble, whose presence or absence, or whose wall state, constitutes a bit of information.[13-18] In a typical "field-access" device, a structure of permalloy elements, consisting for example of periodic T and I shapes (e.g., see Fig. 4.4), is photolithographically patterned on top of a bubble material. An externally applied in-plane field polarizes the permalloy elements, creating localized $z$-field wells in which a bubble can sit, as illustrated schematically in Fig. 6.1. The well may be viewed as representing the average $z$-field acting on the bubble wall, and it is characterized by inflection points where the maximum gradient occurs that can act on a bubble as it moves from side to side in the well. It is also important to note that the stray field of the permalloy inevitably has in-plane components that act on the domain wall as shown in Fig. 6.1. Potential wells for propagating bubbles can also be created by ion-implanted patterns[75,194] in which the magnetization lies in the plane of the film and that can be polarized by an in-plane field. Such devices are called contiguous disk or IIPP devices (see Section 2,E).

As the direction of the in-plane field is rotated by 360°, the T and I shapes are such that the bias field well is caused to move a full period in the structure. A simple example is afforded by the permalloy bar of Fig. 6.1. If the direction of the in-plane field is reversed from $+x$ to $-x$ in the figure, the well moves to the other end of the bar, dragging the bubble with it. Such moving field wells can also be created by an appropriately pulsed array of current lines, giving rise to another class of devices called "current-access" devices.[13-18,195-197] For further details on device design and modeling of the field well, we refer the reader to the literature.[198-201]

Let us now consider a well translating at a velocity $V_d$ determined by the

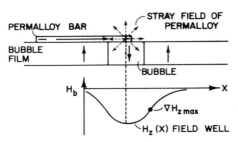

**Fig. 6.1.**  Schematic representation of bubble in a bias-field well created by the stray field emanating from a permalloy bar polarized by an in-plane field.

operating frequency of the device. The bubble will lag the bottom of the well until it experiences a gradient sufficient to cause it to move at $V_d$. Given $\mu$ and $H_{cb}$, the size of this gradient is determined by setting Eq. (6.3) equal to $V_d$. If the required gradient exceeds the maximum gradient that the well can sustain, then the bubble will escape the well and the device will fail. Because the shape of the well is imperfectly known, it is a difficult task to predict high-frequency device operation from a knowledge of wall-dynamics parameters such as $\mu$ and $H_c$. The present status of such predictions is described in Section 21. It is equally questionable to attempt to use high-frequency device operation to deduce information about the wall-dynamics parameters. In principle this could be done by measuring the bubble lag distance or the frequency at which the device fails, but only if the well shape were known. In one approach,[201a] the well depth at various points in the circuit was measured empirically by determining the extra bias field needed to bring a bubble under a permalloy element to the same diameter as it has away from the element. Then a burst of in-plane field can be used to determine the time required for the bubble to move between two such well positions. However because this is a pulse method, it suffers from uncertainty if ballistic overshoot occurs, as will be discussed in the next section.

One case in which the field well shape is known with some certainty is the round permalloy disk.[202,203] An in-plane field induces a pole distribution of the type illustrated in Fig. 6.2. The bubble seeks the point under the edge of the disk where the bias field is minimum. The field well in this case is like a long trough wrapping around the edge of the permalloy disk with a minimum at $\phi = 0$ where $\phi$ is the angular position measured from the point of maximum positive pole density. From the symmetry of the pole distribution, it is clear that the net bias field in the trough may be approximately represented by

$$H = H_b - H_{pd} \cos \phi \qquad (6.6)$$

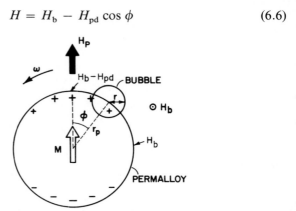

**Fig. 6.2.**   Schematic permalloy disk with bubble and rotating in-plane field.

where $H_b - H_{pd}$ is the bias field at the bottom of the well and $H_b$ is the bias field at $\phi = \pm\pi/2$, which is also the bias field in the absence of the permalloy disk. $H_{pd}$ is measured by subtracting the static collapse field at the bottom of the well from its value without the permalloy disk. If the bubble's stray field has a negligible back reaction on the permalloy poles, then Eq. (6.6) holds even in the presence of the bubble and the gradient for a bubble located at $\phi$ is $\nabla H_z = r_p^{-1} H_p \sin \phi$, where $r_p$ is the radius of the disk. Let us assume $r_p$ is large and ignore the bubble's centrifugal effect as it circulates synchronously with the in-plane field at a circular frequency $\omega$. Assuming steady-state motion and ignoring $H_c$ in Eq. (6.1), we find

$$b\dot{X} = b\omega r_p = 2Mrr_p^{-1} H_p \sin \phi. \tag{6.7}$$

Thus the drag coefficient $b$ and the mobility can easily be found from a plot of rotating frequency $\omega$ versus the sine of the bubble's lag angle $\phi$.

## C. EXTERNAL-GRADIENT TECHNIQUES

In the techniques of the previous section a bubble seeks the field gradient it needs to sustain an externally imposed velocity. By contrast, bubble dynamics can also be studied by determining the bubble velocity in response to an externally imposed field gradient, created for example by an appropriate array of current lines. These current lines can be miniature wires or they can be patterned photolithographically and deposited directly on the sample or else on a glass substrate. For large bubbles such as in orthoferrites, one can use a current-line array in which a central conductor, narrow compared to the bubble, provides an oscillating gradient field, and two outer conductors provide a dc field well to keep the bubble centered.[204] The amplitude of the bubble response is measured by observing the blur in the bubble position normal to the exciting conductor.

For smaller bubbles, one may use a single wide conductor under the bubble or a stripline configuration surrounding the bubble.[117,205-207] The configuration of Fig. 6.3 has proven most popular.[207] Consider first the outer two conductors. Currents of magnitude $I_1$ (in amps) in the same direction through each of the conductors give rise to a bias-field distribution of the type sketched schematically in the figure. In the center of the pattern between the striplines, the gradient (in oersteds/centimeter) is given by

$$|\nabla H_z| = 1.59I_1/d_1^2, \tag{6.8}$$

where $d_1$ is the center-to-center spacing of the conductors (measured in centimeters). If the current is ac, the gradient will oscillate and the amplitude of the bubble displacement can be measured from the blur seen by eye in the microscope[117,208] One problem with this approach is that the bubble tends

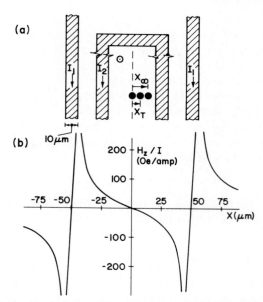

**Fig. 6.3.** (a) Schematic strip-line pattern for pulsed-gradient bubble-propagation experiment. Circles indicate bubble positions before, during, and long after the experiment. (b) Schematic gradient-field distribution created by currents pulsed in the same direction down the two outer lines of part (a).

to drift with time out of the center. Alternatively, one may apply a single gradient pulse of fixed length $T$ and measure the net bubble displacement $X_\infty$ as observed visually in the microscope.[117,207] If one ignores the mass term in Eq. (6.1), the average velocity is just $X_\infty/T$. A typical plot of $V = X_\infty/T$ versus the gradient drive $\Delta H = 2H_g$ is shown in Fig. 6.4 for a LuGd-AlIG film.[208] The linear slope at low drives gives a bubble mobility $\mu = 2M/b$ of 3200 cm/sec Oe and half of the back-extrapolated intercept gives a coercive field of $H_{cb} = 0.5$ Oe. This bubble-propagation experiment has been one of the most popular techniques of bubble-domain dynamics because of its apparent simplicity and applicability to a wide range of device materials. However the technique has many problems that were not apparent in the early years of its use.

One problem is that as the bubble moves in the gradient, the average bias field changes (see Section 2,C) and therefore the radius changes. To compensate for this effect,[207] one may use the inner loop of Fig. 6.3, which, when pulsed with a current $I_2$ (in amperes), gives a bias field (in oersteds) in the center of the striplines of

$$H_z = 0.80 I_2/d_2, \qquad (6.9)$$

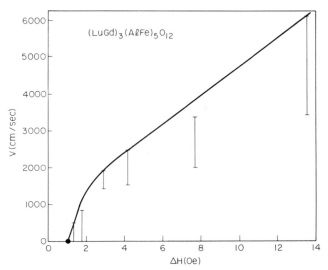

**Fig. 6.4.** Apparent velocity versus drive $\Delta H = |2r\nabla H_z|$ in a conventional pulsed gradient propagation experiment for a bubble in a LuGdAlIG film (after Vella-Coleiro[209]).

where $d_2$ is the center-to-center spacing of the inner conductors (measured in centimeters). The bubble is positioned at the center line of the gradient to start with and the bias field change during the ensuing motion is compensated by applying a triangular pulse that rises gradually but turns off abruptly at the moment the gradient pulse turns off. The required amplitude for the bias compensation pulse is $X_T|\nabla H_z|$, where $X_T$ is the displacement of the bubble at the moment the gradient pulse turns off. In early work it had been assumed that $X_T$ was simply the final displacement of the bubble $X_\infty$, that is, that the bubble stopped moving when the gradient pulse turned off.

High-speed photography has been used to investigate the instantaneous bubble position during the gradient pulse, and it has been found that contrary to initial expectations, there can be large differences between $X_T$ and $X_\infty$ in high mobility films,[210,211] as illustrated for a EuGaYIG film in Fig. 6.5. The results show that the bubble keeps moving a long distance even after the gradient pulse turns off. Such a motion is called "ballistic overshoot" (see Sections 18 and 19). Furthermore the true average velocity $X_T/T$ appears saturated as a function of drive field, suggesting that Eq. (5.5) is a better representation of the bubble motion than Eq. (6.1). These effects invalidate the conventional interpretation of $X_\infty/T$ as a velocity and also complicate the achievement of correct bias compensation. They imply that to correctly interpret bubble propagation experiments, one cannot rely on visual observation but must use some sort of high-speed photography.

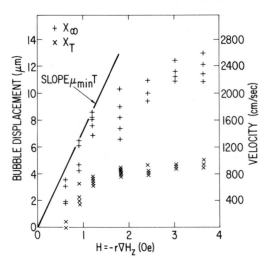

**Fig. 6.5.** Bubble displacement versus gradient drive $H = |r\nabla H_z|$ in a EuGaYIG film in a bubble-propagation experiment with high-speed photography. $X_T$ is the displacement at the end of a pulse of length $T = 0.5$ μsec, and $X_\infty$ is the displacement long after the pulse. The difference $X_\infty - X_T$ indicates ballistic overshoot (after Malozemoff and De Luca[210]).

The single-shot high-speed photography method for observing bubble propagation is quite effective because in addition to determining bubble displacement, it also reveals any transient skew deflection or bubble-shape distortion. However it does not give information as a direct function of time for a single propagation. An alternative method,[130] which does this, is illustrated in Fig. 6.6. A mask with a rectangular hole is placed at the focal plane of a microscope. If the long dimension of the rectangle is approximately the bubble diameter, domain contrast gives a photometric signal proportional to the position of the bubble as it moves under the hole. Sufficient signal-to-noise has been achieved in an 8 μm thick SmCaGeYIG sample to give the single-shot oscilloscope trace indicated in the figure, showing bubble position as a direct function of time. In this case, a delay was observed in the initiation of bubble motion relative to the onset of the gradient field pulse, as will be discussed further in Section 19,B. This method has also been applied to measure phase delays in devices.[211a]

Another complication in the gradient-propagation experiment is the large amount of scatter in bubble displacement $X_\infty$ observed particularly in high mobility films at high drives as illustrated in Figs. 6.4 and 6.5. In uncompensated experiments, some of the scatter can be attributed to bubble runout caused when the bubble moves far enough in the gradient to lower the average bias field below the runout field.[128,212] Scatter also arises if care is not taken

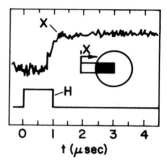

**Fig. 6.6.** Pulsed gradient propagation experiment in a SmCaGeYIG film. The lower trace represents a 1 μsec field pulse of strength $H = |rVH_z| = 1.25$ Oe. The upper trace is a one-shot photometric signal, proportional to bubble position, from the bubble moving under a mask with the rectangular window indicated schematically (after Vella-Coleiro[130]).

to prepare the bubble in a consistent and careful way. This usually means that the bubble must initially be propagated back and forth with weak gradients to insure that its motion is regular and consistent and then it must be brought up, usually in the same direction as the eventual propagation, as close as possible to the exact center of the stripline pattern.[175,213] The reason for this procedure is to avoid inconsistencies in the bubble wall structure, as discussed in Chapters IV and IX. Even if all these precautions are taken, however, scatter still can occur in bubble-propagation experiments, and possible reasons for it are addressed in Section 19,B.

Another bubble-propagation geometry that is particularly useful in characterizing when scatter does or does not occur, is shown in Fig. 5.3. This experiment is called a "rocking experiment" because a bubble is rocked back and forth repetitively by a series of gradient pulses.[165] Thus it is essentially a repetitive bubble-propagation experiment. The bubble is prevented from wandering by a shallow field well created by dc current run in series as a loop around the two inner rectangular offset conductors of the figure. A gradient pulse is applied by means of the outer striplines to propagate the bubble to a new position where it stays because the field-well gradient is weaker than coercivity. Then a fixed sequence of weak reverse gradient pulses is applied to move the bubble back to its starting position. The timing of this sequence is adjusted using a pulse word generator so that the bubble spends most of its time at the two end points. Thus if the sequence of pulses is repeated at a sufficient rate, the eye will only perceive the two end points. This technique permits a very rapid and accurate measurement of $X_\infty$ as a function of the strength and length of the gradient. In principle, if a strobing light source were used, one could also trace out the bubble position as a direct function of time. The averaging inherent in the rocking experiment permits more accurate determination of $X_\infty$ than in the conventional propagation experiment because operator error in initial positioning of the bubble is eliminated. On the other hand if scatter is an intrinsic feature of the motion, then the bubble positions appear blurred,[166] as illustrated in Fig. 5.3b. It has been discovered that the blurring sets in quite abruptly as a function of gradient drive, and the significance of such observations is discussed in Section 19,B.

Yet another complication in gradient propagation experiments is that contrary to initial expectations and the implications of Eq. (6.1), bubbles do not always move parallel to the gradient, i.e., perpendicular to the gradient striplines, as illustrated in Fig. 5.3. Often there is a skew deflection which can be in different directions for different bubbles. If $X_\infty$ is the bubble displacement along the gradient and $Y_\infty$ is the displacement perpendicular to the gradient (parallel to the striplines), then the skew deflection angle $\rho$ is simply

$$\rho = \arctan(Y_\infty/X_\infty). \tag{6.10}$$

Such deflections indicate the existence of distinct bubble wall states associated with distinct wall magnetization distributions (see Chapters IV and VII), and there has been considerable interest in determining the skew angles accurately. An improved method for doing this is the bubble "deflectometer," [187,214] in which a regular array of striplines are placed on top of the bubble material. Let us number these lines 1 through $n$. If the bubble is initially near line 2, then lines 1 and 3 can be activated to apply a gradient to move the bubble to the potential well at the edge of line 3. Then lines 2 and 4 can be activated to move the bubble to the edge of line 4, and so forth. Thus the bubble can be made to translate over a large distance, and the transverse $(Y_\infty)$ and forward $(X_\infty)$ displacements can be measured very accurately.

A related method of performing a gradient-propagation experiment is the rotating gradient ("cyclotron") method[215] illustrated in Fig. 6.7. A combination of static and sinusoidal currents are applied to the array of four conductor lines in such a way as to create a static potential well in the center and a time-varying bias field gradient whose orientation in the plane changes with time at a radial frequency $\omega$:

$$\nabla H_z(t) = |\nabla H_z|(-\hat{x} \sin \omega t + \hat{y} \cos \omega t). \tag{6.11}$$

A bubble responds by moving in a circle, as indicated schematically in the figure. The radius $r_p$ of the trajectory can be measured visually, and the phase lag $\theta_p$ of the bubble motion relative to the rotating gradient drive can be measured photometrically. These quantities are determined by the viscous

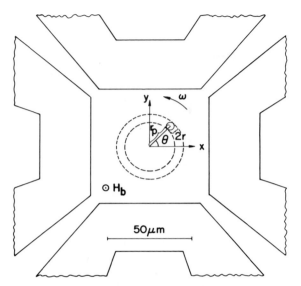

**Fig. 6.7.** Schematic conductor pattern and path of a bubble in a rotating gradient experiment (after Jones et al.[215]).

and coercive drags [Eq. (6.1)], by the skew deflection effect mentioned above and by the centrifugal force $m_T r_p \omega^2$ arising from the total bubble mass $m_T$. The latter is of particular interest because a "steady-state" bubble mass is not easily determined from other experiments. By balancing tangential and radial forces, one obtains the following equations:

$$r|\nabla H_z| \cos \theta_p = \mu^{-1} \omega r_p + H_{cb}, \qquad (6.12)$$

$$r|\nabla H_z| \sin \theta_p + r|dH_r/dr_p| = \pm 2S\omega r_p \gamma^{-1} r^{-1} + (m_T r_p \omega^2 / 4\pi Mhr). \qquad (6.13)$$

Here $dH_r/dr_p$ is the gradient of the static potential well containing the bubble. The term $\pm 2S\omega r_p \gamma^{-1} r^{-1}$ represents the effective gradient of the skew deflection and is discussed more fully in Sections 13 and 14. The sign depends on the direction of propagation. Clearly if the bubble skews to the right, the orbit will be larger for counterclockwise than for clockwise rotation. It is essential to perform the experiment with both senses of rotation, so as to separate the skew deflection term from the mass term in Eq. (6.13), since both terms contribute to the radial force. Thus from measurements of the size of the orbit $r_p$ and the phase lag $\theta_p$, and knowing $\nabla H_z$, $\omega$, $dH_r/dr_p$ and $r$, one can in principle extract $\mu$, $H_{cb}$, $S$, and $m_T$.

## D. DOMAIN DRAG EFFECT

A novel method of applying a drive field to a domain wall is by means of a current flow in a conducting bubble material or in a conducting material placed on top of a nonconducting bubble material.[216–221b] A current can of course give rise to a field distribution that acts on a domain wall. However, the current will also be perturbed by the magnetic fields associated with the wall, for example, through the Hall effect (Lorentz force) or magneto-resistance. It turns out that this perturbation can cause additional drive fields on the wall. Such drive fields are "self-induced" in the sense that currents that are uniform in the absence of a wall will be deflected in such a way as to create drive fields once the wall is present. In the case of the Hall effect, the self-induced drive has been called a "domain drag" effect because the current acts to drag the bubble in the same direction as the net current-carrier motion.

To give a simple model for the drive force arising from the Hall effect,[216] let us consider a regular array of parallel plane domain walls with spacing $d$ small compared to other dimensions, as shown in Fig. 6.8. In this case surface demagnetizing effects can be neglected, so that $B = M$ (we use mks units only in this section). Further, consider an average current density $j_x$ flowing in the material perpendicular to the walls in the $x$ direction. The Hall effect causes the current to deflect in a sense depending on the direction of the magnetization, as indicated in the figure. Thus there is a component of current in the $+y$ direction for up domains but in the $-y$ direction for down domains (assuming electron carriers). If $j_y$ is the *magnitude* of this component, it is clear from Ampere's law that the magnitude of the drive field at the wall is

$$H = j_y d/2 \tag{6.14}$$

and that it alternates in sign from wall to wall in such a way as to exert a force in the $-x$ direction on all the walls.

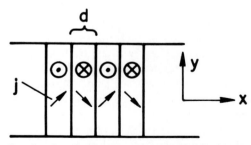

**Fig. 6.8.** Schematic stripe-domain pattern and current distribution in a strip of conducting bubble material with a Hall effect (after Berger[217]).

To estimate the size of this force, we must determine $j_y$ from Maxwell's equations and Ohm's law. We assume a resistivity $\rho$ and a tangent $\beta$ of the Hall angle, which, for the ordinary Hall effect, is given by

$$\beta = M/ne\rho, \tag{6.15}$$

where $n$ is the current carrier density and $e$ is the electronic charge. We also define $v_e$ as the average drift velocity of the carriers in the $x$ direction, such that

$$j_x = nev_e. \tag{6.16}$$

Ohm's law (including the Hall effect) determines $j_y$ in terms of $j_x$ and the electric field $E_y$ by

$$E_{y1,2} = \pm\rho(j_y + \beta j_x) \tag{6.17}$$

where 1,2 and $\pm$ refer to up and down magnetic domains, respectively. If the domain walls are not moving, then $E_y$ must be continuous across the domain wall, and so $j_y = -\beta j_x$. More generally, if the domain wall is moving at a velocity $v_x$,

$$E_{y1} - E_{y2} = 2Mv_x, \tag{6.18}$$

where $2Mv_x$ is the rate of change of flux across the wall. Combining Eqs. (6.14)–(6.18), we obtain the result

$$H = H_{do}(1 - v_x v_e^{-1}) \qquad H_{do} = v_e Md/2\rho = \beta dj_x/2. \tag{6.19}$$

While this formula applies to a parallel array of domain walls, a similar result has been derived for the case of an isolated bubble.[217] Furthermore, because of the long-range nature of the magnetic fields, there is no need for the current to flow in the magnetic medium itself. Considering a separate conducting layer lying on top of an insulating bubble material, one finds once again a result similar to Eq. (6.19), provided $d$ is taken to be the thickness of the conducting layer and is less than either the domain spacing or magnetic film thickness.[218,220]

The magnitude of the domain-drag drive field $H_{do}$ in Eq. (6.19) can be compared to the characteristic field $H = dj_x/2$ parallel to the surface of a current sheet of thickness $d$, which is a rough measure of the bias field variation created across a bubble of diameter $d$ by the current $j_x$. According to Eq. (6.19), $H_{do}$ is always smaller than this characteristic field by the factor $\beta$, which is usually considerably less than 1.

Assuming a semiconducting layer on top of a garnet film, we use the typical parameters $\rho = 10 \ \mu\Omega$ cm, $\mu \equiv (ne\rho)^{-1} = 2.6 \ \mathrm{m^2/V}$ sec, $M = 0.02 \ \mathrm{Wb/m^2}$ ($4\pi M = 200$ G), $d = 5 \ \mu$m, and $j = 5 \times 10^8 \ \mathrm{A/m^2}$, and we find $H_{do} = 0.8$ Oe and $v_e = 130$ m/sec. In this case $\beta = 0.05$. If no forces opposed the domain wall motion ($H = 0$), a wall velocity equal to $v_e$ would be required for dynamic equilibrium in Eq. (6.19). However, for a coercive field $H_c$ and domain wall mobility $\mu$, the field of Eq. (6.19) must equal $H_c + \mu^{-1}v_x$, which implies[220]

$$v_x = \mu(H_{do} - H_c)(1 + \mu H_{do}v_e^{-1}) \approx \mu(H_{do} - H_c), \qquad (6.20)$$

since usually $\mu H_{do} \ll v_e$. Thus one should see a domain wall velocity determined primarily by $H_{do}$. The difficulty in observing this effect is that $H_{do}$ is generally small because $\beta$ is small, and the domain pattern can all too easily be destabilized by Joule heating and the large transverse gradient from $j_x$ (see above).

The above discussion is appropriate to the case of a semiconductor on top of an insulating bubble material, but so far a domain-drag effect in this geometry has not been reported. Recently however, domain drag has been reported in amorphous GdCoMo films by passing the current directly through the bubble film itself.[221a,221b] To apply Eq. (6.19) to this case, $\beta$ must be interpreted as the anomalous Hall angle, which is of order $5 \times 10^{-3}$ for GdCoMo. Current was pulsed down photolithographically patterned strips of the amorphous material, using extremely short pulse times ($\lesssim 50$ nsec), so that very high current densities (up to $3 \times 10^{10}$ A/m$^2$) could be achieved without excessive heating. Motion of bubble rafts ($d \sim 1$ $\mu$m) was observed along the length of the strip at velocities in rough agreement with Eq. (6.19) and (6.20). The direction of motion reverses for films on either side of magnetic compensation because the sign of $\beta$ reverses. It is too early to say whether this novel effect might have application in devices. In a related phenomenon based on magnetoresistance rather than the Hall effect, a bubble can be dragged perpendicular to the current[219,221]; however this effect is even more difficult to verify because it is also small and in this case the effect is indistinguishable from that of a transverse gradient. Another related effect is a magnetoresistance arising from the presence of domain walls, which has been detected in a study on cobalt metal at 4.2 K.[222]

E.   SUMMARY OF DYNAMICAL TECHNIQUES

To conclude this chapter we note that this plethora of experimental techniques has provided a rich diet of results for the physicist to digest. The remaining chapters of this review will describe some of the theories that have been proposed for this purpose. Nevertheless, this variety leaves the materials researcher and device engineer somewhat at a loss for choosing the most effective technique for evaluating their materials in a simple way. It turns out that for low-mobility materials (e.g., $\mu < 500$ cm(sec Oe) most of the techniques described in this chapter will give consistent values of mobility and the choice should be dictated by convenience. Pulsed field-gradient bubble-propagation experiments usually require photolithographically patterned striplines, but if these are available, the technique is very straight-forward. In the absence of striplines, bubble collapse using small hand-wound coils is to be recommended. If rf circuitry and fast photodetection apparatus is available, then magnetooptic magnetic susceptibility techniques are convenient for thin films, while inductive techniques are better for bulk samples.

Unfortunately, most device interest centers on high-mobility materials ($\mu > 500$ cm/sec Oe), and for these the interpretation of the experiments is difficult. The linear mobility becomes hard to determine by any technique because of the early onset of nonlinearity, and the most reliable measure-ments have been done either in the presence of very large in-plane fields or on special materials having large in-plane anisotropy, high $g$-factors, or Dzialoshinski exchange, all of which suppress nonlinearity. In such cases, bubble propagation, bubble collapse, and photometric techniques are equally easy and the choice can again be dictated by convenience. On the other hand for high-mobility materials without these special conditions, the most important parameter for devices is the bubble saturation velocity at moderate drives ($< 10$ Oe), because this, rather than the mobility, limits bubble device operation (see Section 21). One method for determining the saturation velocity in device materials is by a bubble propagation experiment with high-speed photography. Bubble propagation around a permalloy disk should also be possible and has the virtue of simulating the device en-vironment most closely, particularly by the presence of an in-plane field. This method also avoids high-speed photography. However it has not yet been demonstrated for measuring the saturation velocity. The saturation velocities of collapsing or expanding bubbles and stripes are not necessarily the same as those of bubbles translating at low drives; so techniques involving these geometries are mostly of interest for physics experiments rather than device material characterization. Of course the ultimate test of a material is operation in the device itself. Stripe expansion in chevron expanders has given particularly straightforward data on domain-wall velocity.

# IV

## Domain–Wall Statics

In Section 2 we briefly reviewed domain statics, treating domain walls simply as surfaces carrying surface energy. In this chapter we investigate in detail the static structure of the domain wall, breaking up the discussion into three parts: the one-dimensional model, multidimensional structures involving Bloch lines, and finally transitions between structures and capping layer effects. By and large our knowledge is a complex pyramid of theoretical deductions because the static wall structure is usually not directly accessible experimentally. Experimental justification for the deduced structures has mostly come from dynamic experiments described in more detail in Chapters V–IX.

### 7. One-Dimensional Model

We start by reviewing the conventional theory for a static one-dimensional domain wall. We ignore the finite thickness of the film and consider an infinite space with the coordinate axes $xyz$ illustrated in Fig. 7.1. The space is filled with the bubble material of magnetization $4\pi M$ and uniaxial anisotropy $K$ with the easy axis along the $z$ direction. In applying this model to a bubble material, the $xy$ plane would represent the film plane. The space is partitioned by a domain wall centered at $y = 0$ and extending in the $xz$ plane. The wall separates two semi-infinite spaces, the one on the left $(y < 0)$ representing a domain with magnetization pointing up along positive $z$, and the one on the right $(y > 0)$ representing a domain with magnetization pointing down. The problem is to determine the structure of the wall, that is, how the magnetization rotates from $+z$ at $y = -\infty$ to $-z$ at $y = +\infty$.

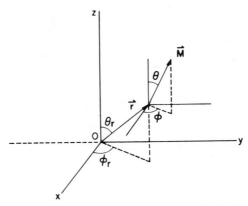

**Fig. 7.1.**  Coordinate axes for one-dimensional wall model (Section 7) and Bloch point (Section 9,A).

Clearly the uniaxial anisotropy favors an abrupt change of direction at $y = 0$ because in that way no spin need point in the hard anisotropy direction in the plane. On the other hand exchange stiffness favors a gradual change of orientation. Thus one may expect that the width of the transition region will be some compromise between these two competing effects.

To solve the problem analytically, we begin by ignoring demagnetizing energy and write the local exchange energy [Eq. (1.11)] and anisotropy energy density [Eq. (1.14)] in terms of polar coordinates $(\theta, \phi)$ defined relative to the $z$ axis as shown in Fig. 7.1:

$$w = w_A + w_K = A[(\partial\theta/\partial y)^2 + (\sin\theta\ \partial\phi/\partial y)^2] + K\sin^2\theta. \qquad (7.1)$$

The form of the exchange energy here is specific to the one-dimensional model in which $\theta$ and $\phi$ vary only along the $y$ axis. The physical requirement for static equilibrium is that the torques on every spin arising from the anisotropy and exchange be zero. The equivalent mathematical condition is that $\int w\,dy$ be stationary with respect to variations of the functions $\theta(y)$ and $\phi(y)$. This condition is expressed by the Euler equations

$$\delta w/\delta\phi = \delta w/\delta\theta = 0, \qquad (7.2)$$

where $\delta w/\delta\phi$ and $\delta w/\delta\theta$ are "functional derivatives," defined by

$$\delta w/\delta\theta = \partial w/\partial\theta - \mathbf{V}\cdot\partial w/\partial\mathbf{V}\theta. \qquad (7.3)$$

Thus, substituting Eq. (7.1) in (7.2), we find the differential equations

$$(\partial^2\phi/\partial y^2)\sin^2\theta + (\partial\phi/\partial y)(\partial\theta/\partial y)\sin 2\theta = 0, \qquad (7.4)$$

$$2A(\partial^2\theta/\partial y^2) - [K + A(\partial\phi/\partial y)^2]\sin 2\theta = 0. \qquad (7.5)$$

A solution of these equations satisfying the boundary conditions $\theta(\pm\infty) = (\pi, 0)$ is

$$\phi(y) = \psi = \text{const} \tag{7.6}$$

$$\theta(y) = \pm 2 \arctan \exp(y/\Delta_0), \tag{7.7}$$

where

$$\Delta_0 = (A/K)^{1/2}. \tag{7.8}$$

This is the well-known solution for the structure of a 180° domain wall, illustrated in Figs. 7.2 and 7.3. $\theta$ varies in a smooth fashion as **M** rotates from up to down. Most of the rotation is concentrated in a region of width $\pi\Delta_0$ around the center. Thus $\Delta_0$ is called the wall width parameter; it is a measure of the wall width and shows the expected competition between $A$ tending to increase the wall width and $K$ tending to decrease it. If we substitute the solutions (7.6) and (7.7) in the energy expression (7.1) and integrate over all space, we find the total energy per unit area of the wall

$$\sigma_0 = 4(AK)^{1/2}. \tag{7.9}$$

Exchange and anisotropy contribute equally to this total. Another useful relation is obtained from the derivative of Eq. (7.7):

$$\Delta_0 \partial\theta/\partial y = \sin\theta. \tag{7.10}$$

We note that in the above solution for the wall structure, the value of the azimuthal angle $\phi(y) = \psi$ is arbitrary because so far we have ignored all energy terms except uniaxial anisotropy and exchange.

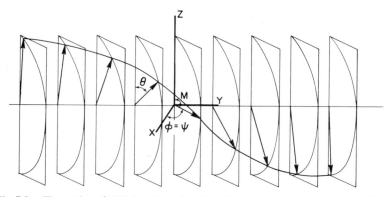

**Fig. 7.2.** Illustration of 180° domain wall with arbitrarily constrained constant wall-moment angle $\psi$. The plane of the wall is $x$–$z$. The spontaneous magnetization **M** lies parallel to a plane inclined at the constant angle $\psi$ to the $x$–$z$ plane. The polar angle $\theta(y)$ increases monotonically from $\theta(-\infty) = 0$ to $\theta(\infty) = \pi$.

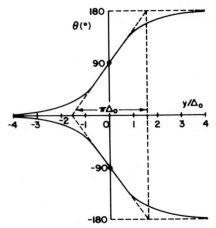

**Fig. 7.3.** Solution $\theta(y)$ for one-dimensional static wall structure (Section 7). The substitution $\theta \to \psi$, $y \to x$, and $\Delta_0 \to \Lambda_0$ gives the structure of an isolated Bloch line (Section 8,A).

Next we consider the additional effects of (1) local magnetostatic energy arising from magnetic charges within the wall when $\phi = \psi \neq 0$, (2) an "in-plane" field $H_p$ oriented in the $xy$ plane at an angle $\psi_H$ to the wall plane, and (3) an "in-plane" anisotropy $K_p$ whose easy axis lies at an angle $\psi_p$ to the wall plane. By "in-plane" we refer to the plane of the bubble film that corresponds to the $xy$ plane of our model. The total local energy density is now [Eqs. (1.6), (1.9), and (1.17)]

$$w = A[(\partial\theta/\partial y)^2 + (\sin\theta \partial\phi/\partial y)^2]$$
$$+ [K + 2\pi M^2 \sin^2\phi + K_p \sin^2(\phi - \psi_p)]\sin^2\theta$$
$$- MH_p \cos(\phi - \psi_H)\sin\theta. \tag{7.11}$$

We note that if $MH_p$ is small compared to $K + 2\pi M^2 + K_p$, Eq. (7.11) tends to a form like Eq. (7.1). Therefore to first order in $H_p$ we may assume $\phi = \psi$ and the wall satisfies equations of the form (7.7) or (7.10) with $\Delta_0$ replaced by a quantity $\Delta$ to be determined. Using Eqs. (7.10) and (7.11) one can now evaluate the local wall energy

$$\sigma = \int_{-\infty}^{+\infty} w \, dy = \int_0^\pi w(\partial\theta/\partial y)^{-1} \, d\theta$$
$$= 2A\Delta^{-1} + 2\kappa\Delta, \tag{7.12}$$

where

$$\kappa = K + 2\pi M^2 \sin^2\psi + K_p \sin^2(\psi - \psi_p) - (\pi MH_p/2)\cos(\psi - \psi_H). \tag{7.13}$$

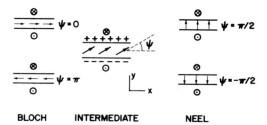

BLOCH      INTERMEDIATE      NEEL

**Fig. 7.4.** Schematic Bloch and Néel wall structures as seen from $+z$ direction.

Minimizing Eq. (7.12) with respect to $\Delta$ one finds

$$\sigma = 4(A\kappa)^{1/2} \qquad \Delta = (A/\kappa)^{1/2} = 4A/\sigma. \qquad (7.14)$$

Comparing Eqs. (7.13–7.14) with (7.8–7.9), we see that the energy terms in $\psi$ simulate a contribution to $K$ in their effects on $\sigma$ and $\Delta$. We emphasize that Eqs. (7.13) and (7.14) are valid only to first order in $H_p$. Furthermore in bubble materials $K$ usually exceeds both $2\pi M^2$ and $K_p$. Therefore it is useful to expand Eqs. (7.14) to first order:

$$\Delta = \Delta_0[1 - (1/2)Q^{-1}\sin^2\psi - (K_p/2K)\sin^2(\psi - \psi_p)$$
$$+ (\pi MH_p/4K)\cos(\psi - \psi_H)] \qquad (7.15)$$

$$\sigma = \sigma_0 + 4\pi M^2\Delta_0\sin^2\psi + 2K_p\Delta_0\sin^2(\psi - \psi_p)$$
$$- \pi\Delta_0 MH_p\cos(\psi - \psi_H), \qquad (7.16)$$

where

$$Q = K/2\pi M^2.$$

To interpret these results let us first ignore $K_p$ and $H_p$ and consider the magnetostatic term in Eq. (7.16). This wall energy is equivalent to that of a plate of thickness $2\Delta_0$ with an average in-plane wall magnetization whose orientation is specified by the angle $\psi$ as indicated schematically in Fig. 7.4. The component of wall magnetization $M_n = M\sin\psi$ perpendicular to the wall gives rise to magnetic charge at the surfaces, which gives rise to a demagnetizing field $H_d = 4\pi M\sin\psi$ and hence a demagnetizing energy $\frac{1}{2}M_nH_d = 2\pi M^2\sin^2\psi$ per unit volume of the plate [see Eq. (1.9)]. Equation (7.15) shows that as demagnetizing energy increases, the wall width squeezes down. Clearly the minimum wall energy occurs for $\sin^2\psi = 0$, i.e., for $\psi = 0$ or $\pi$, in which case the wall magnetization stays in the plane of the wall, thus avoiding any surface charges. This means that as the spins rotate through the wall, they exhibit what might be likened to a screw rotation. Such walls are called "Bloch walls" and they are the most common type of wall in bubble materials because they minimize demagnetizing energy. Symmetry permits two possible senses of screw rotation, right- and left-handed, corresponding to $\psi = 0$ and $\pi$, respectively, both types having equal energy (see Fig. 7.4).

By contrast, the magnetostatic energy in (7.16) is maximum if $\psi = \pm\pi/2$. In this case the spins rotate through the wall in a "head-on" fashion as illustrated schematically in Fig. 7.4. Such walls are called "Néel walls" and also exist in two types. Néel walls or other intermediate configurations with $\psi \neq 0$ or $\pi$ may become favored over Bloch walls if we include the in-plane field and anisotropy terms of Eq. (7.16). Consider for example the case of no in-plane field ($H_p = 0$) but an in-plane anisotropy with its easy axis oriented perpendicular to the wall plane ($\psi_p = \pi/2$). In this case energy minimization of $\sigma(\psi)$ shows straightforwardly that for $K_p < 2\pi M^2$, the Bloch orientation is preferred, but for $K_p > 2\pi M^2$, the Néel orientation ($\psi = \pm\pi/2$) is preferred. A remarkable situation occurs when $K_p = 2\pi M^2$: the wall energy is independent of $\psi$, so that Bloch and Néel walls or any intermediate orientation are equally favored. In this case the wall magnetization may be termed "labile," i.e., it can turn freely. Consider next the case of no in-plane anisotropy but an in-plane field oriented perpendicular to the wall ($\psi_H = \pi/2$, $H_p = H_y$). Minimizing $\sigma(\psi)$, we find

$$\psi = \begin{cases} \arcsin(H_y/8M), & |H_y| \le 8M, \\ \pm\pi/2, & |H_y| \ge 8M. \end{cases} \tag{7.17}$$

Thus as $|H_y|$ increases, the wall magnetization is pulled away from the Bloch orientation, and for $|H_y|$ greater than $8M$ the Néel orientation is preferred. If $H_y = 8M$ and $\psi = \pm\pi/2$, there is a "labile" spin condition because $d^2\sigma/d\psi^2 = 0$ under these conditions. A field $H_x$ in the plane of the wall ($\psi_H = 0$) stabilizes the parallel Bloch orientation ($\psi = 0$), but it destabilizes an antiparallel orientation ($\psi = \pi$) above the critical field

$$H_{x1} = 8M, \tag{7.18}$$

above which the magnetization spontaneously flops into the direction of $H_x$.

The effect of cubic anisotropy $K_1$ on wall energy in the garnets is negligible for usual epitaxial (111)-oriented garnet films.[40] The reason is that $K_1$ is usually an order of magnitude smaller than $K$. Moreover, we see from the energy expression (1.15) that the torques $\partial w_{K_1}/\partial\phi$ are odd about the $xy$ plane of a bubble film ($\theta = \pi/2$). Thus the first-order correction obtained by integrating $\partial w_{K_1}/\partial\phi$ over the wall structure vanishes.

A special situation occurs in the orthoferrites because of Dzialoshinski exchange $\mathbf{D} \cdot \mathbf{m}_1 \times \mathbf{m}_2$ described in Section 1,B. For small canting angle $\beta$

and small deviation $(\phi - \phi_D)$ of the net magnetization from the ortho-rhombic ac plane denoted by $\phi_D = \psi_D$, one finds an approximate energy density term[223]

$$w_D = D\beta(\phi - \phi_D)^2 \sin^2 \theta. \tag{7.19}$$

Combining this with the usual energy terms considered earlier, one finds for the wall energy of the orthoferrites

$$\sigma = \sigma_0 + 4\pi M^2 \Delta_0 \sin^2 \psi + \pi \Delta_0 M H_p \cos(\psi - \psi_H) + 2D\beta\Delta_0(\psi - \psi_D)^2, \tag{7.20}$$

where $\psi$ denotes the "in-plane" angle of wall magnetization as measured from the $x$ axis in the wall plane, as before. Because of the size of $D$, the Dzialoshinski term dominates the other terms in determining the orientation of $\psi$, and so $(\psi - \psi_D)$ is small as required in the derivation. In this case we can visualize the rotation of magnetization through the wall as in Fig. 1.1b. Whether a Bloch or a Néel wall is preferred depends primarily on whether the domain-wall normal lies along the $b$ or $a$ orthorhombic axis, respectively, for the Dzialoshinski exchange effectively forces the magnetization to rotate in the ac plane. The torques due to the Dzialoshinski term may be likened to a large in-plane anisotropy of magnitude $D\beta$ with its easy axis along $\psi_D$. However the Dzialoshinski term differs from an effective in-plane anisotropy in one important respect: as $\psi - \psi_D$ increases during wall motion (see Section 11,C) or in a Bloch line (see Section 8,A), the wall-moment plane no longer lies perpendicular to the Dzialoshinski vector $\mathbf{D}$, and thus the canting angle is reduced. At $\psi - \psi_D = \pi/2$ the net magnetization disappears altogether.

## 8.    Bloch-Line Statics

Once one goes beyond the one-dimensional model of wall structure, to a two- or three-dimensional model, the number of possible structures is vastly increased. The first experimental evidence for more complex structures came from Lorentz micrographs,[143-145] as illustrated in Figs. 4.5 and 4.6. Two different kinds of wall contrast can be discerned in these figures, corresponding to the two polarities of Bloch wall described in the previous section. However in certain places there is a reversal of contrast, as for example in Fig. 4.5c, indicating a changeover from one Bloch-wall polarity to another. Presumably the line separating the two different Bloch-wall polarities runs vertically through the film. Therefore this structure has been termed a "vertical Bloch line." Chains of such structures are visible in Fig.

4.6. Such collections of large numbers of vertical Bloch lines are believed to account for the properties of "hard" bubbles, "dumbbells," and other abnormal bubble domains. Bloch lines can also run horizontally along the wall of a bubble or in a variety of curved configurations. In this section we describe the statics of such structures. The implications for dynamics are discussed in Chapters VI–IX.

## A. ISOLATED BLOCH LINES

Here we derive the static structure of an isolated Bloch line in an infinite wall.[143,224–226] The results are of great utility in practical problems involving bubbles whenever the dimensions of the wall surface are large compared to the Bloch-line width. This condition, often called the "Bloch-line approxima-tion," is well fulfilled in typical bubble films. Calculations going beyond this approximation are described in Section 17. The structure of a static Bloch line can be understood in analogy to the structure of the static Bloch wall. Let us consider a bubble material in an $xyz$ coordinate frame as in Fig. 8.1 with a wall in the $xz$ plane. We assume a right-handed Bloch wall ($\psi = 0$) to the left ($-x$) and a left-handed Bloch wall ($\psi = \pi$) to the right ($+x$) with a transition region or "Bloch line" running along the $z$ axis in the center. One way the in-plane wall magnetization can rotate from $\psi = 0$ to $\psi = \pi$ to accomplish this transition is illustrated in the figure: by a gradual rotation of $\psi$, such that at the center of the transition there is a small Néel wall segment.

Two types of magnetic pole distributions occur within this Bloch-line transition region.[227] One type is due to the $y$ component of magnetization and gives rise to the usual local demagnetizing energy in this region. This part of the pole distribution is called a $\pi$-charge because, in analogy to $\pi$-orbitals of molecules, it has odd symmetry about the midplane of the wall. There are also poles, called $\sigma$-charge, arising from the convergence into the Bloch-line region of the $x$ components of magnetization occurring in the

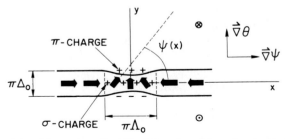

**Fig. 8.1.**   Schematic vertical Bloch line viewed from the $+z$ direction.

flanking Bloch-wall regions. The $\sigma$-charge gives rise to a nonlocal demagnetizing energy in the sense that this energy arises from the interaction of poles at different positions on the wall. The $\sigma$-charge gives rise to local demagnetizing energy as well. For an isolated Bloch line, the local demagnetizing energy from $\pi$-charge is dominant over that from $\sigma$-charge in the high-$Q$ limit because, as will be seen below, the wall width then becomes small relative to the Bloch-line width. The local energy tends to narrow the Bloch-line width, much as uniaxial anisotropy tends to narrow a Bloch wall. However exchange tends to widen the transition just as it widens the wall. The equilibrium width of the Bloch line is expected to be a balance of the two competing effects. We note that uniaxial anisotropy plays no role in the Bloch-line width in the high-$Q$ limit because everywhere along the $x$ axis, the in-plane wall magnetization has equal anisotropy energy.

To solve for the Bloch-line structure analytically, we write the local energy density including anisotropy and demagnetizing energy and also exchange energy for variations of magnetization direction along both $x$ and $y$:

$$w = A(\partial\theta/\partial y)^2 + [K + A(\partial\phi/\partial x)^2 + 2\pi M^2 \sin^2 \phi]\sin^2 \theta. \tag{8.1}$$

Following the treatment of Section 7, we minimize $\int w \, dy$ with respect to the function $\theta(x,y)$ for arbitrary $\psi(x)$ to find a wall width parameter

$$\Delta = \Delta_0[1 + \Delta_0^2(\partial\psi/\partial x)^2 + Q^{-1} \sin^2 \psi]^{-1/2}, \tag{8.2}$$

and a wall energy

$$\sigma = \sigma_0[1 + \Delta_0^2(\partial\psi/\partial x)^2 + Q^{-1} \sin^2 \psi]^{+1/2}, \tag{8.3}$$

where $\psi$ is the azimuthal angle of the wall magnetization measured from the wall plane. Assuming the $\psi$-dependent terms in (8.3) are small, we can expand to first order to obtain

$$\sigma = \sigma_0 + 2A\Delta_0(\partial\psi/\partial x)^2 + 4\pi\Delta_0 M^2 \sin^2 \psi. \tag{8.4}$$

Now the Euler equation for the equilibrium static Bloch-line structure is $\delta\sigma/\delta\psi = 0$ where once again $\delta/\delta\psi$ represents a functional derivative defined by Eq. (7.3). Substituting Eq. (8.4) we find the differential equation for $\psi(x)$:

$$2A(\partial^2\psi/\partial x^2) - 2\pi M^2 \sin 2\psi = 0. \tag{8.5}$$

This equation has the same form as the $\theta(y)$ equation (7.5) for a Bloch wall. Thus we find

$$\psi(x) = \pm 2 \arctan \exp(x/\Lambda_0), \tag{8.6}$$

where

$$\Lambda_0 = (A/2\pi M^2)^{1/2}. \tag{8.7}$$

Equation (8.6) is illustrated in Fig. 7.3 provided we read $\psi$ for $\theta$, $x$ for $y$, and $\Lambda_0$ for $\Delta_0$. We see that $\psi$ varies in a smooth fashion as it rotates from one Bloch-wall orientation to the other. However, most of the rotation is concentrated in a region of width $\pi\Lambda_0$ around the center. Thus we define $\Lambda_0$ as the isolated Bloch-line width parameter. In Eq. (8.7) it shows the expected competition between $A$ tending to increase the line width and $2\pi M^2$ tending to decrease it. If we substitute Eq. (8.6) in (8.4) and integrate over the wall, we find an energy per unit length of the isolated Bloch line

$$E_{\mathrm{LO}} = 8AQ^{-1/2}. \tag{8.8}$$

Equation (8.6) and Fig. 7.3 indicate there are two possible types of Bloch line corresponding to the two different senses of twist or handedness, analogous to the two different types of Bloch wall.

Bloch lines can be expected to occur wherever there is a dividing line between wall regions of opposite Bloch-wall polarities. They can run vertically along the $z$ axis, perpendicular to the film, as assumed in the above model and as shown in Figs. 8.1 and 8.2a. These are called "vertical Bloch lines." They may also run horizontally, along the $x$ axis of the model; such lines are called "horizontal Bloch lines," as shown in Fig. 8.2b. In complex dynamical processes (Chapters VIII and IX) one expects generally shaped "Bloch curves" bounding regions of different Bloch-wall polarities, with the wall magnetization rotating smoothly from one Bloch-wall polarity to another across the Bloch line. The topological possibilities can become exceedingly complex. In all these cases Eq. (8.6) still applies provided one considers a high-$Q$ limit and provided in-plane field, Bloch-line curvature or other perturbing energy terms can be neglected. One merely replaces the coordinate $x$ of Eq. (8.6) by the coordinate normal to the Bloch line in the plane of the wall. For example, a horizontal Bloch line centered at $z = 0$ in an infinite medium and running along the $x$ axis is described by

$$\psi(z) = \pm 2 \arctan \exp(z/\Lambda_0). \tag{8.9}$$

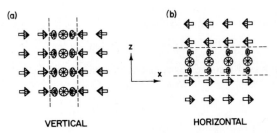

**Fig. 8.2.** Schematic vertical (a) and horizontal (b) Bloch-line spin structures at the midplane of a wall viewed from the $-y$ direction.

Figure 8.2 contrasts vertical and horizontal Bloch-line structure and illustrates an important difference between the two, namely that while both kinds have $\pi$-charge, only the vertical Bloch line has $\sigma$-charge. This difference becomes important in considering Bloch-line interactions (see Section 8,B).

The sense of twist of any Bloch line can be characterized by the two possible directions of the unit vector $\mathbf{t}$ tangent to the Bloch line. $\mathbf{t}$ is defined by

$$\mathbf{t} = (\nabla\phi \times \nabla\theta)/|\nabla\phi \times \nabla\theta| = \mathbf{g}/|\mathbf{g}|, \tag{8.10}$$

evaluated at the geometrical center of the Bloch line. Here $\theta$ and $\phi$ are the usual polar coordinates of the magnetization, shown in Fig. 7.1 and defined relative to the $z$ direction of uniaxial anisotropy. For future reference, $\mathbf{g}$ is the "gyrovector" defined in Sections 9,A and 12,F. ($\mathbf{t}$ is independent of whether $z$ is chosen up or down provided only that the polar coordinates are chosen to be right-handed.) For example, in the special case of the vertical Bloch line shown in Fig. 8.1, $\nabla\phi$ and $\nabla\theta$ point along $+x$ and $+y$, so $\mathbf{t}$ points along $+z$.

Let us next consider the effect of in-plane fields on the isolated Bloch-line structure: Clearly any component of in-plane field $H_x$ parallel to the wall favors one Bloch-wall polarity at the expense of another. According to Eq. (7.16), a displacement of the Bloch line by $\xi$ along a coordinate axis perpendicular to the Bloch line must lead to an energy change per unit length of $-2\pi\Delta_0 MH_x\xi$. Therefore the force per unit length tending to displace the Bloch line is

$$F_{\mathrm{L}} = 2\pi\Delta_0 MH_x. \tag{8.11}$$

The Bloch line will continue to move until it gets to a point on the wall where the in-plane field is perpendicular to the wall or where it runs in to other Bloch lines. Such effects have been observed in electron microscopy experiments, where vertical Bloch lines are observed to sit at opposite ends of bubbles, on a diameter parallel to the in-plane field.[143]

Turning now to the component of in-plane field perpendicular to the wall, we find that in this case the problem is a purely static one, requiring the solution of $\delta\sigma/\delta\psi = 0$ with $\sigma$ given by the usual exchange and magnetostatic terms [Eq. (8.4)] plus an in-plane field term $-\pi\Delta_0 MH_y \sin\psi$. We assume the Bloch-line and field polarities shown in Fig. 8.3, with positive $H_y$ defined to point along the Néel wall segment of the Bloch line. We quote the results of this calculation without a detailed proof.[224] There is no Bloch line at all if $H_y > 8M$, for the entire wall becomes a Néel wall. For $|H_y| \leq 8M$, the Bloch-wall magnetization is canted toward the field direction and the Bloch line must represent a spin rotation by $\Phi$, where

$$\Phi = 2|\cos^{-1}(H_y/8M)| \tag{8.12}$$

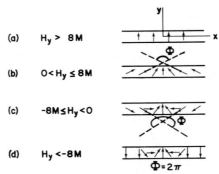

**Fig. 8.3.** Schematic wall structures with a vertical Bloch line in an applied in-plane field $H_y$. (d) Illustrates a $2\pi$ line.

is the angle of Bloch-line rotation (see Fig. 8.3). In this case the extra Bloch-line energy per length is found to be[224]

$$E_L = 8AQ^{-1/2}[\sin(\Phi/2) - (\Phi/2)\cos(\Phi/2)], \qquad |H_y| \le 8M, \quad (8.13)$$

which reduces to $8AQ^{-1/2}$ as found earlier for $H_y = 0$. For $H_y \le -8M$, the in-plane field creates a Néel wall with a $2\pi$-Bloch line "locked in" (see 8.3d). In this case $\Phi = 2\pi$, and one finds[226]

$$E_L = 16AQ^{-1/2}\{[(H_y/8M) - 1]^{1/2} + (H_y/8M)\sin^{-1}(8M/H_y)^{1/2}\}, \quad (8.1$$

$$H_y \le -8M.$$

These equations show that the Bloch-line energy increases with $\Phi$ and $-H_y$. In the presence of an in-plane field, $\psi(x)$ no longer has the simple form of Eq. (8.6). For this case we define the Bloch-line width as $\pi\Lambda \equiv \Phi/\psi'$, where $\Phi$ is given by Eq. (8.12) and $\psi'$ is the derivative of $\psi(x)$ at the center of the Bloch line. Then one finds

$$\Lambda = 2\pi^{-1}\Lambda_0[1 - (H_y/8M)]^{-1}\cos^{-1}(H_y/8M), \qquad |H_y| \le 8M. \quad (8.15)$$

This result shows that the Bloch-line width increases for $H_y$ applied along the direction of magnetization of the Bloch line, and it diverges when $H_y \to 8M$, where the Bloch line disappears. All the above results apply equally to straight vertical or horizontal Bloch lines in the high-$Q$ limit.

Similar effects on Bloch-line energy and structure arise with in-plane anisotropy. For the easy axis parallel and perpendicular to the wall plane, respectively, we find the same form of $\psi(x)$ as in Eq. (8.6), but with a width parameter

$$\Lambda = (A/|2\pi M^2 \pm K_p|)^{1/2} \quad (8.16)$$

and the energy per unit length

$$E_L = 8AQ^{-1/2}(|1 \pm (2\pi M^2/K_p)|)^{1/2}. \tag{8.17}$$

At $K_p = 2\pi M^2$ the Bloch-line width is seen to diverge and the energy goes to zero. For directions of the easy axis intermediate between parallel and perpendicular, the Bloch-line energy scales monotonically. Thus we find that if the wall is curved, as in a bubble, there can be a force on a vertical Bloch line. The Bloch line will seek the point on the bubble wall where its energy is minimized, which occurs where the in-plane easy axis is perpendicular to the bubble wall. However, there is no force on the Bloch line if the wall is straight, in contrast to the in-plane field case where a field $H_x$ in the plane of the wall can cause a Bloch line to move.

It is appropriate to note at this point that Bloch lines of the type considered above cannot exist in orthoferrites, for the magnetization disappears at $\psi - \psi_D = \pi/2$ (Section 7). One must visualize a line of zero magnetization separating two Bloch or Néel wall polarities, which would be the analog of the Bloch line for the orthoferrite case. Such a line should have a high energy and there is as yet no experimental evidence for its existence.

## B. INTERACTIONS BETWEEN BLOCH LINES

Two principal interaction mechanisms between Bloch lines arise from magnetostatic and exchange energies, respectively.[227] With respect to magnetostatic energy, which dominates at long distances, Fig. 8.4 makes clear that any arrangement of vertical Bloch lines at large distances comprises a set of alternating (north–south) line pole distributions arising from the $\sigma$-charges of the alternating Bloch-wall orientation. Therefore, a pair of neighboring vertical Bloch lines at a large distance $s$ always attract. In the limit $h \gg s$ the attractive force is $8(\pi\Delta_0 M)^2/s$ per unit length.

The effect of exchange energy can also be visualized from Fig. 8.4a. If two neighboring Bloch lines of the same handedness approach each other, they clearly repel each other because exchange energy increases. The combination is like a double twist in a rubber band in that there is no continuous way the spin structure can unwind unless the boundary conditions, that is, the spin orientations at the ends of the segment shown, are released. The net rotation of wall magnetization is $2\pi$. On the other hand, if the two Bloch lines have opposite handedness, then it is apparent from Fig. 8.4b that the combination can unwind continuously and thus attain the lower energy state of a pure Bloch wall. The net rotation is zero. Whether or not they annihilate spontaneously in practice depends on how close the Bloch lines are to each other

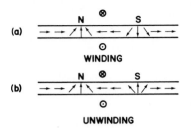

(a)

**Fig. 8.4.**  Schematic wall structure with winding (a) or unwinding (b) Bloch-line pair viewed from the $+z$ direction. N and S denote north and south poles of the $\sigma$-charge distribution (see Fig. 8.1).

initially and on the presence of other influences such as an external field which may hold them apart.

An array of vertical Bloch lines having a common twist sense evidently may be stable. Consider the periodic Bloch-line array in the uniform-twist model described by the expression

$$\psi(x) = \pi x/s, \tag{8.18}$$

with the separation $s$ to be determined. Curiously, the magnetostatic energy contribution arising from $\sigma$-charges just cancels that from the long range part of $\pi$-charge.[227] Thus one needs to consider only the local contribution from $\pi$-charge, which is correctly included in Eqs. (8.2) and (8.3) with the substitution of Eq. (8.18). To first order in $Q^{-1}$ the average energy density from Eq. (8.3) is written

$$\bar{\sigma} = \sigma_0[1 + (\pi \Delta_0/s)^2 + (2Q)^{-1}]^{1/2}. \tag{8.19}$$

The excess energy per Bloch line compared to the energy of the simple Bloch-wall structure is $s(\bar{\sigma} - \sigma_0)$. Minimization of this quantity with respect to $s$ provides the equilibrium separation $s = s_{eq}$, with

$$s_{eq} = \sqrt{2}\,\pi[1 + (2Q)^{-1}]^{-1/2}\Lambda_0 \tag{8.20}$$

for a large finite cluster of vertical Bloch lines in an infinite plane wall.[227]

The structure of a periodic array of vertical Bloch lines has also been computed by means of the Ritz procedure utilizing a highly general ansatz function.[227] It agrees well with Eq. (8.18) for $s \leq s_{eq}$. Since the case $s > s_{eq}$ is anyway unstable with respect to segregation of the Bloch lines into multiple clusters, it follows that the uniform twist model of the Bloch-line array is adequate in all realistic applications. The case $s < s_{eq}$ arises when geometrical constraints squeeze the Bloch lines together and prevent $s(\bar{\sigma} - \sigma_0)$ from attaining its minimum value, as in the hard bubbles and dumbbells described in Section 8,D. When $s$ lies in the even smaller range $s < \pi\Delta_0$, then Bloch-line exchange energy dominates over anisotropy and stray-field energies,[228] and the average wall thickness diminishes since the average of Eq. (8.2) reduces to

$$\bar{\Delta} = \Delta_0[1 + (\pi\Delta_0/s)^2 + (2Q)^{-1}]^{-1/2}. \tag{8.21}$$

## C. BUBBLE STATES WITH SMALL NUMBERS OF VERTICAL BLOCH LINES

We consider a closed domain such as the bubble shown in Fig. 2.5, stabilized by an upward $(+z)$ bias field. There are many possible static equilibrium spin structures or "states" for such a domain,[179] as described by the wall magnetization angle $\psi(\beta,z)$, defined relative to some given direction in the film plane, called the $x$ direction. In general $\psi(\beta,z)$ will be a function of $\beta$, the angle around the perimeter of the domain as defined in Fig. 2.5, and $z$, the position through the thickness. In the remainder of this section we shall enumerate some of these possible states. Because of the great variety of Bloch-line structures that can be accommodated in a bubble wall, there are a great variety of wall states to consider. To a reader unfamiliar with the field, the variety may initially seem staggering, but in fact almost all the states to be described here have been experimentally identified. In addition to their fundamental interest, the states have an added practical significance as candidates for wall-state coding in bubble-lattice devices, and also in their effect on the dynamic stability of more conventional isolated-bubble devices.

For convenience in describing the states we introduce a shorthand notation $(S, l, p)^\alpha$, where $S$ is the "winding" or "revolution" number, $l$ the number of vertical Bloch lines, $p$ the number of Bloch points, and $\alpha$ an index $\pm 1$ describing complementary pairs of states (as explained below). Next we will discuss the winding number $S$ and its relation to $l$ and $\alpha$. A description of the index $p$ and the phenomenon of Bloch points is given in Section 9.

The concept of the winding number [125] is based on the simple fact that if the spin structure is continuous along the perimeter of the closed domain, then $\psi(\beta)$ must change by an integral multiple of $2\pi$ as $\beta$ varies from 0 to $2\pi$. We call this integral multiple $S$. It can be defined as[179]

$$S = (2\pi)^{-1} \int_{\beta=0}^{2\pi} d\psi \tag{8.22}$$

$$= (2\pi)^{-1} \oint (d\psi/ds)\, ds,$$

where $s$ is the arc-length coordinate along the perimeter (see Fig. 2.5) and $\oint ds$ is the contour integral taken counterclockwise around the perimeter. For example, let us consider the case of a bubble with a pure Bloch wall of a single handedness all the way around the circumference as indicated in

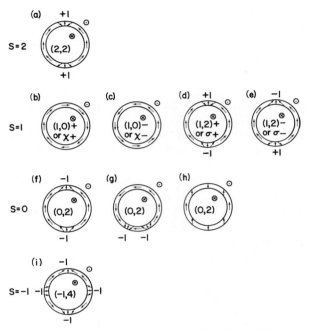

**Fig. 8.5.** Schematic illustrations of selected bubble wall states viewed from the $+z$ direction (bias-field direction). Conventional state notation is indicated inside the bubble, and the winding number $S$ [Eq. (8.22)] is given on the left. $\pm 1$ beside each Bloch line is the handedness index [Eq. (8.23)].

Figs. 8.5b and 8.5c. In this case $\psi(\beta) = \beta \mp \pi/2$ for right-handed and left-handed Bloch walls, respectively. Therefore $d\psi = d\beta$ and substitution in Eq. (8.22) gives $S = 1$ for both cases. This result means the magnetization rotates *one* full turn around the wall of the bubble. We call these two states the "chiral states" and refer to the two possibilities as right or left-handed "chiralities." In terms of our notation, $S = 1$, $l = 0$, $p = 0$, and by convention $\alpha = \pm$ for left- and right-handed chiralities, respectively, so that the two states are denoted $(1,0,0)^+$ and $(1,0,0)^-$ or $1^+$ and $1^-$ for short. (The Bloch-point number $p$ is consistently omitted in Fig. 8.5.) Alternatively the notation $\chi^+$ and $\chi^-$ has been used. (The convention for $\alpha$ assumes the handedness is determined with the thumb in the direction of the magnetization inside the domain. This convention, adopted earlier,[229] means a positive-chirality bubble has a left-handed wall according to the definition of Section 7.) It is obvious that in the absence of in-plane fields or in-plane anisotropy, these are the lowest energy wall states of the bubble. Experimentally these states are observed in Lorentz microscopy as the two cases of light and dark bubbles in which the contrast is unvarying around the bubble (see Fig. 4.5).

If vertical Bloch lines are present in the bubble, we may define a "handedness index" or "signature" for the $i$th Bloch line

$$n_i = \mathbf{t}_i \cdot \mathbf{z}_0, \tag{8.23}$$

where $\mathbf{t}_i$ is the Bloch-line tangent unit vector defined in Eq. (8.10) and where the unit vector $\mathbf{z}_0$ represents the direction of magnetization *outside* the bubble. Since $n_i = \pm 1$, vertical Bloch lines in a closed domain may be termed either "positive" or "negative." Note that this definition relies on the existence of a closed domain with an "outside" and an "inside"; otherwise $n_i$ has no meaning and only $\mathbf{t}_i$ may be used to characterize the Bloch-line handedness. Examples of the signs of vertical Bloch lines are given in Fig. 8.5. The total number of Bloch lines $l$ is given by

$$l = \sum_i |n_i|. \tag{8.24}$$

Because of the extra twists in $\psi(\beta)$ that the Bloch lines represent, it is obvious that the Bloch lines can affect the winding number $S$. The Bloch lines each represent a twist of $\psi$ by $\pi$ while the net winding of $\psi$ around the bubble must be an integral number times $2\pi$. Therefore the number of Bloch lines in a bubble must be even, that is, the Bloch lines come in pairs. One finds

$$S = 1 + \tfrac{1}{2}\sum n_i. \tag{8.25}$$

For example, if a pair of Bloch lines are both of negative sense as illustrated in Fig. 8.5f, then $S = 0$. Such a state is denoted (0,2,0) or (0,2) for short and carries no $\alpha$ index. It is obvious that the $S$-state is independent of the location of the Bloch lines around the wall of the bubble, e.g., whether they are on opposite sides as in Fig. 8.5f or on the same side as in Fig. 8.5g. It is interesting to note that while the chiral states considered above have cylindrical symmetry, the (0,2) state has mirror symmetry along the Bloch-line axis when the Bloch lines lie on opposite sides of the bubble (see Fig. 8.5f). This fact is useful in understanding why $S = 0$ bubbles have no gyrotropic deflection tendency in gradient-propagation experiments (see Section 14,B). The $S = 0$ state is expected to be the lowest energy state in the presence of sufficiently large in-plane fields or in materials with large in-plane anisotropy such as orthoferrites or (110)-garnets, in which case the wall moments align with the preferred direction as shown in Fig. 8.5h, and $\psi(\beta)$ is constant.

Another type of state can occur in a bubble with two Bloch lines of opposite senses, as illustrated in Figs. 8.5d and 8.5e. Since $n_1 = -n_2$, it is clear that $S = 1$. The states can be denoted $(1,2,0)^+$ and $(1,2,0)^-$ or $(1,2)^+$ and $(1,2)^-$ for short, where the index refers to the complementary cases in which the Bloch-line polarities point inward (Fig. 8.5d) or outward (Fig. 8.5e). These states have also been denoted $\sigma^+$ and $\sigma^-$. Such states are energetically favored over the chiral states in an in-plane field because of the considerable Bloch-wall energy reduction: one side of a chiral bubble has wall magnetization lying antiparallel to the in-plane field (see Section 8,E for more details) while both sides of the (1,2) bubbles can have a parallel orientation. However the energy of the (1,2) bubble is still a little higher than that of a (0,2) in an in-plane field because of the antiparallel orientation of one of its Bloch lines. It turns out that (1,2) bubbles tend to be formed from (1,0)'s during gradient propagation and they have a remarkable automotion property (see Section 20,B). We shall also have occasion to consider bubbles with larger numbers of such unwinding Bloch-line pairs ($l$ up to 20), all of which have the same $S$ number (see Sections 18 and 19).

Next consider a state with two Bloch lines, both of positive sense, as illustrated in Fig. 8.5a. From Eqs. (8.22) and (8.24) we can denote the state (2,2). Larger numbers of positive Bloch lines can lead to larger positive $S$ values. Similarly large numbers of negative Bloch lines can lead to large negative $S$ numbers. $|S|$ values of 50 or more, corresponding to 100 or more Bloch lines, have been reported (see Section 8,D). Bubbles with two Bloch lines have been observed by Lorentz microscopy,[143,144] as indicated in Fig. 4.5c. Whether this bubble is a (0,2), (2,2), or (1,2) cannot easily be determined from the micrograph alone because its resolution is insufficient to distinguish the signatures of Bloch lines. Evidence of larger Bloch-line accumulations is shown in Fig. 4.6. In principle these Bloch-line clumps have Bloch lines all of the same handedness, or they would spontaneously unwind and disappear.

Isolated Bloch lines or small groups of lines appear to have essentially no effect on the static properties of bubbles in high-$Q$ materials. The reason is that their energy contributes only a constant independent of radius to the wall energy of the bubble, while the bubble radius is affected by only those energy terms which depend on radius. Typically several tens of Bloch lines can be accommodated on a 5 $\mu$m garnet bubble before the Bloch lines begin to interact sufficiently to affect the radius. The critical $S$ number is determined approximately by $2\sqrt{2\pi\Lambda_0}|S - 1| = 2\pi r$ according to Eq. (8.20).

However Bloch lines have a dramatic effect on bubble dynamics, and this will be the principal topic of Chapters VI–IX. It is useful to anticipate this discussion and mention the key experiments that allow one to identify the bubble state. In the bubble-propagation experiment, bubbles are sometimes found to propagate at a skew angle $\rho$ to an applied field gradient (e.g., see Fig. 5.3). Theory shows (see Sections 13,B and 14,B) that the skew angle is directly proportional to the winding number $S$ according to

$$\tan \rho = 2S\Lambda_0/r\alpha \qquad (8.26)$$

in the limit of zero coercivity. Thus measurement of skew angles has been the main experimental method of determining $S$. Furthermore bubbles are sometimes found to displace in response to a uniform bias pulse (see Section 19,C). Such a displacement $X$ can be related to the number of pairs $n$ of unwinding Bloch lines ($n = n_+ = n_- > 0$) initially located on opposite sides of the bubble:

$$X \sim 2n\Delta_0/\alpha. \tag{8.27}$$

In the presence of an in-plane field, such a displacement can be repeated ad infinitum in a process called "automotion" (see Section 20).

## D.  HARD BUBBLES AND DUMBBELLS

Statically observable experimental effects arise when the net winding number becomes so large that the vertical Bloch lines can no longer be accommodated around the circumference at greater than their equilibrium spacing [see Eq. (8.20)].[125,228,230-236] The Bloch-line structure in this case is schematically illustrated in Figs. 8.6b and 8.7. The circumference of the bubble now forms a geometrical limitation providing the boundary condition which forces the Bloch lines to squeeze to a separation

$$s = \pi r/(S - 1) \approx \pi r/S, \tag{8.28}$$

where $r$ is the radius of the bubble and $S$ ($\gg 1$) is the winding number. In this case the wall energy from Eq. (8.19) is

$$\bar{\sigma} = \sigma_0[1 + (S\Delta_0/r)^2 + (1/2Q)]^{1/2}. \tag{8.29}$$

Since this energy decreases as $r$ increases, corresponding to a loosening of the tension in the Bloch-line chain, this term gives rise to a pressure causing the bubble to expand in size as compared to a normal bubble and causing

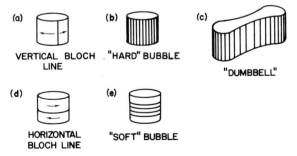

**Fig. 8.6.**  Schematic bubble wall structures with Bloch lines.

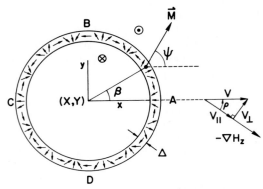

**Fig. 8.7.** Schematic wall magnetization distribution for a hard bubble ($S \gg 1$) (after Slon-czewski *et al.*[125]).

the static bubble collapse field to be increased. To calculate these effects one uses the modified normalized force equation (see Eq. 2.9)[233–235]:

$$(\sigma_0/4\pi M^2 h)[1 + (2S\Delta_0/d)^2 + (1/2Q)]^{1/2} + (d/h)(H_b/4\pi M) - F(d/h) = 0.$$

$$(8.30)$$

An example of data on the diameter versus bias field of various bubbles in a EuErGaIG sample and a fit to Eq. (8.30) is given in Fig. 8.8.[232] The various curves are labeled by a winding number $N \equiv S - 1$, which is determined from the fit. Values of $S$ up to 90 are found, corresponding to 180 Bloch lines in a bubble whose size can range from 4 to 13 $\mu$m in diameter. It may seem incredible that such large numbers of Bloch lines, all of the same polarity, can be obtained. The mechanism of their formation is considered in Section 13,C. Because of their higher collapse fields and larger sizes, these bubbles have been termed "hard bubbles."[230] They have also been called "Bloch-line bubbles,"[231] "extraordinary bubbles,"[234] "anomalous bubbles,"[233] and, if they have less than the maximum collapse field, "intermediate bubbles."[230]

Equation (8.30) shows that in hard bubbles there are *two* forces, namely the Bloch-line repulsion force and the magnetostatic force, which can balance the contracting forces of Bloch-wall surface tension and bias field to give a statically stable bubble. In thin films these forces have sufficiently different diameter dependences that two separate stable solutions can be obtained at a given bias field, both having the same number of Bloch lines.[237] Such bistable bubble states have been confirmed experimentally in 1 $\mu$m thick films of EuLuCaGeYIG.[237] The larger bubble had a diameter of roughly 5 $\mu$m, while the small bubble had a diameter of roughly 0.5 $\mu$m. It should be

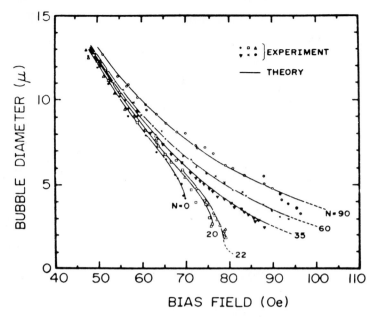

**Fig. 8.8.**   Bubble diameter versus bias field for hard and normal ($N \approx 0$) bubbles in a EuEr-GaIG film. Solid line gives the theory according to Eq. (8.30) (after Kobayashi *et al.*[234]).

emphasized that such bistable bubbles have the same number of Bloch lines, hence the same wall state, in contrast to the different bubble wall states arising from different numbers or dispositions of Bloch lines, as described in the previous section.

If the number of Bloch lines becomes even larger, the bubble diameter may become so large that the bubble will distort into a stripe or "dumbbell" shape illustrated in Fig. 8.6c. Now a long isolated stripe without any Bloch lines has an energy of approximately $2(H_b - H_{ri})MwLh$, where $w(H_b)$ is the width of the stripe, $L$ is its length, and $H_{ri}$ is its run-in field [see Eq. (2.17)]. Since the net Bloch-line exchange energy, using Eq. (8.4), is approximately

$$W_A = 4\pi^2 A\Delta_0 hS^2/L, \tag{8.31}$$

we can minimize the total energy to find the relation

$$L = \pi S\Lambda_0[4\pi M\Delta_0/w(H_b - H_{ri})]^{1/2}, \tag{8.32}$$

which is valid when $H_b$ is close to $H_{ri}$. Thus the stripe length depends linearly upon the number of Bloch lines (twice the winding number $S$) and inversely on the square root of $(H_b - H_{ri})$, as found experimentally.[125,231]

Although the theory suggests that these hard-bubble and dumbbell shapes are "quantized" into states specified by the index $S$, one finds that for typical parameters of garnet or amorphous films, the difference in dimensions between different states is of order 0.1 $\mu$m or less. Thus in practice no convincing quantization of static properties of hard bubble or dumbbell states has been observed. The mechanism of collapse of hard bubbles appears to be different from the purely magnetostatic instability of normal bubbles discussed in Section 2,B.[232,235,238] As the bias field is increased above the normal collapse field, the Bloch lines become squeezed together, increasing the wall energy far above its normal value $\sigma_0 = 4(AK)^{1/2}$. If it is assumed that spontaneous annihilation of Bloch lines can occur at a critical local energy density $w_c$, the observed hard bubble collapse fields can be accounted for consistently, using $w_c$ of order $10^5$ ergs/cm$^3$, which is almost an order of magnitude larger than anisotropy energy in typical 5 $\mu$m garnet films. Bubble collapse occurs because once the Bloch lines annihilate, the bubble is no longer stable above its normal collapse field. Further discussion of Bloch-line annihilation is given in Section 9,B. The remarkable dynamic properties of hard bubbles and dumbbells are described in Section 13.

### E.  STRAY FIELDS AND HORIZONTAL BLOCH LINES

In our discussion of vertical Bloch lines, we have ignored any variations in wall structure through the thickness of the film. Here we consider such variations, which arise even in films of uniform composition because of stray demagnetizing fields near the film surfaces.[224,239–245] Such stray fields are illustrated schematically by the dashed lines in Fig. 8.9. The fields point perpendicular to the wall plane, and for an isolated straight wall they have the approximate analytic form[224]

$$H_y(z) = 4M \ln[z/(h - z)], \tag{8.33}$$

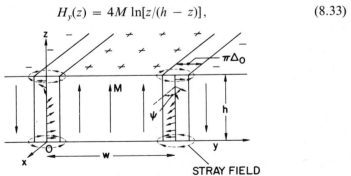

**Fig. 8.9.**  Schematic stray field and wall magnetization distribution in walls of a stripe array (after Slonczewski[255]).

where $z$ is taken to be zero at a film interface as indicated in the figure. This form is appropriate for the high-$Q$ or "thin wall" limit. Equation (8.33) is modified in the presence of neighboring walls, as for instance in a stripe array of stripe width $w$, for which[242]

$$H_y(z) = 4M \ln\{\tanh(\pi z/2w)/\tanh[\pi(h - z)/2w]\}. \qquad (8.34)$$

In both cases the stray field is zero at the center of the wall and diverges logarithmically in opposite directions ($+$ or $-y$) near the two surfaces. The logarithmic divergence at the surfaces in Eq. (8.33) no longer applies within a wall width $\pi\Delta$ of the surface.[226] It is important to recognize that these stray fields are an unavoidable concomitant of bubble-material domain structure. If one attempted to eliminate them, for example by means of highly permeable in-plane layers on both surfaces, the result would be to destabilize the domain structure.

What is the effect of the stray fields on the wall structure? Let us first ignore the effect of exchange and consider only stray-field and demagnetizing energy. The stray fields tend to twist the wall magnetization away from the Bloch-wall direction as shown schematically in Fig. 8.9. The twisting is maximum near the surfaces. The equilibrium positions for $\psi(z)$, calculated from Eqs. (8.33) and (7.17), are shown by the solid lines in Fig. 8.10. The periodicity of this figure along the $\psi$ axis reflects the fact that $\psi + 2\pi$ is equivalent to $\psi$. Let us consider the two possible $\psi(z)$ contours indicated by $A$ and $B$. These correspond roughly to the two Bloch-wall polarities of the one-dimensional model. However, the figure shows that the wall structure is "twisted" near the surfaces by the stray field, and the pure Bloch-wall orientation ($\psi = 0, \pi$) occurs only in the midplane of the film. For stray fields above $8\,M$, the Néel orientation ($\psi = \pm\pi/2$) is preferred, and there is thus a band of Néel wall at the surfaces of the film. The orientation of the Néel sections is such as to provide a certain amount of "flux closure," as illustrated schematically in Fig. 8.9. The points $z = h/(1 + e^2)$ and $he^2/(1 + e^2)$ (where $e = 2.718$) are determined by setting $|H_y| = 8M$ in Eq. (8.33). These are "critical points" where, as mentioned earlier with reference to Eq. (7.17), the wall magnetization is labile because small changes in $\psi$ from $\pm\pi/2$ cause no change in energy. In a bubble such points correspond to "critical circles," which ring the bubble near the surface and interface of the film. The critical points are of particular importance as Bloch-line nucleation centers in bubble dynamics (see Section 15).

Because $\psi$ varies as a function of $z$, the wall exchange energy $2A\Delta(\partial\psi/\partial z)^2$ must also be taken into account. Intuitively, this causes the $\psi(z)$ contour to smooth out, as indicated by the dotted lines in Fig. 8.10a. The degree to which this happens depends on the strength of the exchange, or alternatively on the thickness of the film. A measure of the characteristic maximum rate

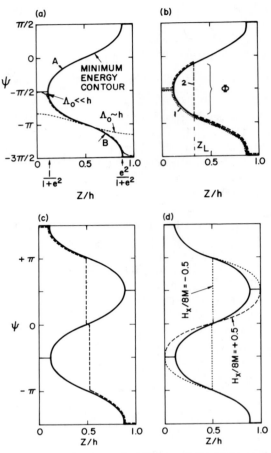

**Fig. 8.10.** Wall-magnetization twist angle $\psi(z)$ through the thickness of a film, assuming a single isolated domain wall with the stray field of Eq. (8.33). (a) Static "Bloch" wall, in the thick and thin film limits. (b) Dynamic wall containing one horizontal Bloch line (2). (c) "Anomalous" static wall containing a $2\pi$ horizontal line. (d) Bloch wall with in-plane fields parallel ($H_x/8M = +0.5$) or antiparallel ($-0.5$) to the average wall magnetization in the wall plane. The antiparallel case causes nucleation of a $2\pi$ horizontal line. Numerical calculations of some of these contours are shown in Fig. 17.5.

of rotation of $\psi$ with $z$ is given by $\Lambda_0 = (A/2\pi M^2)^{1/2}$, the Bloch-line width parameter. If $h < \Lambda_0$, then clearly there can be little wall twisting, and the Néel wall sections at the surface are effectively suppressed. However if $h \gg \Lambda_0$, as is usually the case for most garnet films, then the minimum magnetostatic energy contour is followed quite closely, with only a small amount of rounding at the critical points, as shown in the figure. This is the limit we assume in what follows. We refer the reader to the literature where the minimum energy contour has been calculated in detail and the ensuing corrections to the wall energy reported for the condition of moderate $Q$.[239-245] Direct experimental evidence for the twisted wall structure has come recently from high voltage Lorentz microscopy on cobalt foils tilted about an axis normal to the wall plane.[245a]

Horizontal Bloch lines consist of transitions between the various minimum energy contours of Fig. 8.10a. For example, consider the structures of Fig. 8.10b, which correspond to a horizontal Bloch line lying parallel to the film plane and moving up through the thickness of the film. Such a Bloch line would form a ring around a bubble, as illustrated in Fig. 8.6d. If it sweeps all the way from one critical point to the opposite surface, it effectively switches the polarity of the wall from one average Bloch orientation to the other. The angle of Bloch-line rotation during this process is given by the separation of the end points of the vertical dashed section of contour 2 in Fig. 8.10b. The Bloch-line rotation angle $\Phi(z)$ is given by $2 \cos^{-1}(H_y/8M)$ as in the vertical Bloch line case [Eq. (8.12)]. As the Bloch line moves in the $+z$ direction, the Bloch-line rotation angle increases, because of the dependence of $H_y$ on $z$. At the center of the film $(z = h/2)$, $\Phi = \pi$, and the Bloch line is a $\pi$-line. The horizontal Bloch-line energy can be easily calculated if we assume $h \gg \Lambda_0$, for in this case, which is often called the "Bloch-line approximation," the Bloch line can be considered to lie in a region of constant in-plane field. Then Eqs. (8.13) and (8.14) apply, as for the vertical Bloch-line case considered earlier. The energy is plotted for the two possible Bloch-line polarities in Fig. 8.11, assuming the particular stray field of Eq. (8.33). Clearly such a Bloch line is not statically stable because its energy changes with position and therefore a force acts on the Bloch line pushing it to contract back to a critical point, where it would disappear.[246] Such Bloch lines are, however, of importance in dynamic processes (see Chapter VIII). Numerical calculations of such structures, not relying on the Bloch-line approximation, are described in Section 17,A.

Another horizontal Bloch-line structure that can be statically stable is the $2\pi$ line shown in Fig. 8.10c.[245,247] Larger angles of twist, consisting of multiples of $2\pi$, are also possible. Such structures are stabilized by the fact that the Néel sections at the surfaces act as boundary conditions to pin the wall magnetization. The Bloch lines tend to congregate in the midplane of the film because here the stray field is minimum. The Bloch-line energy can be estimated from Eq. (8.13) for each $\pi$-segment. The interaction forces between horizontal Bloch lines differ from those of vertical Bloch lines in that there are no "$\sigma$-charges" and therefore no magnetostatic attractive

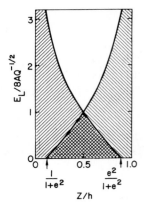

**Fig. 8.11.** Energy per length of right- or left-handed vertical Bloch lines as a function of position $z$ through the film thickness $h$ for the stray field of an isolated plane wall [Eq. (8.33)]. Shaded area under the lines represents total Bloch-line energy. The cross hatched area is the energy of a Bloch line with a Bloch point located at $z/h = 0.5$, ignoring the excess energy of the Bloch point itself.

forces in the absence of stray field. However the exchange repulsion forces are the same.

Thus one can visualize multiple horizontal Bloch-line structures analogous to hard bubbles and illustrated schematically in Fig. 8.6e. An important difference between the two cases is that it is impossible for vertical Bloch-line states to unwind without the creation of a magnetic singularity, whereas the only barrier to unwinding horizontal Bloch lines is the stray-field pinning at the surface. Now the logarithmic divergence of the stray field at the surface [e.g., see Eqs. (8.33) and (8.34)] is mathematically a "weak" divergence in the sense that integration of the wall energy gives a finite result in spite of the very high local fields. Detailed calculations[245] have shown that the surface pinning fields are insufficiently strong to squeeze horizontal Bloch lines tighter together than their natural width $\pi \Lambda_0$. As a crude estimate, the maximum number of horizontal Bloch lines that can be accommodated is $h/\pi \Lambda_0$. The line tension of such Bloch lines tends to squeeze the bubble, and downward shifts in collapse fields of up to several oersteds for typical garnet materials can be estimated.[248] However no experimental identification of such "soft" bubbles has yet been reported.

In addition to its effect on horizontal Bloch lines, the stray field $H_y(z)$ affects the structure of an isolated vertical Bloch line. Using Eqs. (8.13) and (8.14) and ignoring exchange energy in the $z$ direction, we find that the energy per length of the Bloch line is a function of $z$ as shown in Fig. 8.11 for the two cases of right- and left-handed Bloch lines. We can understand this result from Fig. 8.3. For a right-handed Bloch line, the line structure merges with the Néel section at the upper surface and the Bloch-line energy is therefore zero there. Progressing down through the film, the stray field decreases and becomes negative, and therefore the energy increases according to Eq. (8.13). The opposite occurs for the left-handed Bloch line. The total energy of the Bloch line is the area under the curve of Fig. 8.11, which is finite because the divergence in $E_L$ is logarithmic. It is interesting to note that the Bloch-line energy would be significantly reduced as indicated by the cross-hatching if the top half of the Bloch line were right-handed and the bottom half left-handed. How such a Bloch line of mixed handedness can occur is discussed in Section 9,B.

Next we consider the effect of applied in-plane fields on a wall subject to stray fields. Consider a straight wall. If the in-plane field is applied parallel to the average wall magnetization, the $\psi(z)$ contour becomes "straightened out" as shown in Fig. 8.10d for the case $H_x/8M = 0.5$. However if the in-plane field is applied antiparallel to the average magnetization, the field twists the magnetization at the two surfaces in its direction and for sufficiently large field, a Bloch line of twist $2\pi$ will result in the center of the film, as shown in Fig. 8.10d for the case $H_x/8M = -0.5$. To estimate the field $H_{2\pi}$ at which such a $2\pi$ Bloch line forms, one balances the in-plane field force on the Bloch line at the center of the film $2\pi\Delta M H_x$ [see Eq. (8.11)] against the force caused by the internal Bloch line energy. The latter force is given by the derivative of Eq. (8.13) with respect to $z$, where $\Phi$ is a function of $z$ according to Eqs. (8.12) and (8.33). Evaluating the Bloch line force at $z = h/2$, one finds[245,249-250a]

$$H_{2\pi} = 4(2\pi A)^{1/2}/h, \tag{8.35}$$

which is of order 10 Oe for typical 5 $\mu$m films. This field is to be compared to the in-plane field $H_{x1} = 8M$ required to destabilize and switch the wall moment in the one-dimensional model [see Eq. (7.18)]. $H_{2\pi}$ is an order of magnitude less than $H_{x1}$, showing the important effect that stray fields have in determining the static equilibrium wall configuration. Horizontal Bloch-line nucleation by in-plane field has also been verified by detailed micro-magnetic calculations.[245,251]

If a chiral bubble is placed in an external in-plane field, one side of the bubble has magnetization parallel to the field while the other has it anti-parallel. Thus one can expect nucleation of horizontal Bloch lines on one side of the bubble. At those points along the circumference where the applied field component in the wall plane drops to zero, the Bloch lines are expected to recede back to the "critical circles" from which they were nucleated. Thus one arrives at the structure shown in Fig. 8.12a for a chiral bubble in

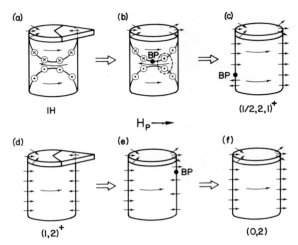

**Fig. 8.12.**   Schematic capping layer and bubble wall states. (a)–(c) represent the "cap-switch" transition from a 1H to a $(\frac{1}{2}, 2, 1)^+$ state in an in-plane field $H_p$. (d)–(f) represent the cap-switch transition from $(1, 2)^+$ to $(0, 2)$ (after Beaulieu *et al.*[250]).

an in-plane field $H_p > H_{2\pi}$. Such a bubble has been termed a 1H bubble, where the H stands for "horizontal" and where the shape of the letter reminds one of the presumed disposition of the Bloch lines.[229,249,250]

## 9.   Bloch Points, State Transitions, and Capping Layers

In micromagnetism, one usually assumes that the magnetization vector $\mathbf{M}(\mathbf{x})$ varies continuously with position $\mathbf{x}$, and this assumption underlies most of the theory in this review. However some magnetic configurations demand topologically the presence of singularities called "Bloch points."[252-256] A defining property of a Bloch point is that all possible directions of $\mathbf{M}$ are found on a sphere of infinitesimally small radius centered on it.

In Section 9,A we describe some general topological properties of Bloch points. A reader solely concerned with the role of Bloch points in bubbles may turn directly to Section 9,B, where their predicted effects on bubble dynamics are compared to experiment. Bloch points are believed to be agents of *S*-state transition of bubble domains. One class of such transitions involves elimination of hard bubbles, and we review "hard bubble suppression" techniques in Section 9,C. Finally in Section 9,D we describe the control of *S*-states in films with a capping layer.

## A. BLOCH POINTS—BASIC PROPERTIES

The definition of a Bloch point implies that the magnitude of the gradient of $\mathbf{M}(\mathbf{r})$, where $\mathbf{r}$ is the position vector, tends to infinity in its neighborhood. Therefore, the exchange energy density $w_{ex} = (A/\mathbf{M}^2)(\nabla\mathbf{M})^2$ [Eq. (1.11)] dominates all other energy terms at points sufficiently close to a Bloch point. In static equilibrium, the energy must be a minimum, and thus the distribution $\mathbf{M}(\mathbf{r})$ satisfies the Euler–Lagrange equation

$$[\delta(\nabla\mathbf{M})^2/\delta\mathbf{M}] - \lambda\mathbf{M} \equiv -(2\nabla^2 + \lambda)\mathbf{M} = 0, \tag{9.1}$$

where the scalar function $\lambda(\mathbf{r})$ is the Lagrangian multiplier function corresponding to the condition that $\mathbf{M}^2(\mathbf{r})$ is constant. By referring to the standard expression for $\nabla^2$ in spherical coordinates, centered on the Bloch point, one readily confirms one special pair of spherically symmetric particular solutions

$$\mathbf{M} = \pm M\hat{\mathbf{r}} \qquad \lambda = \pm 4/r^2, \tag{9.2}$$

where $\hat{\mathbf{r}} \equiv \mathbf{r}/r$ is the unit radial vector. The plus($+$) and minus($-$) signs correspond to what may be called, respectively, divergent and convergent Bloch points located at $r = 0$. They are depicted schematically in Figs. 9.1a and 9.1b, respectively. These solutions may also be written in the forms

$$\theta = \theta_r, \qquad \phi = \phi_r \qquad \text{(divergent)} \tag{9.3a}$$

and

$$\theta = \pi - \theta_r, \qquad \phi = \phi_r + \pi \qquad \text{(convergent)} \tag{9.3b}$$

where $\theta_r$ and $\phi_r$ are the polar angles for $\mathbf{r}$ while $\theta$ and $\phi$ are the polar angles for $\mathbf{M}$ (Fig. 7.1).

There exists in fact an infinite set of distinct Bloch-point configurations. To see this, consider any linear transformation $\mathbf{R}$, independent of $\mathbf{r}$, such that $M_i'(\mathbf{r}) = (\mathbf{R}\cdot\mathbf{M})_i \equiv \sum_j R_{ij}M_j(\mathbf{r})$ is a new distribution generated from $\mathbf{M}(\mathbf{r})$. Here $M_i(i = x,y,z)$ is the $i$th Cartesian component of $\mathbf{M}$. Since Eq. (9.1) is linear, $\mathbf{M}'(\mathbf{r})$ also satisfies it. To preserve the constancy of $\mathbf{M}^2$ we must restrict $\mathbf{R}$ to the real unitary orthogonal-axis transformations. Since the two solutions (9.2) are already mutual inverses, it is sufficient to consider only the infinte group of three-dimen-

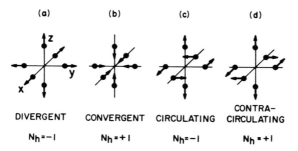

(a)     (b)     (c)     (d)

DIVERGENT    CONVERGENT    CIRCULATING    CONTRA-CIRCULATING

$N_h = -1$     $N_h = +1$     $N_h = -1$     $N_h = +1$

**Fig. 9.1.** Schematic illustrations of some of the possible micromagnetic configurations in the infinitesimal neighborhood of a Bloch point. $\mathbf{M}$ rotates uniformly on any plane circle centered on the origin. $N_h$ indicates Bloch point signature or "homotopy number" [see Eq. (9.12)].

sional proper (i.e., those that preserve the right-handedness of the *xyz* coordinate system) rotations **R**. Accordingly, we have the infinite family

$$\mathbf{M} = \pm M\mathbf{R} \cdot \hat{\mathbf{r}}/r \tag{9.4}$$

of configurations allowed in the infinitesimal neighborhood of a Bloch point.

Consider the example $\mathbf{R} = \mathbf{R}_{z90°}$, which represents a rotation through 90° about the *z* axis in the right-hand screw sense. Application of $\mathbf{R}_{z90°}$ to the divergent point (Fig. 9.1a) yields the Bloch point with "*z*-circulation" represented by

$$\theta = \theta_r, \quad \phi = \phi_r + 90° \quad \text{(z-circulation)} \tag{9.5}$$

and shown in Fig. 9.1c. Reflection of the latter distribution in the *xz* plane produces the further example of a "*z*-contracirculating" distribution

$$\theta = \theta_r, \quad \phi = -(\phi_r + 90°) \quad \text{(z-contracirculation)} \tag{9.6}$$

shown in Fig. 9.1d.

Topological discussion illuminates the properties of such defects in magnetic and other ordered media, such as liquid crystals.[251a,251b] Considerable insight into Bloch points is provided by visualizing a "map" in **M** space of all the vectors **M(r)** corresponding to all the points **r** lying on a simple closed surface *S* (nonpathologically deformed sphere) in the magnetic medium.[252] One distinguishes multiply mapped **M** values by providing slight artificial variations in the moment magnitude *M*. Assuming that the field distribution **M(r)** is continuous for **r** on *S*, then the map **M(S)** cannot have any edges, but must also be closed. A simple map occurs if, for example, **M(r)** deviates little from a common direction for all **r** on *S*; then the map resembles a deflated (but closed) balloon that is pressed, with any number of folds, against one side of a solid sphere of radius *M*, as illustrated by the sectional view in Fig. 9.2a. Note that in this particular case each direction of **M** occurs an even number of times or not at all (except at bends), and that the map cannot enclose the origin.

If, however, *S* is a small sphere enclosing a Bloch point, it is clear from Eq. (9.4) that the mapped surface corresponds to a balloon stretched to *enclose* the solid sphere, as illustrated in Fig. 9.2b. In Fig. 9.2c, the sphere is doubly enclosed. The three mappings shown in Fig. 9.2 may be called *heterotopic* because one map cannot deform into another without its being torn to allow the solid sphere to pass. (The balloon analogy is not complete because the map may have to intersect itself along a curve as indicated by the sectional view in Fig. 9.2c.)

Whenever two maps *can* be deformed, one into the other, without tearing, they may be called *homotopic*. A homotopy number $N_h$ may be assigned to any class of homotopic maps according to the number of "layers" of the map that remain covering the sphere when the map is shrunken to a minimimum total surface area, without tearing, by smoothing all folds. Examples of

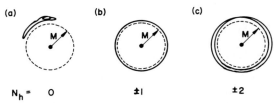

**Fig. 9.2.** Sections of **M**-map for a closed surface which (a) may exclude Bloch points (there may be pairs of Bloch points of canceling signature), (b) must include at least one Bloch point, and (c) must include at least two Bloch points.

homotopy number $N_h = 0$, $\pm 1$, and $\pm 2$ are illustrated schematically in Fig. 9.2. (The significance of the signs will be discussed below.)

Return now to the case in which $S$ is a small sphere centered on a Bloch point ($N_h = \pm 1$)
lying within a static magnetic medium. Imagine that this surface expands continuously, enlarging the enclosed volume. Suppose that $S$ encounters no Bloch points in the process of expansion.
Since a Bloch point is the only physically allowable discontinuity, $\mathbf{M}$ is therefore always continuous on $S$ in this case. It follows that the map cannot tear during the expansion in this case
and the homotopy number ($N_h = \pm 1$) cannot change. Since a surface not enclosing Bloch
points has $N_h = 0$, it follows that the allowable distribution of $\mathbf{M}(\mathbf{r})$ for $\mathbf{r}$ on any closed surface,
however large its dimensions, depends on whether or not it contains a Bloch point. For ex-
ample, if $S$ contains a single Bloch point, then $\mathbf{M}$ cannot be constant on $S$. Thus we see that the
effects of Bloch points are not limited to their immediate neighborhoods, but that Bloch points
have an ineluctable relationship to the magnetization distribution at great distances.

Discussion of complicated configurations involving Bloch points is systematized by the concept of gyrovector density $\mathbf{g}$, which is defined by[257]

$$\mathbf{g} = (\nabla \cos \theta) \times \nabla\phi, \tag{9.7}$$

and which plays a significant role in dynamics (Section 12,F) even when Bloch points are not
present. Application of identities from vector analysis shows that

$$\nabla \cdot \mathbf{g} = 0 \tag{9.8}$$

identically by virtue of the definition (9.7). $\mathbf{g}$ is undefined at a Bloch point where $\nabla\theta$ and $\nabla\phi$
do not exist. One can easily verify the relation

$$\mathbf{g} = \mp \mathbf{r}/r^2, \tag{9.9}$$

which holds in the infinitesimal neighborhood of a Bloch point. The $\mp$ signs correspond with
$\pm$ of Eq. (9.4), respectively. Thus the plus sign in Eq. (9.9) holds for the convergent Bloch point
or any Bloch point obtained from it by a proper rotation; we henceforth refer to such Bloch
points as positive and all others negative. The signs are indicated as $N_h$ in Fig. 9.1. Integrating
over a sphere of radius $r$ we find

$$\int_{BP}\mathbf{g} \cdot d\mathbf{S} = \pm 4\pi, \tag{9.10}$$

with plus($+$) and minus($-$) corresponding to positive (convergent) and negative (divergent)
Bloch points ($d\mathbf{S}$ assumed to point *outward*).

It is clear then that $\mathbf{g}$ is a kind of "flux density" for which Bloch points are quantized sources
and the only sources, just as electrons are unit sources of electric field. $\mathbf{g}$ may have any value at
the free surface of a magnetic film. Consider then the surface integral of $\mathbf{g}$ carried over any closed
surface $S$ that does not pass through any Bloch points. It has the value

$$\int \mathbf{g} \cdot d\mathbf{S} = 4\pi(n_+ - n_-), \tag{9.11}$$

where $n_+(n_-)$ is the number of positive (negative) Bloch points enclosed. We see therefore that
the homotopy number is best regarded algebraically according to

$$N_h = n_+ - n_- \tag{9.12}$$

and that it can be identified with the number of gyrovector "flux quanta" emanating from $S$ as
well as with the net Bloch-point content.

The above connection between topological properties of the mapping and the gyrovector
may be explained thusly: The gyrovector density, which may also be written $\mathbf{g} = (\nabla\phi)\sin\theta \times
(\nabla\theta)$, represents $M^{-1}$ times the algebraic area of the map of $S$ in $\mathbf{M}$ space per unit directed

area on the surface $S$ in position space. The word "algebraic" in this statement refers to the fact that whenever the map is multiple because of folding, as illustrated in Fig. 9.2a, contributions to $\int \mathbf{g} \cdot \mathbf{dS}$ from pairs of superimposed layers cancel. One can thus visualize how the integral $(4\pi M)^{-1} \int \mathbf{g} \cdot \mathbf{dS}$ counts the number of times the $M$-sphere is enclosed by the map, whilst discounting folds which could be removed without tearing.

Higher order singularities, satisfying $|N_h| \geq 2$ on an infinitesimal surrounding sphere, can be constructed mathematically,[253] but are not believed to correspond to Nature. The role of atomic-lattice discreteness in Bloch-point structure has been considered.[254]

## B. BLOCH POINTS IN BUBBLE-DOMAIN WALLS

The role of Bloch points in domain-wall theory can be illustrated in the following way.[255,256] We may clearly regard a domain wall as a boundary region separating domains of opposite magnetic orientation ($\theta = 0$ and $\pi$). Proceeding further, the magnetostatic energy within the wall region is minimized by either of two wall-moment orientations $\phi = \psi = \psi_1$ and $\psi_2$ whose values depend on the external field [Eq. (7.17)] and that correspond to the two chiral alternatives of twist within the wall. Next we consider a Bloch line separating two such different domain wall regions. The direction of magnetization at the center of the line is specified by the bisector of the angle difference $\psi_1 - \psi_2$ between the two domain-wall regions. Thus the Bloch line is defined merely by the *polarity* of the wall magnetization, corresponding to the two types of Bloch-line handedness described in Section 8,A. Finally by analogy we can now consider a Bloch point separating two sections of Bloch line with different magnetic polarity or handedness. Such a structure is drawn in Fig. 9.3. The structure is fully specified by energy minimization consistent with the surrounding domain-, wall-, and line-magnetization directions indicated by arrows. By inspection of the **M** arrows close to the origin in Fig. 9.3 one sees by continuity that all directions occur, thus verifying the Bloch-point definition given earlier. The Bloch point effectively "pinches" the wall and the Bloch-line thicknesses to zero at the origin, as indicated by the constant angle contours shown in the figure. At the origin there must be a singularity where the magnetization abruptly changes direction from one magnetic site of the crystal lattice to the next. The particular Bloch point shown in Fig. 9.3 is of the contracirculating positive type shown in Fig. 9.1d, except that $x$ rather than $z$ is the preferred axis. One can easily see that if a series of Bloch points exist on a vertical $\pi$-Bloch line, they must alternate between circulating negative and contracirculating positive types.

Noteworthy is the singular character of a Bloch point, which sets it apart conceptually from the superficially similar concepts of Bloch line and Bloch wall, which have strictly continuous $\mathbf{M(r)}$ configurations. Although the energy density diverges near a Bloch point the total energy of a region in-

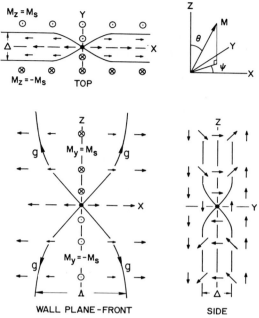

**Fig. 9.3.** Principal orthogonal sections of a micromagnetic configuration including one Bloch wall, one Bloch line, and one contracirculating ($N_h = -1$) Bloch point (after Slonczewski[255]). In the wall-plane cross section ($xz$) gyrovectors are indicated by **g**.

cluding it is finite. The exchange energy density approaches the divergent function $Ar^{-2}$ in the neighborhood of a Bloch point. However the total exchange energy in a sphere of small radius $\rho$ is

$$\int w_A \, dV = \int_0^\rho Ar^{-2}4\pi r^2 \, dr = 4\pi A\rho, \tag{9.13}$$

which is finite. Since the region whose energy density is most affected by the presence of the Bloch point is of order $\Delta_0 = (A/K)^{1/2}$ in radius, the total energy of a Bloch point is of order $4\pi A^{3/2}K^{-1/2}$.

To discuss the energetics more precisely, we define the Bloch-point "insertion" energy as the difference between the energy of a mixed-handedness Bloch line carrying a Bloch point and the energy of a single-handedness Bloch line without a Bloch point. In the limit of large $Q$, this difference is found by numerical computation to be [256]

$$W_{bp} = 2\pi A^{3/2}K^{-1/2}(\ln Q + 1.90), \tag{9.14}$$

in substantial agreement with the above estimate. For a thick film this energy makes only a small contribution to the total energy $8AQ^{-1/2}h$ for a vertical $\pi$ line in a film of thickness $h$, ignoring stray fields.

If we include stray fields the energy per unit length of a vertical Bloch line changes with position through the thickness of the film as shown in Fig. 8.11 and the previous section. A split Bloch line would therefore gain an energy, proportional to one of the singly shaded areas in Fig. 8.11, over a Bloch line of a single handedness. In comparison, the extra energy of the Bloch point is small in thick films as discussed above. This means that a split Bloch line with a Bloch point is actually the lowest energy Bloch-line structure in sufficiently thick films. Numerical computation shows that in thinner films satisfying $h < 7.3(A/2\pi)^{1/2}M^{-1}$ the split Bloch line has a *higher* energy.[258]

A Bloch point in Fig. 8.11 would be located at the position through the thickness of the film where the stray field vanishes (i.e., on the midplane). More generally one can argue that to minimize energy, a static Bloch point must be located at a point of vanishing field (combining applied, demagnetizing, and stray field). Consider the geometry of Fig. 9.3, assuming straight walls and lines. If $H_z \neq 0$, then the system would gain domain energy by wall displacement. Or if $H_x \neq 0$, wall energy would be gained by Bloch-line displacement in the $x$ direction. Finally if $H_y \neq 0$, Bloch line energy could be gained by Bloch-point displacement in the $z$ direction, assuming that the field-dependence of the Bloch-point energy itself is negligible. This conclusion may be modified, however, if the Bloch point can lower its energy at a defect in the material. In case the defect is a small spherical void or nonmagnetic inclusion, then a Bloch point is attracted to it with a binding energy given by Eq. (9.13) with $\rho$ now the radius of the sphere.

Consider now a bubble in a film without in-plane anisotropy but with an applied in-plane field $H_p$ and two Bloch lines lying on opposite ends of the bubble on the diameter parallel to $H_p$ as shown in Fig. 9.4. The Bloch-line energy is minimized if the Bloch lines have Bloch points sitting at the points $z_i$, where[255]

$$H_p + H_r(z_i) = 0, \tag{9.15}$$

and $H_r$ is the component of stray field normal to the wall. Since $H_r(z)$ is known from magnetostatic calculations [e.g., Eqs. (8.33)–(8.34)], the Bloch-point positions can easily be determined and controlled by external in-plane fields.

In connection with Bloch points it is useful to write the surface gyrovector for a wall

$$\mathbf{g_s} = \int_{-\infty}^{\infty} \mathbf{g}(\xi)\, d\xi = -2\mathbf{n} \times \nabla\psi. \tag{9.16}$$

Here $\xi$ is a position coordinate normal to the wall, $\mathbf{n}$ is a unit normal vector directed from the $\theta = 0$ to the $\theta = \pi$ side, and $\nabla\psi$ is the two-dimensional

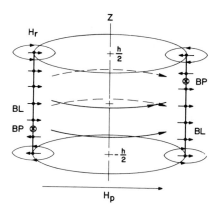

**Fig. 9.4.** A bubble domain with two Bloch lines and two Bloch points in the presence of an in-plane field $H_p$.

surface gradient of the wall-moment angle $\psi$. The large arrowheads in the $xz$ section of Fig. 9.3 indicate schematically the disposition of $\mathbf{g}_s$, which follows contours of constant $\psi$ exactly.

The presence of Bloch points in a bubble wall affects the bubble's winding number $S$, as discussed fully in Section 12,F. It is useful for discussion of dynamic experiments (see below) to define a local winding integer as a function of position $z$ through the thickness of the film. To give a formal expression for the winding integer, let $ds$ be an increment of arc on the intersection of the plane $z = $ constant with the surface of a closed domain of arbitrary shape. Let us write the closed contour integral

$$I(z) = (4\pi)^{-1} \oint g_{sz} \, ds = (2\pi)^{-1} \oint (\partial \psi / \partial s) \, ds. \tag{9.17}$$

The last equality, obtained from Eq. (9.16), shows that $I(z)$ is the "winding integer" specifying the total number of wall-moment twists in the domain. That is, $I(z)$ is the number of rotations of the wall magnetization angle $\psi$ as one moves along a locus defined by the intersection of the bubble wall and a plane at position $z$ within the film. There will be steps in $I(z)$ at any $z$ that defines a plane containing one or more Bloch points. The $S$-state for the entire bubble is the average $I$ over $z$:

$$S = h^{-1} \int I(z) \, dz \equiv I(h) \pm h^{-1} \sum_i N_{hi} z_i. \tag{9.18}$$

Here $N_{hi}$ is the signature of the $i$th Bloch point with $z$ coordinate $z_i$ [see Eq. (9.10)], where $z$ is defined to be 0 at the bottom surface and $h$ at the top surface. The $\pm$ sign depends on the magnetization polarity, being negative if the magnetization outside the domain is up and positive if it is down.

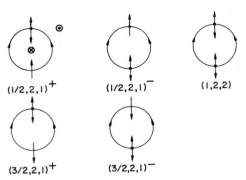

**Fig. 9.5.** Schematic illustration of the five statically stable bubble states with two Bloch lines, one or both of which contain a Bloch point. The double-headed arrow indicates a Bloch line containing a Bloch point at the midplane of the film; in each case, the inward pointing part of the Bloch line lies above the outward pointing part, following the stray field.

Using Eq. (9.18) we can now summarize the various possible bubble states having two Bloch lines and one or two Bloch points as in Fig. 9.5. There are two complementary states with $S = \frac{1}{2}$ (in zero in-plane field), denoted $(\frac{1}{2}, 2, 1)^+$ and $(\frac{1}{2}, 2, 1)^-$ according to the notation introduced in Section 8,C, except now the superscript $+$ or $-$ refers to the sign of the Bloch point according to the convention (9.10). These states have also been called, respectively, $\frac{1}{2}$ and $\frac{1}{2}*$ for brevity.[250] There is one $S = 1$ state, denoted (1, 2, 2), with a Bloch point on each of its Bloch lines. And there are two $S = \frac{3}{2}$ states, denoted $(\frac{3}{2}, 2, 1)^+$ and $(\frac{3}{2}, 2, 1)^-$. As discussed above, an applied in-plane field acts to shift the Bloch point away from the midplane of the film and in such a direction as to lengthen the part of the Bloch line oriented along the applied field direction. This leads to a reduction in the $S$ number. Thus the in-plane field lifts the $S$-degeneracy of the (1, 0, 0) and (1, 2, 2) bubbles.

Experimental evidence for the existence of Bloch points in bubbles has primarily come from measurements of skew deflection in bubble propagation measurements according to the equation (8.26).[259-261,250] For example, experimental $S$ values determined in this way in ion-implanted GdTmGa films[259] were $S = 0, 0.5, 1.02, 1.42$, and $1.82$. The most common state was $S = 1.42$, but it is not understood why. These values are close to half-integral, supporting the Bloch-point picture. The dependence of experimental deflection angles of (1, 2, 2) and the two $S = \frac{1}{2}$ states on external in-plane field $H_p$ is shown in Fig. 9.6a for a detailed study of a SmCaGeYIG film.[250] The angle of the (1, 2, 2) state is observed to deviate rapidly with $H_p$ from the characteristic angle for $S = 1$. The $\frac{1}{2}$ state follows that field dependence at just half the angle of the (1, 2, 2) state. This data agrees very well with the theory based on Eqs. (8.26), (9.15), and (9.18).

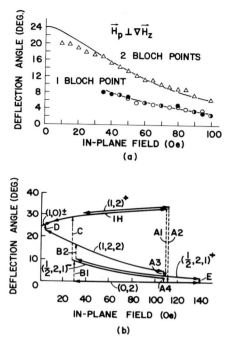

**Fig. 9.6.** Bubble deflection angle, determined in a gradient-propagation experiment, versus in-plane field for different bubble states in an ion-implanted SmCaGeYIG film (after Beaulieu *et al.*[250]). The in-plane field is perpendicular to the gradient, which gives a drive of ~3.4 Oe across the 6 μm diameter bubble. In (a) is shown a comparison of experiment and theory for $(1, 2, 2)$ and $(\frac{1}{2}, 2, 1)$ bubbles. In (b) are shown the experimentally observed state transitions labeled A through E.

The stability of any of the bubble states of Fig. 9.5 depends on the presence of an in-plane field or in-plane anisotropy to keep the two Bloch lines separated. If such forces are not present, the Bloch lines are free to move around the perimeter of the bubble. Thus, although they may be statically metastable in the positions of Fig. 9.5, a dynamic perturbation could cause the Bloch lines to approach each other. In the case of two Bloch lines, one of which has a Bloch point, as shown in Fig. 9.7, half of the Bloch line structure can unwind directly. The line tension of the remaining Bloch-line loop is then expected to cause the loop to retract to the surface and annihilate. It has been confirmed experimentally that as in-plane fields are reduced to zero in a bubble propagation experiment, $(1, 2, 2)$ and $\frac{1}{2}$ states convert irreversibly to $S = 1$, presumably by such a process (see $C$ and $D$ transitions of Fig. 9.6b).[250]

If a very large in-plane field is applied, Eq. (9.15) implies that a Bloch point on a Bloch line will be pushed close to a surface. In the SmCaGeYIG

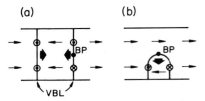

**Fig. 9.7.** Schematic annihilation of two vertical Bloch lines, one containing a Bloch point (BP).

film of Fig. 9.6b (see $E$ transition), an in-plane field of 143 Oe was sufficient to transform a $\frac{1}{2}$ state irreversibly into a $(0, 2)$ state (as determined by the deflection angle). For this field, Eq. (9.15) implies that the Bloch point lies within a Bloch-line width of the surface, so that a spontaneous annihilation of the Bloch point structure is plausible.[250] Furthermore, it is observed that a sufficiently large in-plane field (a few hundred oersteds in a typical 5 μm garnet film) is sufficient to clean out any excess Bloch-line structure and reduce the bubble to a $(0, 2)$ state.[262,263] This means that it is possible to field-nucleate Bloch points onto Bloch lines which oppose the field. Consider for example the bubble of Fig. 8.5e with an in-plane field applied up in the plane of the figure. Bloch-point nucleation is most likely to occur at the top surface, on the lower Bloch line of the figure, where the applied field adds to the stray field. The precise field threshold for such a process is not known experimentally or theoretically. Once nucleated, the Bloch point is expected to run through the thickness and be forced out at the bottom by the process described above, thus converting a $(1, 2)$ bubble into $(0, 2)$. A related mechanism for Bloch-point generation involves the use of in-plane surface layers; this will be discussed in Section 9,D.

Certain proposed bubble-lattice storage schemes in which information is coded by means of the winding number $S$ rely on the long-term stability of vertical Bloch lines to prevent coding errors. Thus the question of Bloch-line stability against thermal activation of Bloch points arises. One easily concludes that preserving a large capacity store free of errors for a period of years at room temperature requires a minimum activation-energy barrier $E_a$ of the order of 1 eV.[255] This condition is of course comparable to that for the stability of chemical compounds. Two alternative processes for thermal Bloch-point nucleation suggest themselves: 1) nucleation of a single Bloch point at the film surface and 2) nucleation of a pair of Bloch points having opposite signature within the interior of the film. Up to the present time, estimates have been given only for processes 2) in the absence of in-plane field, thus applying only to nucleation at the midplane of the film. One sets $W_{bp} = \frac{1}{2}E_a \approx \frac{1}{2}$ eV in Eq. (9.14) to find for optimal values of $A$ and $Q$ the corresponding bubble diameter of order 0.1 μm. For diameters greater than 0.1 $\mu$m, information should be secure with respect to this particular process.

Experimental evidence for the phenomenon of thermal Bloch point nucleation comes from the observation of the "self-collapse" of hard bubbles, to which Eq. (9.14) unfortunately does not apply. When hard bubbles are biased just under their static collapse field, the bubble diameter is observed to decrease spontaneously with time, even in the absence of any external perturbations. After several minutes the bubble collapses, presumably because thermally nucleated Bloch points cause annihilation of the Bloch lines that are necessary for hard bubble stability in high bias fields.[264] Bloch points may also be generated dynamically by driving bubbles in large gradient fields. Experimental evidence has come from bubble-propagation experiments in which skew angles changed or hard bubbles became normal.[166,265–266a] Further discussion of the mechanism of Bloch-point generation will be given in Section 18,D.

For completeness we also quote here one result on the dynamics of Bloch points that allows us to estimate the speed of Bloch-point motion along a Bloch line.[256] One can define a mobility $\mu_{BP}$ for the motion of a Bloch point along the Bloch line. In the geometry of Fig. 9.3, the component $M_y$ at points on the Bloch line reverses sign as the Bloch point moves in the $z$ direction. Therefore the Bloch point is driven in the $z$ direction by a field component $H_y$, such as an external field or the radial stray-field component in a cylindrical bubble. For small $H_y$,

$$V_z = \mu_{BP} H_y. \qquad (9.19)$$

In the high-$Q$ limit $\mu_{BP}$ has been computed numerically as

$$\mu_{BP} = 2\pi\gamma\Lambda_0/\alpha(\ln Q + 1.93) \qquad (9.20)$$
$$= 2\pi Q^{1/2}\mu_0/(\ln Q + 1.93),$$

where $\mu_0$ is the conventional Bloch-line mobility $\gamma\Lambda_0/\alpha$ (see Section 11,A). Taking typical values, $\mu_0 = 1000$ cm/sec Oe, $Q = 9$, and $h = 6$ $\mu$m, one finds that on the order of 1 nsec is required for the Bloch point to relax from a film surface to its midplane, under the influence of a stray field of $\sim 100$ Oe.

## C. HARD-BUBBLE SUPPRESSION

Devices operated in as-grown high-mobility films often fail with the concomitant appearance of hard bubbles off the device propagation track.[230] Assuming hard bubbles are somehow generated in the device (see Section 21) and cause it to fail, one wants to treat device films in such a way that hard bubbles are suppressed. This presumably means making vertical Bloch lines energetically less favorable and/or providing a mechanism whereby they can spontaneously annihilate. In this section we review the various techniques for hard-bubble suppression that have been proposed in terms of what is known about the mechanism by which they act. The most common test for hard-bubble suppression is to measure the spread in collapse fields of a large number of bubbles generated in a variety of ways. A small collapse-field range ($\Delta H < 0.01 \times 4\pi M$) is taken to indicate effective hard-bubble suppression and has been found to correlate with satisfactory device operation.[64,267] A more sensitive test is to propagate bubbles in a field gradient so as to determine the actual bubble $S$-state or net number of Bloch lines in the wall.

One may divide the hard-bubble suppression techniques into two categories depending on whether or not magnetic surface or interface layers are used. First we consider films without such layers. Generally, even in a uniform film, there is a critical temperature, depending on sample composition, above which hard bubbles cannot be generated.[268,269] However, hard bubbles generated at a lower temperature remain stable when raised above the critical temperature.[268] This observation appears to rule out the possibility of thermal Bloch-point activation as a mechanism for hard-bubble suppression (see Section 9,B). A correlation with in-plane or cubic anisotropy has been suggested.[268-270] In one study on a series of SmGaYIG films on (111) samples with different amounts of crystallographic misorientation, hard bubbles were no longer found once the misorientation exceeded $7°$.[270] This effect may be understood in terms of an effective in-plane field or in-plane anisotropy tending to favor the $S = 0$ state of Fig. 8.5h. The effective in-plane anisotropy of orthoferrites or (110)-garnets can similarly account for the lack of hard bubbles in these materials.[45]

Another observation was that hard bubbles could not be generated in materials with sufficiently low $Q$ (e.g., less than $\sim 5$ at room temperature).[271-272] The effect of the $Q$ factor, which usually decreases with temperature, may also explain the critical temperature. To understand the $Q$-effect, one may speculate that low $Q$ permits a larger wall width at the surface, thus giving a greater energy gain there for Bloch-point nucleation on energetically unfavored Bloch lines. The low-$Q$ technique has not found widespread use because of the requirement of large $Q$ for preventing bubble nucleation in field-access devices.[273]

Historically the first and still the most widely used techniques for hard-bubble suppression involve the use of magnetic surface or interface layers in which the magnetization is exchange-coupled to the bubble film. These techniques can be divided into two groups, depending on whether the magnetization in the layer remains perpendicular to the film or whether it lies in-plane (see Section 2,D). Two perpendicular anisotropy layers can be obtained by growing two different films one on top of the other.[64,66] Similar effects can be achieved in films grown with a nonuniform growth process, as in the case of the LaGaYIG films grown by vertical dipping without rotation.[274] Alternatively, annealing a garnet surface coated with silicon lowers the magnetization, presumably by creating oxygen vacancies which permit gallium redistribution between tetrahedral and octahedral sites.[275] Annealing in an inert atmosphere could cause diffusion of gallium from a GGG substrate into a garnet film, lowering the magnetization at the interface.[276] Sometimes even device-processing steps such as rf sputtering of silicon oxide can cause surface gradients in LaGaYIG films.[70] All of these methods are reported to suppress hard bubbles.

The presumed mechanism of hard-bubble suppression in these cases follows from domain structures for Type 1 and Type 2 double layers shown in Fig. 2.6. In the case of a Type 1 structure, a domain wall caps the bubble at the interface between the two layers. If such a capping wall has a continuous distribution of magnetization with no Bloch points, then it follows from topological arguments that the bubble wall it connects to by exchange coupling must have a winding number of zero.[65] A hypothetical spin distribution for this case is shown in Fig. 2.6. Alternatively the bubble stray field, which points radially, could polarize the cap and create the spin structure labeled $S = 1$ in Fig. 2.6, with a Bloch point at the center. Such a configuration must connect with an $S = 1$ bubble wall.[67] Theoretical calculations have not been attempted to determine which configuration is more stable. However, different bubble states are seen, including some having diffuse boundaries.[66] This observation suggests that the wall structure is distorting and perhaps the wall is penetrating through the second layer to the surface to avoid the magnetic singularity of the $S = 1$ case. It is reasonable to suppose that, because of exchange coupling, hard-bubble walls would demand topologically even more complex spin configurations in the capping layer, with significantly higher energy than either the $S = 0$ or $S = 1$ states, thus explaining why hard bubbles are strongly disfavored.

A similar argument applies in the case of Type 2 films. Here the bubble domain wall connects with a compensation wall which spreads out away from the bubble (see Fig. 2.6). The compensation wall is similar to a Bloch wall in that its width is of order $\Delta_0 = (A/K)^{1/2}$ and its energy of order $\sigma_0 = 4(AK)^{1/2}$, assuming that the exchange and anisotropy constants of the two layers are the same and that only their magnetizations are different.[68] If the sublattice magnetization in the compensation wall is all in the same direction in the region around the bubble as shown in Fig. 2.6, then the bubble should have an $S = 0$ state. The $S = 1$ state is not likely to be as strongly stabilized by stray fields as in the Type 1 case, because the net magnetization in the compensation wall is zero.

The most widely used method for hard-bubble suppression involves the use of a layer of in-plane magnetization. Such a layer can be achieved in garnets by ion-implantation, which is believed to cause a lattice strain that acts through magnetostriction to lower the anisotropy.[267,277] A similar effect could be achieved by growing a second garnet layer with low anisotropy[268] or by depositing a permalloy layer exchange-coupled to the garnet bubble film.[278-280] At least for the first two or the above methods, there is a critical temperature above which the layer is effective in preventing hard-bubble generation, much as in the case of uniform films.[268] However the layer does not prevent the existence of hard bubbles, for hard bubbles generated below the critical temperature remain stable even when the temperature is then raised above it. This result indicates that although layers can change the relative energy of different types of bubbles, some additional dynamic mechanism, such as Bloch-point nucleation, is needed to permit a transition between the different states. Another poorly understood feature of the results is that the collapse range changes smoothly rather than abruptly as a function of in-plane layer thickness.[277,281] In-plane layers also occur in GdCo amorphous films from oxidation of the surface.[102]

The mechanism of hard-bubble suppression for in-plane magnetic layers is presumably similar to that for Type 1 and 2 double layers discussed above. Because of exchange coupling, a hard bubble demands a complex, high-energy structure in the cap that is strongly disfavored. A more detailed discussion of bubble states favored by in-plane caps follows in the next subsection.

## D. WALL STATES IN CAPPED FILMS

Considerable work has been done to investigate the correlation of in-plane domain structures of capping layers with the underlying bubble states and the transitions between various states. For example in a study on GdTmGa-YIG films overcoated with permalloy,[76] two types of in-plane domain structures, called $X$ and $Y$, were observed around bubbles (Fig. 2.7b), with presumed magnetic distributions illustrated in Figs. 2.8c and 2.8d. It was discovered that $X$ bubbles propagate at an angle to the gradient, indicating an $S = 1$ state, while $Y$ bubbles propagate straight, indicating $S = 0$. These results can be understood from Fig. 2.8. For example the $X$ bubble (Fig. 2.8c) has a radial magnetization distribution in the cap above the domain wall, giving a unit winding number, which connects to the bubble state by exchange coupling.

An even more complete picture has emerged of the wall states in ion-implanted garnet films and of the various transitions between them.[67,79,]

[250,259,261,266a,282] We describe in detail here one study on an ion-implanted SmCaGeYIG film.[250] Our description serves to bring together the various concepts introduced in earlier sections of this chapter. The results are of great practical interest for wall-state coding in bubble lattice devices. A plot of observed deflection angles determined in a "deflectometer" (see Section 6,C) is shown in Fig. 9.6b for in-plane field $H_p$ perpendicular to the bias field gradient.[250] The tangent of the deflection angle is proportional to $S$ according to Eq. (8.26). At $H_p = 0$, the predominant state is found to be $S = 1$. A similar result is found for most other ion-implanted garnet compositions.[67,261,282] The reason for this result is the same as in the permalloy case: The stray field favors a radial magnetization distribution in the cap above the domain wall (see Fig. 2.8a) and this distribution, with winding number $S = 1$, is exchange coupled to the bubble wall underneath. This "ground state" of the bubble is assumed to be the simple chiral state $(1, 0, 0)^{\pm}$ (see Section 8,C).

As $H_p$ increases, the deflection angle $\rho$ and apparent $S$-number of the $S = 1$ state increase.[187,250,261] This effect has been explained as due simply to the static bubble ellipticity induced by the in-plane field; the bubble wall structure does not in fact change in any qualitative way. However, above about 10 Oe, a noticeable change in mobility was observed, and evidence to be described below indicates that Bloch lines are then present in the bubble. Three candidates are the 1H bubble described in Section 8,E and the $(1, 2)^{\pm}$ bubbles described in Section 8,C. But the $(1, 2)^{-}$ bubble can be ruled out because its magnetization distribution is easily seen to be incompatible with the capping layer distribution by comparing Figs. 8.5e and 2.8a.[283] Of the two remaining candidates, it turns out, for reasons to be described in Section 19,C, that the choice is determined by the direction of the bubble propagation. Turning to the other states of Fig. 9.6b we have already identified the states having $\rho$ decreasing with $H_p$ as $(1, 2, 2)$ and $(1/2, 2, 1)^{\pm}$ in Section 9,B. The state with no deflection is of course the $(0, 2)$ state.

The transitions between these many different states are labeled in Fig. 9.6b as $A1$, $A2$, etc. The mechanisms of the $C$, $D$, and $E$ transitions have been discussed in Section 9,B. To explain the $A$ and $B$ transitions, we invoke the concept of a "cap switch" described in Section 2,E. The cap has two characteristic states, one containing the closure domain and the other fully saturated, as shown in Figs. 2.8a and 2.8b. The cap state is assumed to be primarily controlled by the strength of the in-plane field, not by the nature of the wall state underneath.[67,76,250] Thus at a sufficiently large in-plane field the closure state switches to the saturated state and this causes $A$-type transitions at 110 Oe in Fig. 9.6b. As the in-plane field is reduced, the reverse cap switch back to the closure state occurs at a somewhat lower in-plane field, because of a "hysteresis" arising from the nucleation barrier characteristic of a first-order transition. Presumably, the reverse cap switch causes the $B$-type transitions at 30 Oe in the figure. As further confirmation of this interpretation, the precise fields for the cap-switch processes are observed to be sensitive to the characteristics of the ion-implanted layer, such as the implant dosage and energy.[250]

Now, as discussed earlier, the in-plane capping layer is exchange coupled to the bubble wall underneath and therefore the bubble winding number $I(z)$ just under the capping layer ($z = h$) is controlled by the cap. As apparent from Figs. 2.8a and 2.8b, the closure state requires $I(z = h) = 1$, while the saturated state requires $I(z = h) = 0$. Now the net $S$-number of the bubble is given by the average of $I(z)$ through the thickness of the film [Eq. (9.18)], and as discussed in Section 9,B, Bloch points can cause $I(z)$ to vary through the thickness. Thus the cap does not in fact determine the net $S$-number uniquely but only $I(z)$ just under the surface.

At the in-plane fields for which cap switching occurs, a Bloch line is usually located on the side of the bubble domain just under the closure domain. For example, consider the 1H and $(1, 2)^+$ bubbles illustrated in Figs. 8.12a and 8.12d. Since $I(z)$ just under the capping layer must change from 0 to 1 or vice versa during the two cap-switch processes, a Bloch point must thereby be "injected" into the Bloch line under the closure domain. The effects of such a Bloch-point injection are shown in Fig. 8.12. For the 1H bubble, the Bloch point runs down the Bloch line because the line is gaining in-plane field energy. Next, unwinding of the $2\pi$ Bloch-line structure occurs to reduce exchange energy and the Bloch lines retract under the influence of line tension to form a $(\frac{1}{2}, 2, 1)^+$ bubble (Fig. 8.12c). This process explains the $A1$ transition of Fig. 9.6b. For the case of the $(1, 2)^+$ bubble, Bloch-point injection is followed by motion of the Bloch point down the Bloch line (see Fig. 8.12e). Since the in-plane field increases as the Bloch point moves down, the Bloch point is presumably pushed all the way to the bottom surface, leaving a $(0, 2)$ state as observed in the $A2$ transition. When the in-plane field is lowered and the cap-switch process reverses, a new Bloch point is generated on the Bloch line under the closure domain. In a similar way, this process converts the $(0, 2)$ state into $(\frac{1}{2}, 2, 1)^-$ and the $(\frac{1}{2}, 2, 1)^+$ state into a $(1, 2, 2)$, thus explaining transitions $B1$ and $B2$, respectively. It is remarkable that these transitions allow one to put Bloch points controllably on either or both of the two Bloch lines and thus to distinguish experimentally the two complementary $\frac{1}{2}$ states. Bloch-point injection by means of cap switching thus offers a method for controlling bubble states[67] and has been success-fully used for wall-state coding in bubble lattice devices.[283a]

Wall states have also been studied in double-layer films in which the capping layer has perpendicular anisotropy and is saturated, creating a domain wall above the bubble (Type 1, Section 2,E).[283b] The state transitions are similar to those in ion-implanted films, with one important difference: The $S = 0$ state is stable down to zero in-plane field (in contrast to Fig. 9.6b), making stable coding possible without an in-plane bias field. The reason for this difference can be traced to the very different nature of the capping walls in the two cases (compare Figs. 2.6 and 2.8).

# V

# Wall Dynamics in One Dimension

## 10. One-Dimensional Theory

In Section 3,A we introduced the Landau–Lifshitz equation[2] and we showed how it predicts the motion of spins on the basis of two effects—a conservative gyrotropic precession around an effective field and a dissipative viscous damping of the precession. Here we apply the equation to the one-dimensional domain-wall structure derived in Section 7, limiting our discussion to the high-$Q$ limit and referring to the literature for more general treatments.[176,177,284–294d]

As in Section 3,A, let us define $w\{M\}$ to be the local static magnetic energy density arising from magnetostatic energy, exchange, anisotropy, etc. Let us describe the direction of magnetization $\mathbf{M}$ in terms of polar coordinates $(\theta, \phi)$ defined in Fig. 7.1. The Landau–Lifshitz equations prescribe that the static forces or torques $\delta w/\delta\theta$ and $\delta w/\delta\phi$ must be balanced by "dynamic reaction forces" arising from spin precession and viscous damping. The basic idea of this section is to integrate these forces over the thickness of the domain wall, so that instead of dealing with $(\theta, \phi)$ as continuously varying functions through the domain-wall thickness, one can simplify the problem and deal with only two parameters to describe domain-wall dynamics: the normal-displacement coordinate $q$ and the wall-magnetization angle $\psi$. This simplification becomes possible in the high-$Q$ limit because the wall structure is primarily determined by anisotropy and exchange forces, while dynamic reaction forces and other static forces arising from magnetostatic energy or applied fields are first-order corrections. Thus, following the derivation of static wall structure in Section 7, we assume the dynamic wall structure:

$$\theta(y, t) = 2 \tan^{-1} \exp\{[y - q(t)]/\Delta(t)\}, \qquad (10.1)$$

$$\phi(y, t) = \psi(t). \tag{10.2}$$

That is we assume $\theta(y, t)$ has the usual Bloch-wall profile, except that $\Delta$ may be a function of time $t$, as given by Eq. (7.14) or (7.15), and $\phi(y, t)$ is independent of $y$. The following derivation is valid provided $|\dot{\Delta}| \ll |\dot{q}|$. Eq. (10.1) implies that

$$\partial\theta/\partial y = \Delta^{-1} \sin\theta, \tag{10.3}$$

$$\delta\theta = -(\partial\theta/\partial y)dq = -\Delta^{-1}dq \sin\theta, \tag{10.4}$$

$$\dot{\theta} = -\Delta^{-1}\dot{q} \sin\theta. \tag{10.5}$$

Note here that the functional variation $\delta\theta(y)$ is proportional to an ordinary differential $dq$ because $q$ does not depend on $y$.

Now we wish to integrate over the Landau–Lifshitz equations written in component form in Eqs. (3.7) and (3.8). This is conveniently accomplished by writing these equations in their equivalent variational form

$$\delta w = (\delta w/\delta\theta)\delta\theta + (\delta w/\delta\phi)\delta\phi$$
$$= M\gamma^{-1}[(\dot{\phi} \sin\theta - \alpha\dot{\theta})\delta\theta + (-\dot{\theta} \sin\theta - \alpha\dot{\phi} \sin^2\theta)\delta\phi]. \tag{10.6}$$

Integrating $\delta w$ over $y$ gives the corresponding differential in wall energy $d\sigma$. Furthermore an integration over $y$ can be replaced by an integration over $\theta$ by using Eq. (10.3). Substituting Eqs. (10.2)–(10.5), one can directly integrate (10.6) over the wall to give the total differential expression

$$d\sigma = 2M\gamma^{-1}[(-\dot{\psi} - \alpha\Delta^{-1}\dot{q})dq + (\dot{q} - \alpha\Delta\dot{\psi})d\psi], \tag{10.7}$$

which is equivalent to the pair of partial differential equations[287]

$$\partial\sigma/\partial\psi = 2M\gamma^{-1}(\dot{q} - \alpha\Delta\dot{\psi}), \tag{10.8}$$

$$\partial\sigma/\partial q = -2M\gamma^{-1}(\dot{\psi} + \alpha\Delta^{-1}\dot{q}). \tag{10.9}$$

These equations are valid not only for a one-dimensional wall but also for a gently-flexed wall (i.e., large radii of curvature), provided one replaces $\partial\sigma/\partial\psi$ and $\partial\sigma/\partial q$ with functional derivatives (see Section 12,A).[224,225]

These deceptively simple equations form a basis of wall dynamics in bubble materials and it is important to understand them thoroughly. The pattern of these equations corresponds, term by term, to that of Eqs. (3.7) and (3.8): The terms in $\theta$ give rise to terms in $q$. This can be understood intuitively from the fact that as the fixed wall structure moves by $dq$, the spins in the center of the wall rotate by $\partial\theta = -dq/\Delta$, as can be seen from Eq. (10.4). The terms in $\phi$ give rise, of course, to the terms in $\psi$, which describe the precession of the in-plane wall magnetization. Finally the volume-torque terms $\delta w/\delta\phi$ and

$\delta w/\delta\theta$ give rise to the wall-surface derivatives $\delta\sigma/\delta\psi$ and $\partial\sigma/\delta q$, respectively.

Let us now consider Eq. (10.8) in detail. $\partial\sigma/\partial\psi$, a derivative of wall energy per unit area, clearly represents the torque on the wall magnetization in the $xy$ plane. We can evaluate it directly from the one-dimensional wall-energy expression derived earlier in Eqs. (7.16) and (7.20). Thus we find

$$\dot{q} = 2\pi\gamma\Delta M \sin 2\psi + \tfrac{1}{2}\pi\gamma\Delta H_p \sin(\psi - \psi_H) + \gamma\Delta K_p M^{-1} \sin[2(\psi - \psi_p)]$$
$$+ 2\gamma\Delta D\beta M^{-1}(\psi - \psi_D) + \alpha\Delta\dot{\psi}, \tag{10.10}$$

with terms on the right-hand side proportional to torques arising from local demagnetization, in-plane field, in-plane anisotropy, and Dzialoshinski exchange (for the orthoferrite case only), respectively. By analogy we consider $\alpha\Delta\dot{\psi}$ to arise from a "damping torque." Thus we call Eq. (10.8) or Eq. (10.10) the "torque equation of wall motion." It represents a fundamental principle of wall motion, namely that the velocity $\dot{q}$ of a domain wall must be sustained by a torque on the wall magnetization. Physically we understand this principle as follows: $\dot{q}$ represents $\dot{\theta}$ motion, and according to the gyrotropic character of the Landau–Lifshitz equations, $\dot{\theta}$ motion can occur by precession around an in-plane effective field $\mathbf{H}_{\text{eff}}$, which may be construed whenever there is a torque $\mathbf{T} = \mathbf{M} \times \mathbf{H}_{\text{eff}}$. For example, in the case of the local demagnetizing torque, the effective field is simply the field $\mathbf{H}_{\text{eff}} = \mathbf{H}_d$ arising from magnetic charges as illustrated in Fig. 7.4; analogous effective fields can be seen to arise from $H_p$ and $K_p$ according to the prescription $\mathbf{H}_{\text{eff}} = -\delta w/\delta\mathbf{M}$.

Next we consider Eq. (10.9). $\partial\sigma/\partial q$ is the rate of change with position of the energy per unit area of the wall. Thus it is simply the "pressure" on the wall. Such a pressure can arise from any field, effective or real, which lowers the energy of one domain with respect to the other. Clearly such a field must lie along the $z$ direction of Fig. 7.1. For example, if there is a net uniform bias field $H_a$ applied to destabilize the wall from its static equilibrium, the pressure is simply $-2MH_a$. The effect of demagnetizing fields due to surface charges of domains may be represented by a restoring-force field $H_k = -kq/2M$ (see Section 2,A). In addition the phenomenological coercive field $H_c$ sgn $\dot{q}$ [see Eqs. (5.1) and (5.2)], acts like an effective field opposing the prevailing motion. Introducing these terms into Eq. (10.9) we find

$$\dot{\psi} = \gamma[H_a - (kq/2M) - H_c \text{ sgn } \dot{q}] - \alpha(\dot{q}/\Delta). \tag{10.11}$$

By analogy to the static terms, we consider $-\alpha\dot{q}/\Delta$ to arise from a "damping pressure." Thus we call Eq. (10.9) or Eq. (10.11) the "pressure equation of wall motion." This equation represents the second fundamental principle of wall motion, namely that the in-plane wall magnetization precesses in the $xy$ plane at the Larmor frequency $\omega = \gamma H_{net}$ corresponding to the orthogonal net effective $z$-field $H_{net}$, which includes viscous and coercive drag:

$$H_{net} = H_a - (kq/2M) - H_c \operatorname{sgn} \dot{q} - (\alpha \dot{q}/\gamma \Delta). \qquad (10.12)$$

As an aside we mention that Eq. (10.11) does not exhaust all the possible pressure terms. For example, a gradient in in-plane field gives a pressure arising from the dependence of wall energy on position in the gradient.[292] In multidimensional problems (Section 12) an additional term can arise from wall curvature.

In summary we have found that the problem of wall motion is reduced in the high-$Q$ limit to a pair of coupled first-order differential equations in two time-dependent variables, the wall displacement $q$ and the in-plane wall-magnetization precession angle $\psi$. These equations of motion have the same form as Hamilton's equations of motion for two "canonically conjugate" dynamical variables, position and momentum. In our case $q$ is the position and $2M\psi/\gamma$ is formally the "canonical momentum" per unit area. Clearly we are not talking here of a physical momentum in the usual sense of moving matter. Nevertheless, as we shall see, it is useful to think of $\psi$ as proportional to the momentum of the wall because many of the features of wall motion are closely analogous to the properties of moving matter, including the existence of a kinetic energy and hence an effective mass.

Considerable insight into the nature of the possible solutions to the equations of wall motion may be obtained from considering energy conservation.[287] When a wall moves at some velocity $\dot{q}$ in a net uniform drive field $H$, the rate of energy gained per unit area is

$$\dot{\sigma} = -2MH\dot{q}. \qquad (10.13)$$

This energy gain must, according to energy conservation, be either stored in the intrinsic wall energy or else dissipated to the lattice by viscous damping (we ignore coercive loss and wave excitations in the remainder of this section). The rate of energy loss per area $\dot{\sigma}$ can be evaluated from the integral over all space of $(\partial \sigma/\partial q)\dot{q} + (\partial \sigma/\partial \psi)\dot{\psi}$, which, using Eqs. (10.8) and (10.9), gives

$$\dot{\sigma} = -(2M\alpha/\gamma)[(\dot{q}^2/\Delta) + \Delta \dot{\psi}^2]. \qquad (10.14)$$

This expression is the "dissipation function" for wall motion and is analogous to Eq. (3.9), the term in $\dot{q}$ representing $\dot{\theta}$ loss and the term in $\dot{\psi}$ representing $\dot{\phi}$ loss.

Let us consider the limiting case in which relatively little kinetic energy is stored in the wall (compared to $K$), so that we can replace $\Delta$ by the constant $\Delta_0 = (A/K)^{1/2}$ in accordance with Eq. (8.2). Then, if the motion is periodic or steady, there can be no continuing change of energy stored in the wall, and so, averaging over time and equating (10.13) and (10.14), one finds

$$\langle \dot{q} \rangle = (\alpha \Delta_0/\gamma H)[\Delta_0^{-2}\langle \dot{q}^2 \rangle + \langle \dot{\psi}^2 \rangle], \tag{10.15}$$

where $\langle \ \rangle$ represents the time average.

Physically Eq. (10.15) expresses the intuitive fact that the mean velocity is proportional to the mean rate of energy dissipation. However the rate of energy dissipation is itself related to the wall velocity through the term in $\langle \dot{q}^2 \rangle$. If one considers the case $\dot{\psi} = 0$, and $\dot{q}(t) = V$, which is the case of steady-state motion, one finds the important result

$$V = \alpha^{-1}\gamma \Delta_0 H. \tag{10.16}$$

The velocity is in this case inversely proportional to $\alpha$. This is the conventional domain-wall mobility formula first derived by Landau and Lifshitz.[2] For a more general kind of motion, one has the inequalities $\langle \dot{q}^2 \rangle \geq \langle \dot{q} \rangle^2$ and $\langle \dot{\psi}^2 \rangle > 0$, whose combination with Eq. (10.15) yields

$$\langle \dot{q} \rangle < \alpha^{-1}\Delta_0\gamma H, \tag{10.17}$$

showing that the conventional formula puts an upper bound on the velocity for any $H$.

Next we derive a lower bound on the velocity. First we rewrite Eq. (10.14) by eliminating $\dot{q}$ and $\dot{\psi}$ using Eqs. (10.8) and (10.9):

$$\dot{\sigma} = -\gamma[2M(\alpha + \alpha^{-1})]^{-1}[\Delta_0(\partial\sigma/\partial q)^2 + \Delta_0^{-1}(\partial\sigma/\partial\psi)^2]. \tag{10.18}$$

Combining Eq. (10.18) with (10.13) and averaging over periodic motion, we find

$$\langle \dot{q} \rangle = \gamma[4M^2H(\alpha + \alpha^{-1})]^{-1}[\Delta_0\langle(\partial\sigma/\partial q)^2\rangle + \Delta_0^{-1}\langle(\partial\sigma/\partial\psi)^2\rangle]. \tag{10.19}$$

Taking $\langle(\partial\sigma/\partial q)^2\rangle = (2MH)^2$ and noting that $\langle(\partial\sigma/\partial\psi)^2\rangle \geq 0$, we have

$$\langle \dot{q} \rangle \geq (\alpha + \alpha^{-1})^{-1}\Delta_0\gamma H. \tag{10.20}$$

Possible velocities falling between the bounds (10.17) and (10.20) range over a factor of $\alpha^{-2} + 1$, which, for small $\alpha$, can be very great. The lower velocity bound is actually realized by one-dimensional wall motion at high drives (Section 11,C) and in hard-wall motion (Section 13).

In the limit $\alpha \to 0$ one might expect $\langle(\partial\sigma/\partial q)^2\rangle$ and $\langle(\partial\sigma/\partial\psi)^2\rangle$ to have finite limits. If this is true then Eq. (10.20) forces the conclusion $\langle \dot{q} \rangle \propto \alpha$ for $\alpha \to 0$. The fact that $\langle \dot{q} \rangle$ vanishes for $\alpha = 0$ should not be surprising when one considers that for $\alpha = 0$ our model does not provide any long-term drain for the energy supplied by the applied field; hence the wall cannot move forward on the average. In the opposite limit $\alpha \to \infty$, the conventional relation $\langle \dot{q} \rangle = \alpha^{-1}\Delta\gamma H$ is valid, and $\langle \dot{q} \rangle$ again vanishes. Thus on general grounds we expect to optimize $\langle \dot{q} \rangle$ at some $\alpha$, for any given $H$.

## 11.  Applications of the One-Dimensional Theory and Comparison to Experiment

### A.  LINEAR MOBILITY

As we have already seen in the previous section, a linear-mobility formula for wall velocity, $\mu = \dot{q}/H = \gamma\Delta/\alpha$ is valid when $\dot{\psi} = 0$. This result also comes from the pressure equation (10.11), provided we include restoring forces and coercive fields in $H$. $\dot{\psi} = 0$ is the case of "dynamic equilibrium" because from the torque equation (10.10) it is clear that if $\psi$ is constant, the torques are constant and hence the velocity is constant. Thus the wall translates in the $y$ direction with a fixed spin structure. Of course dynamic equilibrium is only possible if the torques are sufficiently large to sustain the required velocity. We will consider the torque limits in Section 11,C. The fact that the mobility is proportional to $\Delta$ is reasonable since $\pi\Delta$ gives the approximate distance over which the wall moves for a $\theta$ precession of $\pi$. It is instructive to visualize the dynamic equilibrium spin configuration corresponding to the given velocity $\dot{q} = \gamma\Delta H/\alpha$. In the simple case of demagnetizing torques only, the statically favored wall is a Bloch wall, but at dynamic equilibrium we find that $\psi$ must be canted out of the wall plane by an angle

$$\psi = \tfrac{1}{2}\arcsin(\dot{q}/2\pi\gamma\Delta M) \equiv \tfrac{1}{2}\arcsin(H/2\pi\alpha M). \tag{11.1}$$

On the other hand if we have a large in-plane field or in-plane anisotropy perpendicular to the wall, we have a static Néel wall (see Section 7), and in this case the dynamic equilibrium $\psi$ is rotated slightly back towards the Bloch configuration.

In all these cases, however, it is important to recognize the generality of the linear mobility formula $\mu = \dot{q}/H = \gamma\Delta/\alpha$. The formula is independent of the nature of the torques in Eq. (10.10) except insofar as there are first order changes in $\Delta$, given by Eq. (7.15). Thus the formula applies equally well to orthoferrites with Dzialoshinski exchange and to garnets and amorphous materials with any combination of in-plane fields and in-plane anisotropy, provided of course that the one-dimensional model is valid. Typical results for $\mu$ in various bubble materials by various experimental techniques are given in Table 11.1 extending the previous tabulation of Hagedorn.[20]

Since $\mu$ is inversely proportional to $\alpha$, we see from the table that by and large the mobilities follow the expected trends discussed in Section 3,C, with materials containing only the spherically symmetrical or nonmagnetic ions such as $Gd^{3+}$, $Eu^{3+}$, $La^{3+}$, $Lu^{3+}$, or $Y^{3+}$ on the rare earth sites showing the highest mobilities. For typical garnet material parameters $\Delta \sim 0.05$ $\mu$m and $\gamma = 1.76 \times 10^7$ $\sec^{-1}Oe^{-1}$, the mobilities imply $\alpha$ values ranging

# TABLE 11.1

Mobilities of Epitaxial Garnet (YIG-Based) and Amorphous (GdCo-Based) Films at Room Temperature[a]

| Composition | Mobility (cm/sec Oe) | Experimental Technique[a] | $4\pi M$ (G) | $Q$ | $h$ ($\mu$m) | $w$, $d$, or $l$ ($\mu$m) | Ref. | Comments |
|---|---|---|---|---|---|---|---|---|
| **YIG-based Garnets** | | | | | | | | |
| **Nonmagnetic rare earths** | | | | | | | | |
| $Ca_{2x}V_x$ | 170 | b | 240 | 14 | 3.1 | 5 (w) | 295 | |
| $Ga_x$ | 1340 | a | 88 | | | 0.53 (l) | 296 | CVD, pre- and post-irradiation |
| $La_{0.1}Ga_{1.2}$ | 30000 | a, d | 60 | 21 | 2.6 | | 95, 213 | $H_p$ |
| $Ga_{1.2}$ | 1500 | a | 175 | 4.1 | 6.3 | 7 (d) | 205 | CVD |
| $Ga_{1.4}$ | 380 | a | | | 5 | | 297 | CVD |
| $LuCaGe$ | 2000 | a | 171 | | 11 | 0.47 (l) | 300 | |
| $LaLuCaGe$ | 2700 | — | 145 | | 4.1 | 0.79 (l) | 298 | II |
| $LuLaCaGe$ | 600 | e | 135 | | 3.9 | 0.81 (l) | 192 | |
| **Eu-based** | | | | | | | | |
| $Eu_{0.65}Ga_{1.2}$ | 1750 | a | 160 | 8.9 | 4.3 | 0.77 (l) | 299 | $H_p$ |
| $Eu_{0.6}Ga_{1.2}$ | 1640 | a | 196 | | 8.1 | 6.2 (d) | 93 | |
| $Eu_{1.45}Ca_{1.1}Si_{0.6}Ge_{0.5}$ | 1400 | a | 218 | 6.9 | 4.2 | 5.18 (w) | 83–4 | High-g |
| $Eu_{1.2}Lu_{1.8}Ga_{0.5}$ | 700 | a | 769 | 4 | 1.1 | 1.05 (w) | 301 | $\rho = 40°$ |
| $Eu_{0.2}Ca_{0.92}Ge_{0.92}$ | 2600 | a | 168 | 4.5 | 7.1 | 6.2 (w) | 302 | |
| $Eu_{1.02}Gd_{0.59}Al_1$ | 470 | a | 200 | 8.2 | | | 206 | Preanneal |
| $Eu_{1.02}Gd_{0.59}Al_1$ | 1070 | a | 200 | 0.8 | | | 206 | Postanneal |
| $Eu_{0.75}Gd_{0.75}Al_{0.5}Ga_{0.1}$ | 1000 | a | 327 | 4.5 | 9.2 | 0.18 (l) | 271 | |
| $Eu_{0.18}Gd_{0.65}Yb_{0.47}Ga_{0.95}$ | 1380 | a | 190 | 2.3 | | | 206 | |
| $Eu_{0.6}$ $Gd_{0.65}Ga_x$ | 640 | a | 224 | | 12.2 | | 93 | |
| $Eu_{1.8}Tm_{0.2}Al_x$ | 470 | a | 210 | | 11.3 | | 303 | |
| $Eu_{0.8}Tm_{2.2}Ga_{0.5}$ | 1000 | a | 708 | 3 | 1.0 | 1.15 (w) | 301 | |
| $Eu_{0.15}Tm_{0.35}CaGe$ | 1300 | a | 300 | 4.8 | 3.7 | 3.28 (w) | 304 | $\rho = 45°$ |

*(Continued)*

TABLE 11.1 (Continued)

| Composition | Mobility (cm/sec Oe) | Experimental Technique[a] | $4\pi M$ (G) | $Q$ | $h$ ($\mu$m) | $w, d,$ or $l$ ($\mu$m) | Ref. | Comments |
|---|---|---|---|---|---|---|---|---|
| **YIG-based Garnets** | | | | | | | | |
| **Eu-based (Continued)** | | | | | | | | |
| $Eu_{0.3}Tm_{0.3}Ca_{0.88}Ge_{0.88}$ | 1400 | — | 218 | 4.1 | 4.6 | 4.7 ($w$) | 302 | |
| $Eu_1Yb_2Ga_{0.4}$ | 430 | a | 663 | 4 | 2.0 | 1.5 ($w$) | 301 | $\rho = 30°$ |
| $Eu_{1.7}Yb_{1.3}$ | 400 | a | 1450 | 2.1 | 2.8 | 1.1 ($w$) | 301 | $\rho = 60°$ |
| $Eu_{1.85}Yb_{0.15}Al_x$ | 650 | a | 194 | | 7.4 | | 93 | Pre- and post-II |
| $Eu_{1.7}Yb_{<0.2}Al_1$ | 600 | a | 250 | | | 0.4 ($l$) | 305 | |
| $Eu_{1.99}Ho_{0.01}Al_x$ | 170 | a | 169 | | 9.0 | | 93 | |
| $Eu_{1.99}Pr_{0.01}Al_x$ | 800 | b | 254 | | 7.2 | | 93 | |
| $Eu_{0.66}Tb_{0.04}Ga_{1.15}$ | 580 | a, b | 179 | 8.0 | 5.25 | 5.6 ($d$) | 175 | $\rho = 13°$ |
| **Gd-Based** | | | | | | | | |
| $Gd_{0.45}Ga_{1.1}$ | 6700 | d | 132 | 2.6 | 4.44 | 0.48 ($l$) | 306 | |
| $Gd_{0.46}Ga_{1.05}$ | 2000 | b | 140 | 0.8 | | 0.42 ($l$) | 307 | |
| $Gd_{2.1}Lu_{0.9}Al_{0.6}$ | 5000 | a | | 4–5 | | 6 ($d$) | 32 | |
| $Gd_{0.72}Tm_{1.2}Ga_{0.8}$ | 630 | a | | | | | 268 | Pre- and post-capping layer |
| $Gd_xTm_yGa_z$ | 1000 | a | 225 | | | | 296 | Pre- and post-irradiation |
| $Gd_{1.09}Tm_{0.93}Ga_{0.68}$ | 600 | b | 168 | 6 | | 0.66 ($l$) | 307 | Pre- and post-irradiation |
| $Gd_1Tm_1Ga_x$ | 410–4500 | a | 200 | | 7.8 | | 93 | |
| $Gd_1Tm_1Ga_{0.7}$ | 3250 | a | 200 | | 5.5 | | 93 | Double layer |
| $Gd_{1.2}Tm_{0.9}Al_{0.1}Ga_{0.3}$ | 2700 | a | 249 | 1.8 | 5.6 | 0.29 ($l$) | 271 | |
| $Gd_xTm_yGa_z$ | 1000 | a | 200 | | 5.9 | 0.67 ($l$) | 308–9 | Pre- and post-80 Å NiFe |
| $Gd_{0.9}Tm_{1.8}Ga_{0.75}$ | 2050 | a | 157 | 8.9 | 3.8 | 0.84 ($l$) | 310 | $H_p$ |
| $Gd_{0.7}Yb_{0.75}Ga_{0.9}$ | 1500 | — | 150 | 4.6 | 3.6 | 0.8 ($l$) | 311 | |
| $Gd_{0.99}Yb_{0.86}Ga_{0.84}$ | 1750 | b | 160 | 1.9 | | 0.45 ($l$) | 307 | Pre- and post-irradiation |
| $Gd_{1.55}Yb_{1.55}Al_{0.7}$ | 1560 | a | 144 | 2.4 | | | 206 | |
| $Gd_{1.55}Yb_{1.55}Al_{0.7}$ | 650 | a | 62 | 9.7 | | | 206 | |
| $Gd_{1.5}Yb_{0.6}Al_{0.5}$ | 600 | a | 186 | 2.5 | 10.4 | 0.36 ($l$) | 271 | Pre- and post-irradiation |

| Material | | Technique | | | | | Ref. | Comments |
|---|---|---|---|---|---|---|---|---|
| **Er-Based** | | | | | | | | |
| $Er_3Ga_{0.6-0.7}$ | 200 | b | 132 | 5 | 4.4 | 1.5 (*l*) | 312 | CVD |
| $Er_xGa_y$ | 70 | a | | | | | 296 | CVD, Pre- and post-irradiation |
| $Er_2Eu_1Ga_x$ | 96 | a | 215 | 5.1 | 6.8 | | 93 | |
| $Er_2Eu_1Ga_{1.7}$ | 272 | b | 194 | 3.3 | 6.3 | | 313 | |
| $Er_1Gd_1Al_{0.1}Ga_{0.4}$ | 440 | b | 190 | 7.5 | | 0.38 (*l*) | 271 | |
| $Er_{2.2}Gd_{0.8}Ga_{0.44}$ | 200 | b | | | | 0.78 (*l*) | 307 | Pre- and post-irradiation |
| **Sm-Based** | | | | | | | | |
| $Sm_{0.4}Ga_{1.2}$ | 200–320 | a | 200 | 4–5 | 5–8 | | 314 | |
| $Sm_{0.1}Ca_{0.98}Ge_{0.98}$ | >1500 | a | 1750 | | | 6 (*d*) | 32 | $\rho = 30°$ |
| $Sm_{1.5}Lu_{1.5}$ | 190 | a | 142 | 1.5 | 2.3 | 1.25 (*w*) | 315 | |
| SmGa | 266 | e | 166 | | 4.4 | 0.75 (*l*) | 192 | |
| SmLuCaGe | 236 | e | | | 4.3 | 0.77 (*l*) | 192 | |
| $Sm_{0.3}Tm_{0.75}CaGe$ | 800 | a | 500 | 3.7 | 1.6 | 1.3 (*w*) | 316 | |
| $Sm_{0.4}Lu_{0.56}CaGe$ | 600 | a | 518 | 3.2 | | 1.66 (*w*) | 304 | |
| **Tm-Based** | | | | | | | | |
| $Tm_1Ga_{0.9}$ | 1130 | a | 215 | 3.5 | 11.2 | 0.37 (*l*) | 271 | |
| **Amorphous** | | | | | | | | |
| GdCoMo | 200 | c | 520 | 4.6 | 2 | 4.5 (*w*) | 317 | |
| GdCoMo | 200 | a | 447 | 2.7 | 1.4 | 2 (*w*) | 191 | |
| GdCoCu | 300 | a | 1750 | 3.4 | 2.3 | 1.2 (*w*) | 318 | |
| GdCoCu | 90 | a | 1600 | 2.8 | 0.9 | 0.9 (*w*) | 318 | |
| GdCoAu | 490 | a | 1000 | — | 1.7 | 0.3 (*l*) | 319 | |

[a] This tabulation extends the previous one by Hagedorn. Experimental technique is coded as (a) pulsed gradient-field propagation, (b) dynamic bubble expansion with high-speed photography, (c) bubble collapse, (d) photometric measurement of straight-wall displacement, (e) stripe run-in or run-out. Domain size is indicated by the zero field stripe width $w$, an average bubble diameter $d$, or the length parameter $l$. In "Comments," CVD refers to garnet films made by chemical vapor deposition, all others being prepared by liquid phase epitaxy (LPE). II means "ion-implanted." $\rho$ gives apparent bubble deflection angle. $H_p$ refers to measurements made in the presence of in-plane field.

from 0.5 down to 0.005 in the case of LaGaYIG. However even in materials grown under carefully controlled conditions, the mobility is often variable from place to place in the film and also from film to film.[314] In such cases an unusually low mobility film can be annealed to regain the more typical mobility value, suggesting that structural defects affect the mobility. On the other hand, surface ion implantation usually does not affect mobility severely even though it presumably introduces defects.[209,267,320] Nuclear irradiation has also been found to have little effect on the mobility in garnet films and orthoferrites.[20,296] Mobility of garnets tends to increase with increasing temperature, and an empirical correlation has been found with the temperature dependence of $M/\sigma$.[321,322] Relatively little has been done so far to interpret physically this behavior or the effects of annealing,[302,323] and we refer the reader to the literature in this area.

Since $\mu$ is proportional to $\Delta$, there can be first-order changes in $\Delta$ according to Eq. (7.15) that affect the mobility. For example, apparent velocities in gradient propagation experiments in garnet films tend to increase with in-plane field.[209,259] The effect is stronger for fields oriented perpendicular rather than parallel to the direction of motion, consistent with Eq. (7.15), but the fact that velocities increase in both cases seems inconsistent. In a careful single-wall experiment, a 6% difference in mobility was detected in $YFeO_3$ for walls moving along the $a$ and $b$ directions of the orthoferrite.[324] Since Dzialoshinski exchange forces the wall moment to lie along the $a$ direction, this experiment compares the mobility of Bloch and Néel wall configurations. According to Eq. (7.15) the difference should be just $\frac{1}{2}Q^{-1}$ or 0.1% for $YFeO_3$. The experimental difference was explained in terms of the different lattice parameters along the $a$ and $b$ directions, which cause an anisotropy in the effective $A$ values, and hence in $\Delta_0$.[324,325]

There is evidence that the mobility increases, and the damping constant decreases, with applied in-plane field in a fashion which cannot be accounted for by the $\Delta$-effect described above.[326,327] For example, in a careful study of a LaGaYIG film, $\mu$ was determined from the decay of underdamped wall oscillations in a photometric single-wall experiment.[327] The damping $\alpha$, extracted from $\mu$, decreased linearly from 0.011 at zero in-plane field to 0.004 at 200 Oe. The explanation of this remarkable effect is not certain. One possibility is that flexural vibrations of the domain wall contribute significantly to the wall damping, for the spectrum of these vibrations moves rapidly to higher frequencies in the presence of in-plane fields (Section 22,A).

It should be noted, however, that particularly for high-mobility garnet materials, doubts have arisen about the experimental interpretation of the linear mobility in many cases. One of the problems is that often two different mobility regions occur as a function of drive in gradient propagation experiments as illustrated in Fig. 6.4. Effective damping parameters can be assigned to each region, but the physical significance of these different parameters is obscure.[328] More recent work has revealed that "ballistic overshoot" and other nonlinearities can occur in both gradient-propagation and bubble-collapse experiments, at very low drive fields in low-damping films. Thus many of the experimental mobility values in Table 11.1 are questionable. The only reliable results for high mobilities ($\mu \geq 500$ cm/sec Oe) are those obtained in the presence of strong in-plane field, in-plane anisotropy or angular-momentum compensation, as will be explained in Section 11,C.

## B.  Döring Mass

In the previous section we have discussed the case of wall motion at dynamic equilibrium. Let us now consider some of the transient effects predicted by Eqs. (10.10) and (10.11). At static equilibrium the direction of $\psi$ is such that the net torques on $\psi$ vanish, and according to Eq. (10.10) the wall velocity is therefore zero. For small deviations of $\psi$ from equilibrium, the torques in general increase linearly with $\psi$ and therefore so does the wall velocity. We consider now what happens if a step-function drive field is applied to a stationary wall. Since intially $\dot{q}$ is zero, $\psi$ starts precessing at the Larmor frequency $\gamma H$ according to Eq. (10.11). As $\psi$ increases, the torque terms of Eq. (10.10) increase and therefore the velocity increases, leading to a drop in the precession rate according to Eq. (10.11). $\dot{\psi}$ will drop to zero provided the torques and hence the velocity increase sufficiently. Thus a dynamic equilibrium is soon reached with the spins in their new canted position and the wall moving at a constant wall velocity given by the linear-mobility formula. The direction of the equilibrium canting of $\psi$ away from its static orientation can easily be found by using the right-hand rule (Section 3,B) to determine the polarity of the transient precession $\dot{\psi} = \gamma H$.

The effect of the transient precession process on wall velocity can be more explicitly evaluated for the case of small drives by linearizing the torque terms in Eqs. (10.8) or (10.10):

$$\dot{q} = (\gamma/2M)(\partial^2\sigma/\partial\psi^2)_{eq}(\psi - \psi_{eq}) + \alpha\Delta\dot{\psi}, \tag{11.2}$$

where $\psi - \psi_{eq}$ is the deviation from the static equilibrium value $\psi_{eq}$, and where $(\partial^2\sigma/\partial\psi^2)_{eq}$ is evaluated at $\psi_{eq}$. If we now eliminate $\psi$ in Eqs. (10.11) and (11.2), we find directly

$$m\ddot{q} + b\dot{q} + kq = 2M(H_a - H_c \operatorname{sgn} \dot{q}), \tag{11.3}$$

where

$$b = 2M/\mu, \qquad \mu = \gamma\Delta/\alpha, \tag{11.4}$$

$$m = (2M/\gamma)^2(1 + \alpha^2)(\partial^2\sigma/\partial\psi^2)_{eq}^{-1}. \tag{11.5}$$

This is of course the oscillator equation considered in Section 5,A. If $k = 0$, we see that in response to a step field $H$, the velocity approaches its equilibrium value $v = 2M(H_a - H_c)b^{-1}$ in an exponential fashion with a rise time $\tau = m/b$. Such acceleration effects are caused by the mass term $m\ddot{q}$, which in turn depends on $\partial^2\sigma/\partial\psi^2$. On the other hand, if the drive is so large that $\psi$ no longer remains small, then this linear approximation for the torques breaks down and Eq. (11.3) is no longer valid. One must then integrate the equations of motion (10.10) and (10.11) directly. This case will be considered in the next section.

Further insight into the nature of the wall mass can be gained by evaluating the additional energy per unit area stored in the wall during its motion, assuming small $\psi - \psi_{eq}$. If we ignore damping terms, this kinetic energy is $\frac{1}{2}(\partial^2\sigma/\partial\psi^2)_{eq}(\psi - \psi_{eq})^2$ as can be derived from Eq. (7.16). Since $2M(\psi - \psi_{eq})/\gamma$ is the canonical momentum $p$ as discussed in Section 10, and the kinetic energy is $E = \frac{1}{2}mV^2 = p^2/2m$, we deduce a mass $(2M/\gamma)^2(\partial^2\sigma/\partial\psi^2)^{-1}$, in agreement with Eq. (11.5) ignoring damping. Again it should be emphasized that this mass is not a real mass but only an effective mass whose properties are merely analogous to real matter because of the analogy between the wall-dynamics equations of motion and Hamilton's equations of motion.

We can now evaluate the mass for a number of different situations by finding $(\partial^2\sigma/\partial\psi^2)_{eq}$ from Eqs. (7.16) or (7.20) or by linearizing Eq. (10.10) directly. For the case of local demagnetizing energy only, $(\partial^2\sigma/\partial\psi^2)_{eq} = 8\pi\Delta_0 M^2$; so

$$m_D = (2\pi\gamma^2\Delta_0)^{-1}(1 + \alpha^2), \qquad \Delta_0 = (A/K)^{1/2}. \tag{11.6}$$

This is the famous Döring mass formula including damping[2,84] and is valid for any $Q > 0$. It implies a velocity rise-time constant $m/b$ of

$$\tau_D = (4\pi\alpha\gamma M)^{-1}(1 + \alpha^2). \tag{11.7}$$

That the mass in Eq. (11.6) is inversely proportional to the wall width can be understood from the fact that the stored energy $\frac{1}{2}mV^2$ must clearly be proportional to the wall width whereas the velocity is also proportional to it. Similarly the $\gamma^{-2}$ dependence can be understood from the fact that the velocity is proportional to $\gamma$ whereas the stored energy is independent of $\gamma$. It is harder to see how a damping correction can contribute to the mass, but this is a result of the $\dot{\psi}$ losses that inevitably occur as the wall magnetization moves to its dynamic transient position. There is no analogous term in the mechanics of real matter.

In the presence of in-plane anisotropy with easy axes parallel or perpendicular to the wall plane, the mass can similarly be shown to be

$$m = (2\pi\gamma^2\Delta_0)^{-1}|1 \pm (K_p/2\pi M^2)|^{-1}(1 + \alpha^2). \tag{11.8}$$

Thus if $K_p = 2\pi M^2$ and the wall is perpendicular to the easy axis [negative sign in Eq. (11.8)], we can have a remarkable situation—an infinite mass, which can be understood as follows: If a drive field is applied, Eq. (10.11) predicts the magnetization will precess. But Eq. (10.10) shows that the de-magnetizing and anisotropy torques cancel each other and so the velocity remains zero if damping is negligible. Thus the wall does not move, as one might expect for a wall with infinite mass.

Now we give some results for wall mass in the presence of an external in-plane field $H_p$, correct in the limit $Q \gg 1$ and $H_p \ll 2K/M$. If the field is applied to a wall whose magnetization initially points along the $x$ direction ($\psi = 0$, see Fig. 11.1), the mass is

$$m = (2\pi\gamma^2\Delta_0)^{-1}[1 + (H_x/8M)]^{-1}(1 + \alpha^2), \qquad H_x > -8M \quad (11.9)$$

$$m = (2\pi\gamma^2\Delta_0)^{-1}[1 - (H_x/8M)]^{-1}(1 + \alpha^2), \qquad H_x < -8M \quad (11.10)$$

for fields $H_x$ in the plane of the wall, and

$$m = (2\pi\gamma^2\Delta_0)^{-1}[1 - (H_y/8M)^2]^{-1}(1 + \alpha^2), \qquad |H_y| < 8M \quad (11.11)$$

$$m = (2\pi\gamma^2\Delta_0)^{-1}[|H_y/8M| - 1]^{-1}(1 + \alpha^2), \qquad |H_y| > 8M \quad (11.12)$$

for fields $H_y$ perpendicular to the wall. These results are plotted in Fig. 11.1. We can understand these different cases in the following way: for $H_x$ applied along the initial wall-magnetization direction ($\psi = \psi_H = 0$), the torques holding the spin in that direction increase. Thus for a given momentum $p = 2M\gamma^{-1}\psi$, the energy $p^2/2m$ is increased, and therefore the mass must decrease and $m^{-1}$ shown in the figure increase. The opposite of course

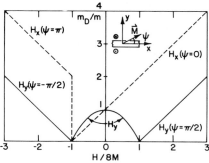

**Fig. 11.1.** Inverse wall mass [normalized to the Döring mass $m_D$ of Eq. (11.6)] versus in-plane field parallel ($H_x$) and perpendicular ($H_y$) to the wall plane in the one-dimensional model with $Q \to \infty$ [see Eqs. (11.9)–(11.12)]. At zero field the wall magnetization is assumed to point initially along the $x$ direction ($\psi = 0$). For $H_x > -8M$, $\psi = 0$, but for $H_x < -8M$, the wall magnetiza-tion is initially stable at $\psi = \pi$. For $|H_y/8M| < 1$, the initial value of $\psi$ varies continuously between $-\pi/2$ and $\pi/2$.

happens for oppositely directed magnetization $-8M < H_x < 0$. Once $H_x < -8M$, an instability occurs and there is no magnetostatic barrier to prevent the magnetization from flopping and lying initially along the field-favored direction ($\psi = \pi$). On the other hand the behavior of the two types of Bloch wall with respect to a field perpendicular to the wall ($\psi_H = \pm\pi/2$) is symmetric, as can be seen in Fig. 11.1. The field gradually pulls the static $\psi$ to the Néel orientation. At $|H_y| = 8M$, the mass becomes infinite ($m^{-1} = 0$) because a critical or labile point is reached (see Section 7).

In the case of the orthoferrites, the torque is dominated by the Dzialo-shinski exchange term and the mass based on Eq. (11.5) is

$$m = (2\pi\gamma^2\Delta_0)^{-1}(2\pi M^2/D\beta)(1 + \alpha^2). \tag{11.13}$$

Clearly the mass is reduced from the Döring value by the fraction $2\pi M^2/D\beta$, which is of order 1/1000. Considering the requirements for observing mass effects discussed in Section 5,B, this result, coupled with low restoring forces, makes it clear why wall-mass effects are not easily observed in ortho-ferrites.[20,162]

We now compare these predictions of wall mass to experiment. The Döring rise time constant of Eq. (11.7) is $\leq 15$ nsec for the typical garnet parameters $\alpha \geq 0.02$, $\gamma = 1.76 \times 10^7$ $Oe^{-1}sec^{-1}$ and $4\pi M = 200$ G. Thus one might expect negligible mass effects in any pulse experiment with pulse times much longer than 15 nsec or any sinusoidal excitation experiment at frequencies much below 70 MHz. For typical bubble films, however, one often finds resonances at frequencies as low as a few megahertz and ballistic overshoot effects lasting hundreds of nanoseconds. These results will be discussed in Chapters VIII and IX. The only cases where good agreement with the Döring

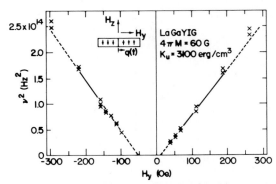

**Fig. 11.2.** Square of wall oscillation frequency versus in-plane field perpendicular to the wall in a LaGaYIG film, determined from photometric measurements on a single wall stabilized in an external field gradient and caused to oscillate by a step in bias field $H_z$ (after de Leeuw and Robertson[95]).

theory has been obtained are 1) in bulk platelets or crystals[160,329-332] or 2) in the presence of large in-plane fields.[94,112,306] Examples of the first case are early rf magnetic susceptibility experiments on polycrystalline ferrites,[329-331] single crystal cobalt ferrite,[332] and, more recently, experiments on GdYIG platelets as shown in Fig. 5.2.[160] An example of the second case is given in Fig. 11.2 where the resonance frequency squared is plotted against in-plane field.[95] Excellent agreement with Eq. (11.12) is obtained at large in-plane fields, except for a curious asymmetry in the behavior for positive and negative in-plane fields, which is discussed further in Section 16,C.

## C.  WALKER BREAKDOWN

Equations (10.8) and (10.10) require that the domain-wall velocity be sustained by a torque on the in-plane wall magnetization. What then, is the maximum velocity that can be sustained, corresponding to the maximum torque that can be supplied? We discuss this question in the limit $Q \gg 1$ or $\Delta = \Delta_0 = (A/K)^{1/2}$. Under conditions of dynamic equilibrium with $\dot\psi = 0$ and $\dot q = $ const., the torque must be supplied by the demagnetizing, applied-field or in-plane anisotropy torques as in Eq. (10.10), and clearly all of these terms have well-defined maxima. For example, if the only torque is a demagnetizing torque, the maximum is achieved for $\sin 2\psi = 1$, i.e., at $\psi = \pi/4$ or $\psi = 5\pi/4$, corresponding to the two static types of Bloch wall (see Section 7), with

$$\dot q_{max} \equiv V_w = 2\pi\gamma\Delta_0 M. \tag{11.14}$$

According to Eq. (10.11) with $\dot\psi = 0$, this velocity occurs at a net drive field $H [= H_a - (kq/2M) - H_c]$ of

$$H_w = 2\pi\alpha M. \tag{11.15}$$

These are the important formulas for the "Walker breakdown velocity" $V_w$ and "Walker critical field" $H_w$, for the high-$Q$ limit.[285] More general results for arbitrary $Q$ are given in the literature.[20,285,288]

Analogous expressions can be derived for the case of demagnetizing energy plus in-plane anisotropy[287]

$$\dot q_{max} = \gamma\Delta_0 K'/M \tag{11.16}$$

where

$$K' = [(2\pi M^2)^2 + K_p^2 + 4\pi M^2 K_p \cos 2\psi_p]^{1/2}. \tag{11.17}$$

This equation is valid under the assumption $K_p \ll K$ as well as $Q \gg 1$. If the in-plane easy axis is perpendicular to the wall plane ($\psi_p = \pi/2$), we once again see a curious anomaly that occurs when $K_p/2\pi M^2 = 1$. Here the critical velocity goes to zero and no velocity at all can be sustained under conditions of dynamic equilibrium. If an in-plane field $H_p \gg 8M$ is applied, then[95, 294a, 294b]

$$\dot{q}_{max} = (\pi/2)\gamma\Delta_0 H_p, \tag{11.18}$$

but for smaller values of $H_p$, the dependence of $\dot{q}_{max}$ on field cannot be expressed in a simple analytical formula. Calculated results for the case $H_p = H_x$ are shown in Fig. 11.3. We see that for opposite directions of the applied field $H_x$ the breakdown velocity can be either reduced or increased. When the velocity is reduced to zero at $H_x = -8M$, an instability occurs and the magnetization flops over to a different direction, causing the breakdown velocity to increase once again. The result for $H_p = H_y$ is the mirror image of Fig. 11.3.

To the critical velocity $\dot{q}_{max}$ in each of the above cases corresponds a critical drive field $H_{max} = \dot{q}_{max}/\mu = \alpha\dot{q}_{max}/\gamma\Delta_0$. Let us now consider what happens when net drive fields $H$ either smaller or larger than $H_{max}$ are applied as a step function to a wall with no restoring force. If $H < H_{max}$, Eqs. (10.10) and (10.11) show that after an initial transient period during which $\psi$ precesses forward and velocity increases, dynamic equilibrium is eventually reached with $\dot{\psi} = 0$ and $\dot{q} = V < \dot{q}_{max}$, as discussed in the previous section.

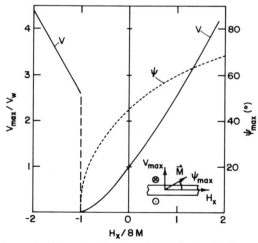

**Fig. 11.3.**   Maximum domain-wall velocity normalized to the Walker velocity $V_w = 2\pi\gamma\Delta M$ (solid line), and critical angle of wall magnetization (dotted line), as a function of in-plane field $H_x$ parallel to the wall, in the one-dimensional model with $Q \to \infty$. A field $H_y$ perpendicular to the wall gives a maximum velocity that is the mirror image of the above plot.

This behavior has been verified by a detailed computer calculation for general $Q$-values.[333]

However if $H > H_{max}$, the viscous drag term in Eq. (10.11) is insufficient to balance the drive term $\gamma H$, and $\psi$ must continue to precess forward. Thus we see that for $H > H_{max}$, no dynamic equilibrium can be established, and this condition is often called "breakdown." From the torque equation it is clear that as $\psi$ continues to precess, $\dot{q}$ must also oscillate because of the periodic nature of the torque terms with $\psi$.[176,177,287,333] In the limit of large driving fields $H \gg H_{max}$, Eq. (10.11) shows that $\psi$ processes essentially as $\dot{\psi} = \gamma H$, and thus the periodic torque terms average out. Under these circumstances, the only torque contributing to the net velocity is the damping torque term $\alpha \Delta_0 \dot{\psi}$, giving a velocity

$$\dot{q} = (\alpha + \alpha^{-1})^{-1} \gamma \Delta_0 H. \tag{11.19}$$

This condition is the "free precession" limit and represents the lower bound on the velocity predicted from energy conservation in Eq. (10.20). Figure 11.4 shows a plot of an analytic expression for the complete field dependence of the velocity averaged over a precessional cycle, for the case of demagnetizing and in-plane anisotropy torques.[287] If the damping is small, the average velocity drops above the breakdown velocity [Eq. (11.16)] and then increases again as the damping torque takes over. The region of decreasing velocity or negative mobility, however, is expected to be unstable, for if one area of the wall begins to lag, it experiences an increasing drive that, in this negative mobility region, will cause it to lag even further. Clearly the one-dimensional model is inadequate to describe such behavior. If the damping is large ($\alpha \gg 1$), the nonlinearity in $V$ versus $H$ is suppressed altogether.

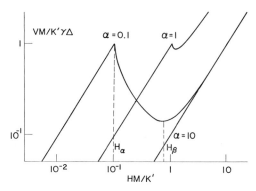

**Fig. 11.4.** Log–log plot of wall velocity normalized to the Walker velocity $V_w = \gamma \Delta K'/M$ [Eqs. (11.16), (11.17)], as a function of drive field $H$ for three values of damping $\alpha$, in the one-dimensional model with $Q \to \infty$. The motion is at dynamic equilibrium for $H < 2\pi\alpha M$ and oscillating otherwise (after Slonczewski[287]).

The precise range of drive fields in which the above two-coordinate $(q, \psi)$ model is valid is not clear. However, a more comprehensive solution of the Landau–Lifshitz equations in one dimension shows that when $H$ exceeds $\frac{1}{2}H_K \equiv K/M$ a great change in wall behavior occurs.[334] For then the fundamental wall-oscillation frequency $2\gamma H$ falls within the band of frequencies propagated by spin waves, thus permitting radiation of excited wall energy into the volume occupied by domains. Even in the range $H < \frac{1}{2}H_K$, harmonics of the fundamental wall oscillation radiate spin waves, but more weakly. The *lower* limit of $H$ for the one-dimensional theory in the presence of inhomogeneous stray fields should be no smaller than the heuristic *upper* limit for quasi-steady two- or three-dimensional behavior discussed in Section 12,E.

How does this model for maximum velocities and the onset of non-linearities beyond Walker breakdown compare to experiment? A vast assortment of experiments, to be discussed in more detail in Chapters VIII and IX, have shown that velocity maxima and nonlinearities do indeed occur, but at velocities generally an order of magnitude lower than Walker's velocity in typical garnet films in the absence of in-plane fields. One example is the photometric experiment on a LaGaYIG film shown in Fig. 11.5.[335] Equations (10.10) and (10.11) predict the solid line for displacement as a function of time of an isolated wall in a gradient, in response to a step drive field. There should be no breakdown, but in fact a strong deviation from the one-dimensional theory occurs, characterized by a roughly constant slope, that is, by a saturation velocity. This and other failures of the one-dimensional theory can only be explained by consideration of multidimensional theories described in the next chapter.

One case in which the one-dimensional theory seems to work is in experiments using drive pulses with rapid rise times.[129,174,336] For example, Fig. 4.3 shows stroboscopic data on a bias-pulsed bubble in a LuGdAlIG film.[129] The dots show experimental points for the radius change as a func-

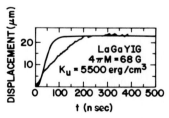

**Fig. 11.5.**  Wall displacement as a function of time determined photometrically for a single wall in a LaGaYIG film stabilized in a field gradient of 2.4 kOe/cm with a 400 Oe in-plane field perpendicular to the wall. The wall responds to a 5.8 Oe step in drive field. Smooth line represents a fit to the one-dimensional theory of Eqs. (10.10) and (10.11) (after de Leeuw[335]).

tion of time, while the solid line is a prediction corresponding approximately to the full integration of Eqs. (10.10) and (10.11).[333] The maximum observed velocity agrees very well with the predicted Walker velocity of 9800 cm/sec, and the oscillation frequency is well reproduced. However after $\sim 15$ nsec, the experimental behavior is observed to deviate radically from the theory and the velocity becomes much slower. A possible reason for the success of the one-dimensional model at short times but not at long times will be discussed in Section 12,E. One difficulty in fitting the theory to the data is that using $\alpha$ as an adjustable parameter, the optimal $\alpha$ may differ considerably from the FMR $\alpha$-value.[98] This problem has been discussed in Section 3,C.

Another case where the one-dimensional theory works tolerably well is in the presence of large in-plane fields $H_K \gg H_p \gg 4\pi M$ or large in-plane anisotropies $H_K \gg 2K_p/M \gg 4\pi M$.[43,44,46,47,94,95,335] For example, in photometric experiments on a single wall in a LaGaYIG film with $4\pi M = 68$ G in an in-plane field of 400 Oe, a dramatic velocity peak was observed, as shown in Fig. 11.6.[94] The peak is a factor of three below the predicted critical velocity of 90,000 cm/sec from Eq. (11.18), but more recent experiments

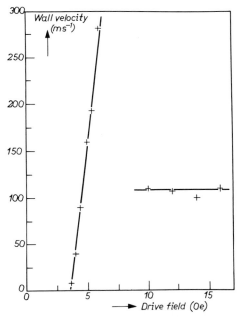

**Fig. 11.6.**  Velocity versus drive field for a photometric plane-wall experiment on a LaGaYIG film in a 400 Oe in-plane field, as in Fig. 11.5 (after de Leeuw[94]).

with more uniform films have shown improved agreement.[95,327] The effect of a large in-plane anisotropy has been studied by bubble collapse in a pair of EuTmGa garnet samples, one a (111)-film, with negligible $K_p$, and the other a bulk platelet cut from a (110) crystal facet with $K_p/2\pi M^2 = 2.6$.[43] The 650 cm/sec saturation velocity of the (111)-film is almost a factor of ten less than the predicted Walker velocity of 4500. By contrast in the platelet, velocities up to 6000 cm/sec were achieved, limited only by experimental considerations, while the expected critical velocity, according to Eq. (11.16) is of order 20,000 cm/sec. In another study of EuLuMnAlIG films grown on (110)-substrates, the critical velocities are high (20,000 cm/sec) but are still a factor of three less than the predicted value (60,000 cm/sec) for this composition.[46,47] Another problem is that there is a factor of three difference in peak velocity between wall motion parallel ($V_p = 7000$ cm/sec) and perpendicular ($V_p = 20,000$ cm/sec) to the in-plane easy axis. According to Eqs. (11.16) and (11.17), there should be no such difference for this film with $K_p/2\pi M^2 = 12$. The explanation for the low velocity in the parallel case may lie in the presence of a horizontal Bloch line in the film (Section 8,E).[337] Even so, use of materials with large in-plane anisotropies would appear to be an attractive way to insure high-velocity device operation. The drawback is in the complexity of materials preparation, although progress has recently been made in the preparation of (110) films with adequate anisotropies.[45-47]

The one-dimensional theory also may be appropriate for high-$g$ materials. A series of EuCaSiGeYIG films were studied in which $g$ ranged from 1 to 30.[86] The corresponding Walker velocities were predicted to range from 6,700 to 100,000 cm/sec. The low $g$ sample showed a velocity saturation in the bubble collapse experiment at 1200 cm/sec, a factor of 6 below the Walker velocity. By contrast the highest $g$ sample showed velocities of up to 60,000 cm/sec, limited only by experimental technique and only a factor of 2 from the predicted Walker velocity. The mobilities of these films were observed to be within 10% of each other, which can be understood from the fact that, as discussed in Section 3,C, $\gamma$ and $\alpha$ diverge together at angular momentum compensation. Thus $\gamma/\alpha$ and hence $\mu = \gamma\Delta/\alpha$ remain finite while the Walker velocity and critical field become very large.

Data on the orthoferrites show the opposite problem from that in the garnets, namely that wall velocities higher than the "Walker velocity" have been observed. The Walker velocity for $YFeO_3$ can be estimated to be 1000 cm/sec on a naive pure magnetostatic model using Eq. (11.14). Taking into account the effective in-plane anisotropy $K_p = D\beta$ arising from Dzialoshinski exchange [see Eq. (7.19)] one might use Eq. (11.16) to estimate a breakdown velocity of $4 \times 10^5$ cm/sec.[20,338] Recently a number of authors using bubble collapse,[339] stroboscopic techniques[340-342] and the Sixtus–

Tonks technique.[97,157] have observed velocities significantly higher than this "breakdown velocity," but with anomalies in this range. It has been suggested that the anomalies are due to magnetoelastic interactions, since the velocity of sound has roughly the same value, rather than to a "Walker breakdown."[157,343] Recently it has been pointed out that use of Eq. (11.16) with $K_p = D\beta$ is in fact a poor approximation except at low velocities, because as $\psi$ increases, the canting angle and net magnetization must decrease [see discussion after Eq. (7.20)].[344,344a] A plausible analysis of this case has not yet been reported.

# VI

# Wall Dynamics in Three Dimensions

In the previous chapter we have seen that the one-dimensional theory of wall motion is inadequate to describe most experimental results in bubble dynamics. The theory fails mostly because of Bloch lines, which demand a two- or three-dimensional treatment. Bloch lines can either be present in the domain wall to start with, as in the various wall states described in Chapter IV, or they can be nucleated and annihilated dynamically. In this chapter we describe the basic equations that are used to treat wall motion with Bloch lines. We also give a rather general physical picture of how Bloch lines affect domain-wall motion. We shall see that the main effects are (1) to provide deflection or distortion forces on moving domains, (2) to reduce the mobility, (3) to reduce the critical ("breakdown") velocities, and (4) to increase the effective mass. Furthermore we shall show how many of these effects can be derived from equations whose form is analogous to those of the one-dimensional model. Explicit examples of these effects are given later.

## 12. General Domain Dynamics

A. Basic Equations of Three-Dimensional Wall Motion

Let us consider a domain of arbitrary shape, illustrated in Fig. 12.1. Any point $P$ on its surface may be specified by the general curvilinear coordinates $\zeta$ and $\eta$. For example in the case of the bubble in Fig. 2.5 these coordinates could be $\zeta = \beta$, the angular position around the bubble circumference, and $\eta = z$, the position through the film thickness. The dynamics of the domain can be described by the evolution with time $t$ of the wall-normal displace-

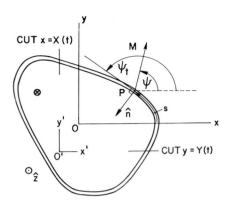

**Fig. 12.1.** Constant-$z$ section of a generally-shaped reversed domain. The cut planes $y = Y(t)$ or $x = X(t)$ mark a discontinuity $2\pi S$ of wall-moment angle $\psi$ around the domain perimeter. The point $P$ on the domain perimeter can be specified by the generalized coordinates $(\zeta, \eta)$.

ment $q(t, \zeta, \eta)$ and the wall-magnetization orientation angle $\psi(t, \zeta, \eta)$. Here $q$ is defined to point along the wall normal $\hat{n}$ and in a direction toward the domain with down magnetization, i.e., inward for a closed domain outside which the magnetization points up along $+z$, as shown in Fig. 12.1. This convention is consistent with that adopted in Chapter V. The angle $\psi(t, \zeta, \eta)$ at any point is measured counterclockwise in the film plane from a fixed axis in the medium. If we define $\psi_t(\zeta, \eta)$ as the angle of the wall tangent $\hat{n} \times \hat{z}$ measured from the same fixed axis, then $\psi - \psi_t$ corresponds to the definition of $\psi$ used in Chapter V.

We base our discussion on the Landau–Lifshitz equations integrated through the wall thickness to give [225,287]

$$\delta\sigma/\delta\psi = 2M\gamma^{-1}(\dot{q} - \alpha\Delta\dot{\psi}), \tag{12.1}$$

$$\delta\sigma/\delta q = -2M\gamma^{-1}(\dot{\psi} + \alpha\Delta^{-1}\dot{q} + \gamma H_c \,\mathrm{sgn}\,\dot{q}). \tag{12.2}$$

They are just like the one-dimensional equations (10.8) and (10.9), but with coercivity included explicitly and with functional partial derivatives $\delta\sigma/\delta\psi$ and $\delta\sigma/\delta q$ in place of ordinary partial derivatives $\partial\sigma/\partial\psi$ and $\partial\sigma/\partial q$. These equations must apply at every point on the curved wall surface. As before, these equations represent the balance of dynamic reaction forces against the torques $\delta\sigma/\delta\psi$ and pressures $\delta\sigma/\delta q$ arising from static internal or external forces such as magnetostatic or applied fields. Employment of functional derivatives, defined in Eq. (7.3), brings to bear additional effects arising from the dependence of $\sigma$ on $\nabla\psi$ and $\nabla q$, where the latter gradient vectors are understood to be tangential to the wall surface.

Consider the case where the wall surface is nearly parallel to the $xz$ plane.

To first order, we may take $q$ parallel to $y$ and we must interpret $\sigma$ to be the energy density projected on the $xz$ plane. Equation (8.3) for the wall energy may be generalized to

$$\sigma = \sigma_0[1 + \Delta_0^2(\nabla\psi)^2 + Q^{-1}\sin^2(\psi - \psi_t)]^{1/2}[1 + (\nabla q)^2]^{1/2}. \quad (12.3)$$

Here the term in $(\nabla\psi)^2$ represents the exchange energy of the $\psi(x, z)$ distribution, and the term in $(\nabla q)^2$ represents the effect of increased wall surface due to curvature. These two contributions were absent in the one-dimensional treatment. Except in the case of very hard walls ($|\nabla\psi| > \Delta_0^{-1}$) where the exchange term can become large, one can expand Eq. (12.3) to first order in the exchange, magnetostatic, and curvature terms. It is also useful sometimes to include other first-order corrections like the effect of in-plane fields $H_x$ and $H_y$ [e.g., see Eq. (7.16)] and the effective surface energy of wall displacement in a drive field $H_z$. Thus we have

$$\sigma = \sigma_0[1 + \tfrac{1}{2}(\nabla q)^2] + 2A\Delta_0(\nabla\psi)^2 + 4\pi\Delta_0 M^2 \sin^2(\psi - \psi_t)$$
$$- \pi\Delta_0 M(H_x \cos\psi + H_y \sin\psi) - 2MH_z q. \quad (12.4)$$

One must bear in mind that $\mathbf{H} = (H_x, H_y, H_y)$ may be a function of $q(x, z)$ if it arises from the magnetic charges on the film and wall surfaces. In particular, we consider $H_z$ to include the restoring force.

Evaluation of the functional derivatives $\delta\sigma/\delta\psi$ and $\delta\sigma/\delta q$ with the approximation $\psi_t = (\pi/2) - (\partial q/\partial x)$ gives

$$\delta\sigma/\delta\psi = 4\pi\Delta_0 M^2 \sin 2(\psi - \psi_t) + \pi\Delta_0 M(H_x \sin\psi - H_y \cos\psi)$$
$$- 4\Delta_0 A\nabla^2\psi \quad (12.5)$$
$$\delta\sigma/\delta q = -2MH_z - \sigma_0\nabla^2 q - 4\pi\Delta_0 M^2\partial[\sin 2(\psi - \psi_t)]/\partial x. \quad (12.6)$$

The partial differential equations (12.1) and (12.2) with the substitutions (12.5) and (12.6) provide a general basis for treating domain-wall movement in more than one dimension.[287]

## B. DYNAMICAL-VARIABLE FORMALISM

Unfortunately the coupled partial differential equations (12.1) and (12.2) cannot be solved in general. One approach is to proceed with a massive computer calculation, and recently there has been some work in this direction.[245,251,345-348] In many cases, however, we can make plausible approximations that allow us to use one or a few time-dependent dynamical variables $X_i(t)$ rather than the entire range of variables $q(t, \zeta, \eta)$ and $\psi(t, \zeta, \eta)$.[225] For example, if we assume that a bubble stays rigidly cylindrical with constant radius during bubble translation, we can describe $q(t, \zeta, \eta)$, with

$\zeta = \beta$ and $\eta = z$ (see Fig. 2.5), completely in terms of coordinates $X(t)$ and $Y(t)$ of the bubble center as

$$q(t, \beta, z) = -X(t) \cos \beta - Y(t) \sin \beta. \tag{12.7}$$

Other possible dynamical variables to be considered depending on the problem are the bubble radius, the average precession angle in the wall or the positions of various Bloch lines.

Let us assume we have specified relations like Eq. (12.7) for $q$ and $\psi$ in terms of each of the dynamical variables $X_i(t)$ $(i = 1, 2, \ldots, n)$ of the given problem:

$$q(t, \zeta, \eta) = q[\zeta, \eta, \{X_i(t)\}], \tag{12.8}$$

$$\psi(t, \zeta, \eta) = \psi[\zeta, \eta, \{X_i(t)\}].$$

Now we can derive equations of motion for the $X_i$'s. For this purpose a variational form of the equations of motion is helpful. Consider the total static energy variation

$$\delta W = \iint dA[(\delta\sigma/\delta\psi)\delta\psi + (\delta\sigma/\delta q)\delta q], \tag{12.9}$$

where $\delta\psi$ and $\delta q$ are arbitrary variations and the indicated surface integral is carried out over the entire wall surface $A$. Substituting Eqs. (12.1) and (12.2) gives us

$$\delta W = (2M/\gamma) \iint dA[(\dot{q} - \alpha\Delta\dot{\psi})\delta\psi - (\dot{\psi} + \alpha\Delta^{-1}\dot{q} + \gamma H_c \,\text{sgn}\, \dot{q})\delta q]. \tag{12.10}$$

This equation gives the increment of the total stored energy due to variations $\delta q$ and $\delta\psi$, which are virtual in the sense that they can be considered regardless of whether or not they actually take place in the motion $q(t)$, $\psi(t)$. Taking the partial derivative with respect to $X_i$, we find the important relation:

$$\partial W/\partial X_i = (2M/\gamma) \iint dA[(\dot{q} - \alpha\Delta\dot{\psi})(\partial\psi/\partial X_i)$$
$$- (\dot{\psi} + \alpha\Delta^{-1}\dot{q} + \gamma H_c \,\text{sgn}\, \dot{q})(\partial q/\partial X_i)]. \tag{12.11}$$

Here $-\partial W/\partial X_i$ is the force arising from the static dependence (hopefully known) of $W$ on $X_i$. A good example is the total force on a bubble in a gradient $\nabla H_z$ [Eq. (2.14)]

$$\mathbf{F} = -\partial W/\partial \mathbf{X} = -2\pi r^2 h M \nabla H_z. \tag{12.12}$$

Of course $W$ can also include internal wall energies such as exchange energy of Bloch lines.

Equation (12.11) shows that the generalized force $-\partial W/\partial X_i$ equals an expression that represents a net dynamic reaction force arising from gyrotropic, viscous damping, and coercive contributions. As we shall see in Section 14, we can often break this expression down into separate dynamic reaction forces from the Bloch wall and from each of the Bloch lines. When the integration in Eq. (12.11) is carried out after substituting Eqs. (12.8), one ends up with a set of $n$ differential equations in the dynamical variables $X_i$, $n$ being the number of dynamical variables. A simple example is the case of a bubble translating in a gradient but with a spin structure assumed to stay constant during the motion and point in a fixed direction everywhere, as in a large constant in-plane field, so that $\dot{\psi} = \partial\psi/\partial X = 0$. Thus the only dynamical variables are $X$ and $Y$ as in Eq. (12.7). Integrating Eq. (12.11) using Eq. (12.7), one finds the well-known result

$$\partial W/\partial \mathbf{X} = -2\pi Mrh\gamma^{-1}(\alpha\Delta^{-1}\dot{\mathbf{X}} + 4\pi^{-1}\gamma H_c\dot{\mathbf{X}}|\dot{\mathbf{X}}|^{-1}), \qquad (12.13)$$

which recovers Eq. (6.3) when combined with Eq. (12.12). Much more interesting results are obtained if one includes different spin structures, as will be described in Sections 13 and 14.

It should be emphasized that to be physically consistent, Eq. (12.11) must use $\psi$ measured from a fixed direction and at a fixed point in the medium. This is a crucial and easily overlooked point. For example consider a hard bubble ($|S| \gg 1$), as shown in Fig. 8.7, translating with a fixed spin structure $v(\beta) = S\beta + v_0$ in the moving coordinate frame. It might be thought that $\partial\psi/\partial X$ is zero becaue the spin structure does not change in the moving reference frame. But in fact, relative to a fixed point in the medium that sees a wall of finite width sweeping past, $\psi$ varies according to

$$d\psi(\beta,z) = r^{-1}S(dX \sin \beta - dY \cos \beta). \qquad (12.14)$$

This equation leads to some remarkable gyrotropic effects discussed in Sections 13 and 14.

## C. PHYSICAL PICTURE OF WALL MOTION WITH BLOCH LINES

Before treating specific examples with the formalism of the preceding section, it is valuable to obtain some physical insight into the general nature of domain-wall motion that emerges when the coupled equations (12.1), (12.2), (12.5), and (12.6) are applied to a problem involving Bloch lines. We start by noting that there is an important analogy between wall motion and precession in the polar angle $\theta$ on the one hand and Bloch-line motion within a wall and precession in the azimuthal angle $\phi$. The wall velocity $\dot{q}$ equals the wall-width parameter $\Delta$ times $\dot{\theta}$ at the center of a wall, and similarly the Bloch-line velocity within the wall $v_{BL}$ equals the Bloch-line width parameter $\Lambda$ times $\dot{\phi}$ at the center of a Bloch line.

Bloch lines play a great role in domain-wall dynamics because spin precession is generally concentrated in them. Consider first the case in which Bloch lines are constrained to fixed positions in the reference frame of a domain moving at velocity **V**, such as the hard bubble of Fig. 8.7. If the spatial rate of spin rotation in the Bloch lines is given by $\mathbf{V}\psi$, the physically relevant time rate of spin precession in the laboratory frame is

$$\dot{\psi} = -\mathbf{V} \cdot \nabla \psi . \tag{12.15}$$

$\dot{\psi}$ is particularly great when vertical Bloch lines are clustered on the flanks of a moving domain (points $B$ and $D$ of Fig. 8.7) because then $\nabla\psi \| \mathbf{V}$. According to Eq. (12.2), this precession must be balanced by a pressure on the flanks of the domains. Since this pressure is orthogonal to **V**, the domain tends to become distorted or deflected off its course. This is essentially the reason for the skew deflection effect to be discussed in Sections 13 and 14. The precession also gives a contribution to the damping through the term $\alpha\Delta\dot{\psi}$ in Eq. (12.1) and thus it affects the mobility, although this effect is small unless the numbers of Bloch lines are very great.

Next consider the case where the Bloch lines are not constrained but are free to move along the domain wall. In this case spins within the Bloch line have the greatest freedom to precess. In fact the Bloch-line spins are characterized by two degrees of freedom, i.e., the polar angles $\theta$ and $\phi$, because both the wall and the Bloch line can in many cases move freely. By contrast spins in the wall away from the Bloch lines are constrained in $\phi$, e.g., by local demagnetizing energy, and spins in the domains away from the walls are constrained in both degrees of freedom by anisotropy energy. The fact that both degrees of freedom are labile in the Bloch lines makes the spins within Bloch lines especially responsive to disturbances. For example, the occurrence of gyrotropic precession of spins requires freedom of two coordinates—one for the disturbing torque to act upon, and the other to vary or precess with time in response to this torque. For example, a wall pressure such as produced by the field component $H_z$ exerts a torque, included in the expression $\delta\sigma/\delta\theta$, and causing changes in $\phi$.

How does this precessional freedom of the Bloch-line regions affect wall motion? Usually it retards wall motion. For consider a wall partly occupied by Bloch lines and partly free of Bloch lines. In the part without Bloch lines we consider the wall moment restrained by wall demagnetization in the classical manner (e.g., see Section 11,A), so that $\dot{\psi} = 0$. Then Eq. (12.2) reduces to

$$\delta\sigma/\delta q = -2M\gamma^{-1}(\alpha\Delta^{-1}\dot{q} + \gamma H_{\mathrm{c}} \operatorname{sgn} \dot{q}) \qquad \text{(without Bloch lines)}. \tag{12.16}$$

But in the part *with* Bloch lines, the aforementioned lack of constraint on $\dot{\psi}$ implies that the torque $\delta\sigma/\delta\psi$ is small. For example, in the case of vertical Bloch lines in collapsing bubbles, $\delta\sigma/\delta\psi = 0$ (see Sections 13 and 14). If we therefore neglect $\delta\sigma/\delta\psi$ in Eq. (12.1), we find

$$\dot{q} = \alpha\Delta\dot{\psi} . \tag{12.17}$$

In such a region the wall velocity $\dot{q}$ is limited by the torque available from the viscous drag on the precession $\dot{\psi}$. This is small, particularly in low-loss materials having small $\alpha$. Thus, the Bloch-line regions tend to retard the wall motion. Indeed, eliminating $\dot{\psi}$ between Eqs. (12.17) and (12.2) results in

$$\delta\sigma/\delta q = -2M\gamma^{-1}[\Delta^{-1}(\alpha + \alpha^{-1})\dot{q} + \gamma H_c \operatorname{sgn} \dot{q}] \qquad \text{(with Bloch lines)}.$$

$$(12.18)$$

This equation for the motion of the Bloch-line containing surface differs from Eq. (12.16) only by the presence of the term with the factor $\alpha^{-1}$, which vastly decreases the effective mobility $\gamma\Delta/(\alpha + \alpha^{-1})$ in this part of the wall when $\alpha$ is small. Physically, much of any pressure exerted externally on the wall and contained in $\delta\sigma/\delta q$ of Eqs. (12.16) and (12.18) is consumed in inducing wall *precession* $\dot{\psi}$, leaving little to overcome the viscous drag $-2M\alpha\gamma^{-1}\Delta^{-1}\dot{q}$ retarding the wall *displacement*.

Because of the disparity in effective mobilities between wall regions with and without Bloch lines, it is obvious that if the external pressure is the same everywhere, the regions without Bloch lines will move farther forward than those with Bloch lines, as shown in Fig. 12.2. Let us assume the wall is sufficiently stiff that the curvature does not become very large and the entire wall moves together, that is, both kinds of wall move with the same mean $\dot{q}$. To satisfy Eqs. (12.16) and (12.18), with the same $\dot{q}$, $\delta\sigma/\delta q$ must be larger in the Bloch-line region than in the region without Bloch lines. This difference is provided by the surface tension term $\sigma_0\nabla^2 q$ of Eq. (12.6) that makes a contribution because of the wall curvature illustrated in Fig. 12.2.[349] The curvature increases the effective drive at the Bloch lines and decreases it away from the Bloch lines. Thus the domain wall billows around the Bloch

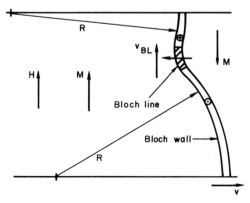

**Fig. 12.2.** Schematic moving domain wall containing a horizontal Bloch line propagating at velocity $V_{BL}$ along the wall. A wall curvature (with radii of curvature $R$) is induced in the non-steady-state motion (after Slonczewski[349]).

lines like the sails of a ship around a mast. According to Eq. (12.2), the effect of the billowing is to enhance the precession rate at the Bloch line because $\delta\sigma/\delta q$ is increased, while the precession rate away from the Bloch line is zero. Precession is therefore effectively concentrated at the Bloch lines, and it corresponds to rapid displacement of the Bloch lines within the wall, as mentioned earlier. Energy loss is concentrated at the Bloch lines, so that energy is effectively "funneled" into the Bloch-line regions from the rest of the domain wall. The net effect of this process is to reduce the mobility of the entire wall, and explicit examples are given in Sections 13 and 14.

In the above discussion we have considered two limiting cases—steady-state motion and completely free Bloch-line motion. Next we consider the basic approximations and formulas that aid in treating quantitatively these cases as well as intermediate cases in which Bloch-line motion occurs and $\delta\sigma/\delta\psi$ is small but not zero. These formulas will give further physical insight into how Bloch lines cause low velocities and large effective masses.

## D. STEADY AND QUASI-STEADY WALL MOTION

The case of nonsteady-state motion of a wall containing unconstrained Bloch lines is so complex that up to now not even the simplest example has ever been solved including the effects of curvature. Instead most problems have been treated using a series of approximations that may be called collectively the "quasi-steady approximation."[224,245,345,349]

To explain this approximation we first must discuss certain aspects of steady-state motion that will also be useful in later discussions (Section 17). We specialize to the case of a plane wall, described by coordinates $x$ and $z$, although similar concepts apply in the case of curved walls (e.g., see Chapter IX). In this case steady-state motion at velocity $V$ corresponds to $\dot\psi = 0$ and $\dot q = V$. Then Eq. (12.1) reduces to

$$2M\gamma^{-1}V = \delta\sigma/\delta\psi, \qquad (12.19)$$

which, taken in conjunction with Eq. (12.4) for $\sigma$, is a differential equation for $\psi(x, z)$. As usual, $\delta/\delta\psi$ represents a functional derivative because $\sigma$ depends on $\partial\psi/\partial x$ and $\partial\psi/\partial z$ as well as on $\psi$. This problem is mathematically equivalent to the Lagrangian problem with the Lagrangian[224]

$$L = A^{-1} \int dA(\sigma - 2M\gamma^{-1}V\psi), \qquad (12.20)$$

where $\int dA$ is a surface integral over an area $A$. The solutions $\psi(x, z)$ are then determined by requiring that $L$ be stationary or $\delta L = 0$ for arbitrary variations of $\psi$, giving a constant $V$. (The Lagrangian formalism has often been applied to dynamical micromagnetic problems.[11]) These solutions represent the physical condition that spins at all points on the wall experience the same torque required to sustain the velocity $V$. In general, for any given $V$, there will be multiple solutions $\psi_i(x, z)$ ($i = 1, \ldots, n$). For reasons to become apparent below, these solutions are conveniently characterized by the surface average

$$\bar{\psi}_i = A^{-1} \int dA \psi_i(x, z). \tag{12.21}$$

For example, a horizontal Bloch line half way through the thickness of a film separating domain-wall regions with $\psi = 0$ and $\psi = \pi$ would give $\bar{\psi} = \pi/2$. Other solutions for the same $V$ might have multiple Bloch lines with different $\bar{\psi}$ values. The solutions are also characterized by their average wall energy

$$\bar{\sigma}_i = A^{-1} \int dA \sigma_i. \tag{12.22}$$

For example, a horizontal Bloch line has an energy which depends on position, as given by Eqs. (8.12)–(8.14) and (8.33), and this energy, divided by film thickness, plus the wall energy without the Bloch line, gives $\bar{\sigma}_i$.

By continuity, one expects that $\psi_i(x, z)$, $\bar{\psi}_i$, and $\bar{\sigma}_i$ vary smoothly with $V$. Thus, as a function of $V$, there may be several branches $\psi_i(x, z, V)$, $\bar{\psi}_i(V)$, and $\bar{\sigma}_i(V)$. We shall see explicit examples of such functions in Section 17, where the different branches correspond to different numbers of Bloch lines. One may also invert the functional relationship and view $V_i(\bar{\psi})$ and $\bar{\sigma}_i(\bar{\psi})$ as functions of $\bar{\psi}$.

Consider a particular pair of values for $i$ and $V$ and the corresponding function $\psi(x, z) = \psi_i(x, z, V)$. Consider further any variation $\delta\psi$ for which $\delta\bar{\psi}$ does not vanish. The fact that $\delta L$ must vanish for an arbitrary $\delta\psi$ implies that $\delta L$ also vanishes for this particular $\delta\psi$. Substituting Eqs. (12.21) and (12.22) into (12.20) one then finds

$$0 = \delta L = \delta\bar{\sigma}_i - 2M\gamma^{-1}V\delta\bar{\psi}_i, \tag{12.23a}$$

which is equivalent to

$$V_i(\bar{\psi}) = (\gamma/2M)\, d\bar{\sigma}_i/d\bar{\psi}. \tag{12.23b}$$

This important steady-state relationship gives the prescription for calculating $V$ if $\psi(x, z)$ is known. It is analogous to the one-dimensional steady-state result that wall velocity is proportional to torque. In this multidimensional case, however the torque is represented by a derivative of average wall energy with average precession angle. The corresponding pressure equation (12.1) is simply

$$\delta\sigma/\delta q = -2M\bar{H} = -2M\gamma^{-1}(\alpha\Delta^{-1}V + \gamma H_c), \qquad V > 0, \tag{12.24}$$

where $\bar{H}$ is the average drive acting on the wall. Equation (12.24) reduces to the usual mobility relation. There is no curvature or wall-bowing effect because $\dot{\psi} = 0$ everywhere.

Now let us turn to the nonsteady-state problem of a moving wall containing moving Bloch lines. Let us assume all the steady-state solutions $V_i(\bar{\psi})$ are already known. We can make use of these in a "quasi-steady approximation" in the following way. From our earlier discussion in the previous subsection, we know that wall bowing must occur because of the different precession rates $\dot{\psi}$ at different points in a wall containing Bloch lines. To preserve the steady-state wall structure, we must assume first of all that the wall bowing is negligible. Secondly we assume that the time-dependent mean wall velocity $\dot{q}$ is also given by Eq. (12.23b), and that the dynamic spin distributions $\psi_i(x, z)$ are the same as the steady-state ones. This is plausible if the drive $H$, the damping $\alpha$, and coercivity $H_c$ are sufficiently small that $\bar{\psi}$ and $\dot{q}$ change gradually enough not to excite the natural oscillations of the domain-wall surface. We discuss more fully the criteria for these assumptions at the end of this section.

Completely apart from the quasi-steady approximation, we may integrate Eq. (12.2) with the quasi-flat wall expression (12.6) over the wall surface. The integral over $\delta\sigma/\delta q$ reduces to $-2M\bar{H}$, where $\bar{H}$ is the average drive. It is noteworthy that a gentle wall curvature gives no contribution to this average, assuming either periodic boundary conditions or the conditions $\partial q/\partial n = 0$ at boundaries such as free surfaces, for then

$$\int dA \nabla^2 q = 0. \tag{12.25}$$

The significance of this result is that although wall curvature is essential in causing rapid Bloch-line precession, it can largely be ignored in subsequent discussion. Therefore one finds

$$\dot{\bar{\psi}} = \gamma\bar{H} - \alpha\Delta^{-1}\dot{q} - \gamma H_c \, \text{sgn} \, \dot{q}, \tag{12.26}$$

where the coercive term is correct providing $\dot{q}$ has only one consistent sign over the whole wall surface. This general relation, which does not rest on the quasi-steady picture, shows that the average precession angle $\bar{\psi}$ is governed by a Larmor precession equation analogous to that of the one-dimensional equation of motion (10.11). The reason for the simplicity of this equation is that it corresponds to the mechanical principle that the rate of change of total momentum of any compound body equals the total external force, no matter how complex the forces of interaction between the body's component parts. The corresponding torque equation, now invoking the quasi-steady approximation, is

$$\dot{q} = V_i(\bar{\psi}), \tag{12.27}$$

provided one can ignore the damping term $\alpha\Delta\dot{\bar{\psi}}$. As before, $V_i(\bar{\psi})$ in Eq.

(12.27) represents the steady-state solutions. Together, Eqs. (12.26) and (12.27) form a basis for treating nonsteady-state motion in the quasi-steady approximation.

The qualitative predictions of Eqs. (12.26) and (12.27) are closely analogous to the behavior of $\psi$ and $q$ in the one-dimensional model of Section 11,C. Generally, but not always, $V_i(\bar{\psi})$ in Eq. (12.27) increases monotonically with $\bar{\psi}$, linearly for small $\bar{\psi}$ but eventually reaching a maximum $V_p$ because of the fact that the maximum torque that the wall structure can sustain is always finite. Thus, if one starts with a static spin structure characterized by $\bar{\psi} = 0$ and $\dot{q} = V_i(\bar{\psi}) = 0$, and if a step field $\bar{H}$ is applied, Eq. (12.26) predicts that $\bar{\psi}$ will initially increase at a rate $\gamma(\bar{H} - H_c)$. However as $\bar{\psi}$ increases, $\dot{q}$ increases, and provided $\bar{H}$ is not too large, the spin structure approaches exponentially a dynamic equilibrium with $\dot{\bar{\psi}} = 0$ as illustrated in Fig. 12.3 and[224]

$$\dot{q}_\infty = V_i(t = \infty) = \Delta\gamma(\bar{H} - H_c)\alpha^{-1}. \qquad (12.28)$$

This is the conventional mobility formula. Clearly if $\bar{H} - H_c > \alpha\Delta^{-1}\gamma^{-1}V_p$, no equilibrium can be reached and a "breakdown" or unstable motion ensues whose nature, often periodic as illustrated in Fig. 12.3, will be discussed at length in Sections 15 and 17,C.

One can similarly define an effective mass $\bar{m}$ analogous to the one-dimensional case

$$\bar{m} = P/\ddot{q}. \qquad (12.29)$$

Here $P$ is the net pressure

$$P = 2M\bar{H} - 2M\gamma^{-1}(\alpha\Delta^{-1}\dot{q} + \gamma H_c \, \text{sgn} \, \dot{q}), \qquad (12.30)$$

which represents the drive pressure minus viscous and coercive terms and which acts to produce acceleration. According to Eq. (12.26)

$$P = 2M\gamma^{-1}\dot{\bar{\psi}}. \qquad (12.31)$$

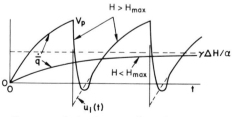

**Fig. 12.3.** Schematic mean velocity $\dot{q}$ versus time $t$ in the presence of a step drive field $H$ initiated at $t = 0$, for two values of $H$, larger and smaller than $H_{max} = \mu^{-1}V_p$ (after Slonczewski[224]). [Note: Depending on assumptions, the curve $q(t)$ might exhibit special features, such as horizontal tangency at points satisfying $\dot{q} = V_p$.]

Furthermore differentiating Eq. (12.23b) with respect to time, one obtains

$$\ddot{q} = (\gamma/2M)(d^2\bar{\sigma}_i/d\bar{\psi}^2)\dot{\bar{\psi}}. \tag{12.32}$$

Combined with Eqs. (12.29) and 12.31), this gives

$$\bar{m}^{-1} = (\gamma/2M)^2 \, d^2\bar{\sigma}_i/d\bar{\psi}^2$$
$$= (\gamma/2M)dV_i/d\bar{\psi}. \tag{12.33}$$

This result is also analogous to the one-dimensional expression Eq. (11.5) for wall mass. Equation (12.33) shows that the inverse mass is proportional to the slope of the $V_i(\bar{\psi})$ curve.

In summary, by writing averaged expressions for $\bar{\psi}$ and $\bar{q}$, one can establish a close analogy of the three-dimensional to the one-dimensional theory in the quasi-steady approximation. The principal difference arises from the fact that when Bloch lines are present the dependence of $\bar{\sigma}$ on $\bar{\psi}$ is much more gradual than in the corresponding one-dimensional case. This is just another way of stating the proposition with which we began this section; namely, that Bloch lines are relatively free to move in a wall, permitting large changes in $\psi$ at the expense of little energy. Therefore, according to Eqs. (12.23b) and (12.33), one may expect average wall velocities to be reduced and effective wall masses to be increased. This is the essential reason for the low saturation velocities and large ballistic effects discussed in Chapters VIII–IX. Although we have considered the case of a planar wall in the above discussion, similar concepts apply to curved walls and moving closed domains, so that Eqs. (12.23), (12.26), (12.27), and (12.33) are also approximately valid, but with appropriately modified definitions for $\bar{\psi}$ and other quantities as discussed further in Chapter IX. The quasi-steady approximation also underlies the use of the dynamical-variable formalism of Section 12,B. In applying this formalism one generally assumes that the domain walls are rigid, with no wall-billowing effects, and that all precession is concentrated at the Bloch lines.

It should be emphasized that Eq. (12.26) is a direct consequence of the general wall-dynamics equations (12.2) and (12.6). Thus it does not rest on the quasi-steady approximation but rather reflects the general principle of momentum conservation. As in our discussion of one-dimensional wall motion, Eq. (12.26) equates the rate of change of total momentum $2M\gamma^{-1}A\bar{\psi}$ to the total force.

For one interesting application of momentum conservation, take the case that $\bar{H}$ represents a rectangular pulse of amplitude $H$ lasting from $t = 0$ to $t = \tau$. Let us neglect $H_c$ and integrate Eq. (12.26) from $t = 0$ to $t = \infty$. Then we have for the net displacement of the wall after it is once again at rest,

$$\bar{q}(\infty) - \bar{q}(0) = \Delta_0\alpha^{-1}[\gamma H\tau - \bar{\psi}(\infty) + \bar{\psi}(0)]. \tag{12.34a}$$

Suppose also that the wall is in a state of absolute minimum energy at $t = 0$ and at $t = \infty$. We note the wall energy (12.4) has the periodic property $\sigma(\psi + 2\pi n) = \sigma(\psi)$ in general [and $\sigma(\psi + \pi n) = \sigma(\psi)$ if $H_x = H_y = 0$]. Thus it follows that the minimum energy states satisfy $\bar{\psi}(\infty) - \bar{\psi}(0) = 2\pi n$ so that (12.34a) becomes[344b]

$$\bar{q}(\infty) = \bar{q}(0) + \Delta\alpha^{-1}(\gamma H \tau - 2\pi n \quad \text{or} \quad -\pi n). \qquad (12.34b)$$

Since the integer $n$ can only increase in steps of unity as either $H$ or $\tau$ is varied, the final position $\bar{q}(\infty)$ should also exhibit steps in the amount $2\pi\Delta\alpha^{-1}$ or $\pi\Delta\alpha^{-1}$. Although such steps have not been observed in motions of plane walls, recent bubble-collapse measurements at high drive field exhibit small steps or fluctuations of $\bar{q}(\infty)$ vs. $\tau$ or $H$ attributable to steps in $\bar{\psi}(\infty)$.[344b] Their magnitudes are of amount $2\pi$ both with and without external in-plane field, although one might expect $\pi$ in the absence of external in-plane field. Additional discussions of periodicity in wall velocity appear in Sections 11,C, 15,D, 16,B, and 17,C.

The picture of wall motion developed in this section has an analogy in the process of magnetization reversal of a large single particle. Just as application of a static field opposite to the initial magnetization causes the magnetization direction to rotate in the particle, so the drive field in the dynamic case causes the magnetization orientation in a wall to rotate from one magnetostatic equilibrium direction to another. And just as a domain wall can facilitate magnetization reversal by reducing the energy barrier, so a Bloch line can facilitate the rotation in the wall by reducing the torque barrier to spin precession.

## E. CRITERIA FOR THE QUASI-STEADY APPROXIMATION

As we have seen in the previous section, one criterion for the quasi-steady approximation is that the wall surface tension $\sigma$ be sufficiently large that the billowing never becomes large enough to appreciably distort the geometric shape of the wall.[349] Under the simplest assumption, the radius of wall curvature $R$ is determined by balancing the pressure due to the wall energy $\sigma_0 = 4(AK)^{1/2}$ against the applied pressure $2MH$ according to the equation

$$2M|H| = \sigma_0/R. \qquad (12.35a)$$

$R$ must clearly be greater than the characteristic spacing $L$ between Bloch lines. Thus we find

$$H < H_{qs1} \equiv 2(2\pi AQ)^{1/2}L^{-1}. \qquad (12.35b)$$

If we take $L = h$, the thickness of the film, for the case of a single horizontal Bloch line, $H_{qs1}$ is typically 10 Oe for 5 $\mu$m garnet films.

A second criterion on the quasi-steady approximation is that the Bloch line move slowly compared to the time required to establish the curvature contour discussed in Section 12,C. Now any membrane, such as a domain wall, which carries a local mass $m$ per unit area and has an energy per unit area (surface tension) $\sigma_0$, transmits flexural waves at the velocity

$$u_f = (\sigma_0/m)^{1/2}. \tag{12.36}$$

Substituting $\sigma_0 = 4(AK)^{1/2}$ and the Döring mass $m_D = (2\pi\gamma^2\Delta_0)^{-1}$, we find[61]

$$u_{fD} = (8\pi A)^{1/2}\gamma. \tag{12.37}$$

For problems involving in-plane fields or in-plane anisotropy, one may derive alternative velocities by substituting the appropriate local mass (see Section 11,B) in Eq. (12.36). [See, however, Section 22 for limitations of Eq. (12.36).] The propagation of such flexural vibrations along the wall is what permits a wall-curvature contour to be established. This can be seen in two ways: First the curvature can always be decomposed into the normal flexural modes of the wall. Secondly, the curvature causes enhancement of precession at the Bloch lines and suppression of precession elsewhere, so that the energy is effectively funneled into the Bloch line regions, and the flexural vibrations are wave disturbances, which transmit that energy.

Therefore, although the question has not been explicitly investigated, a plausible criterion is that the Bloch-line velocity $v_L$ cannot exceed $u_f$ of Eq. (12.37) since the velocity of even a linear disturbance cannot. To estimate $v_L$ let us take $\alpha = H_c = 0$ in Eq. (12.26) and assume that as the Bloch line moves, it switches the wall spins by $\pi$. If the Bloch lines are separated by a characteristic distance $L$, then as $\bar{\psi}$ precesses by $\pi$, the Bloch line must cover the entire distance $L$. Thus we have

$$v_L = \pi^{-1}L\gamma H. \tag{12.38}$$

If $L$ and $H$ are very large, precession rates in the Bloch lines become very high. Combining Eqs. (12.37) and (12.38) we obtain

$$H < H_{qs2} \equiv (2\pi)^{3/2}A^{1/2}L^{-1} \tag{12.39}$$

as a necessary criterion for application of the quasi-steady theory. For typical 5 $\mu$m garnet films, taking $L = h$ as before, $H_{qs2}$ is roughly 10 Oe and of the same order as Eq. (12.35). A related criterion is that the time of the experiment or the rise time of a field pulse should be greater than the time for the wave disturbance to traverse $L$. Using Eq. (12.37), we find

$$t > t_{qs} = L(8\pi A)^{-1/2}\gamma^{-1}. \tag{12.40}$$

For typical 5 $\mu$m garnet films, taking $L = h$, $t_{qs}$ is roughly 25 nsec.

What happens if these criteria are not fulfilled? Physically, regions far from the Bloch line do not "see" it, and wall motion can occur as if the one-dimensional equation were valid in that local region. An example of this kind of behavior is seen in a stroboscopic bubble-expansion experiment on a LuGdAlIG film in Fig. 4.3, where oscillations fitting the one-dimensional theory are observed to last for about 15 nsec compared to $t_{qs} = 25$ nsec for this film.[129] Similar effects have been observed in photometric experiments on stripe arrays.[336,350] For example, in a EuTmGaYIG film, a velocity peak of 2000 cm/sec at drive fields greater than 8 Oe, lasting up to 20 nsec, compares with a Walker velocity of 3000 cm/sec and the estimates for $H_{qs}$ and $t_{qs}$ close to those given above. High initial velocities have also been seen in bubble-collapse experiments in which the bubble was biased very close to its collapse field.[172,350a]

## F. GYROVECTOR AND DISSIPATION DYADIC

The twin concepts of gyrovector and dissipation dyadic provide a penetrating insight into the forces involved in domain dynamics.[257,351] These concepts also provide alternative avenues for the solution of many of the dynamical problems generally attacked in this review with the wall-motion equations (12.1) and (12.2).

To begin, consider statically a nonequilibrium distribution of magnetization, represented by $\mathbf{M}(\mathbf{x} - \mathbf{X})$ or by $\theta(\mathbf{x} - \mathbf{X})$ and $\phi(\mathbf{x} - \mathbf{X})$, in the presence of some externally applied field distribution $\mathbf{H}_{ex}(\mathbf{x})$. Here $\mathbf{x}$ is the usual position vector and $\mathbf{X}$ is a vector representing the "position" of the distribution, which may be visualized as representing a domain-wall, a Bloch line, an entire domain, or the like. In seeking its static equilibrium, the distribution will generally tend to both distort and displace itself. Any true distortion would be described by a change in the functional form $\mathbf{M}(\mathbf{x})$, but a displacement only by a change in $\mathbf{X}$.

The static force $\mathbf{F}_s$ tending to displace any nonequilibrium distribution is obtained by differentiating the total energy $W$ of the distribution with respect to its position:

$$\mathbf{F}_s = -\partial W/\partial \mathbf{X}. \tag{12.41}$$

Since any variation $\delta W$ of the total energy is expressible as a volume integral of local density variations $\delta w(\mathbf{x} - \mathbf{X})$, we can write

$$\mathbf{F}_s = \int \mathbf{f}_s \, dV. \tag{12.42}$$

Here $V$ is the volume and $\mathbf{f}_s$ is the static force density written in the alternative forms

$$\mathbf{f}_s = -\delta w/\delta \mathbf{X} = \delta w/\delta \mathbf{x}, \tag{12.43}$$

where $\delta w/\delta \mathbf{x}$ represents the local static force density. Considering that $\delta w$ depends on $\theta$, $\nabla\theta$, $\phi$, $\nabla\phi$ (Section 1) we employ chain differentiation to write Eq. (12.43) in the form

$$\mathbf{f}_s = (\delta w/\delta\theta)(\nabla\theta) + (\delta w/\delta\phi)(\nabla\phi). \tag{12.44}$$

Although technically expressed as functional derivatives, the quantities $-\delta w/\delta\theta$ and $-(\delta w/\delta\phi)\sin\theta$ are precisely the components of local torque density derived from the total stored energy. Thus Eq. (12.44) relates the force density of an arbitrary instantaneous configuration to local torques and gradients of **M**. In static *equilibrium* of course, $\delta w/\delta\theta$ and $\delta w/\delta\phi$ would vanish at every point.

In a dynamic problem, $\theta$ and $\phi$ depend on time $t$ as well as $\mathbf{x}$; we can still evaluate Eq. (12.44) instantaneously for any given $\theta = \theta(\mathbf{x}, t)$ and $\phi = \phi(\mathbf{x}, t)$ to obtain directly that part of the total force (the "static" part) attributable to stored energy. In dynamics, the local static torques do not vanish but instead give rise to local motion described by the Landau–Lifshitz equations (3.7) and (3.8):

$$\delta w/\delta\theta = \gamma^{-1}(M\dot\phi \sin\theta - \alpha M\dot\theta), \tag{12.45}$$

$$\delta w/\delta\phi = -\gamma^{-1}(M\dot\theta \sin\theta + \alpha M\dot\phi \sin^2\theta). \tag{12.46}$$

Given instantaneous values of $\delta w/\delta\theta$, $\delta w/\delta\phi$, and $\theta$, these equations can be solved for the instantaneous spin precession given by $\dot\theta$ and $\dot\phi$. Upon substituting Eqs. (12.45) and (12.46) into Eq. (12.44) we observe a balance of "static" and "dynamic" force densities represented by

$$\mathbf{f}_s + \mathbf{f}_d = 0, \tag{12.47}$$

where the dynamic force density is

$$\mathbf{f}_d = (M/\gamma)[(-\dot\phi \sin\theta + \alpha\dot\theta)\nabla\theta + (\dot\theta \sin\theta + \alpha\dot\phi \sin^2\theta)\nabla\phi]. \tag{12.48}$$

Now let us consider the special case of steady motion at velocity **v** described by writing

$$\theta = \theta(\mathbf{x} - \mathbf{v}t) \qquad \phi = \phi(\mathbf{x} - \mathbf{v}t). \tag{12.49}$$

These functional forms describe the motion of the **M** distribution as a "rigid wave packet." In this case, we can write the equations

$$\dot\theta = -\mathbf{v}\cdot\nabla\theta \qquad \dot\phi = -\mathbf{v}\cdot\nabla\phi. \tag{12.50}$$

Substitution of Eqs. (12.50) into (12.48), followed by application of the vector identity $\mathbf{a} \times (\mathbf{b} \times \mathbf{c}) = \mathbf{b}(\mathbf{a}\cdot\mathbf{c}) - \mathbf{c}(\mathbf{a}\cdot\mathbf{b})$ gives the result[257]

$$\mathbf{f}_d = (M/\gamma)(\mathbf{g} \times \mathbf{v} + \mathbf{\eth}\cdot\mathbf{v}), \tag{12.51}$$

where **g** is the "gyrovector density"

$$\mathbf{g} = -\sin\theta(\nabla\theta) \times (\nabla\phi) \qquad (12.52)$$

and $\mathfrak{d}$ is the "dissipation dyadic"

$$\mathfrak{d} = -\alpha[(\nabla\theta)(\nabla\theta) + \sin^2\theta(\nabla\phi)(\nabla\phi)]. \qquad (12.53)$$

(The factor $M/\gamma$ has been removed from the original definition of **g** and $\mathfrak{d}$ in the literature.[257]) The total dynamic force on the moving distribution **M** is

$$\mathbf{F}_d = \int \mathbf{f}_d \, dV = (M/\gamma)(\mathbf{G} \times \mathbf{v} + \mathfrak{D} \cdot \mathbf{v}), \qquad (12.54)$$

where

$$\mathbf{G} = \int \mathbf{g} \, dV \qquad (12.55)$$

and

$$\mathfrak{D} = \int \mathfrak{d} \, dV \qquad (12.56)$$

are the total gyrovector and dissipation dyadics for the moving **M** distribution.

One practical impact of these equations (12.51)–(12.56) is that they provide a ready means of solving dynamic micromagnetic problems without actually solving partial differential equations for the time dependence of $\theta$ and $\phi$. Consider some structure, such as a Bloch wall, Bloch line, or entire domain, moving with sufficient slowness that the functional forms $\theta(\mathbf{x} - \mathbf{v}t)$, $\phi(\mathbf{x} - \mathbf{v}t)$ are approximately represented by a solution of the corresponding static problem. One substitutes these forms into the above equations to evaluate $\mathbf{F}_s$, **G**, and $\mathfrak{D}$, thereby getting an explicit equation of motion

$$(\gamma/M)\mathbf{F}_s + \mathbf{G} \times \mathbf{v} + \mathfrak{D} \cdot \mathbf{v} = 0 \qquad (12.57)$$

for the structure under study. This procedure has been used successfully to derive many of the equations of bubble dynamics.[257,351]

Volume integration of **g** is particularly simple. In certain problems only the component $g_z$ normal to the film plane is of interest. According to Eq. (12.52) it may be written in terms of the Jacobian notation $\partial(\cos\theta, \phi)/\partial(x, y) = (\partial\cos\theta/\partial x)(\partial\phi/\partial y) - (\partial\phi/\partial x)(\partial\cos\theta/\partial y)$, thus

$$g_z = \partial(\cos\theta, \phi)/\partial(x, y). \qquad (12.58)$$

From the properties of Jacobians it follows that transforming the integration variables $xyz$ to the coordinates $\cos\theta$, $\phi$, $z$ transforms the $z$-component of Eq. (12.55) into

$$G_z = -\int(\iint \sin\theta \, d\theta \, d\phi)dz. \qquad (12.59)$$

Note that the integral within the parentheses in Eq. (12.59) represents the surface area of the region in the space $(\cos\theta, \phi)$ which maps into the $xy$

region occupied by a given spin structure. In problems of interest $\theta$ and $\phi$ will take on constant values $\theta_1$ and $\theta_2$, and $\phi_a$ and $\phi_b$, at infinity or along certain opposing pairs of boundaries in $xy$ space whose positions are simply related to the micromagnetic structure considered (see Fig. 12.4), so that Eq. (12.59) reduces simply to

$$G_z = \int (\cos\theta_2 - \cos\theta_1)(\phi_b - \phi_a)dz. \tag{12.60}$$

(Note that $\theta_1$, $\phi_a$, $\theta_2$, $\phi_b$ must appear in counterclockwise order around the boundary for $G_z$ to have the correct sign.) The remarkable power and elegance of this relation will be made apparent by considering the applications given below.

As an example of the use of the gyrovector, first consider a Bloch line in the presence of an in-plane field $H_p$, as illustrated in Fig. 12.4. In this case $\cos\theta$ approaches $\mp 1$ at $y = \pm\infty$, assuming $H_p \ll 2K/M$, and $\phi$ approaches $(\pi \pm \Phi)/2$ at $x = \pm\infty$, where $\Phi = \frac{1}{2}|\cos^{-1}(H_y/8M)|$ is the full twist of the Bloch line. Then Eq. (12.60) reduces to[257]

$$G_z = 2\Phi \tag{12.61}$$

for a unit length of the Bloch line. For a generally oriented line one similarly calculates the tangential component of $\mathbf{G}$ and finds that the components of $\mathbf{G}$ orthogonal to the line vanish. Thus it follows that the gyrotropic force per unit length of a Bloch line is

$$\mathbf{F}_{gL} \equiv M\gamma^{-1}\mathbf{G} \times v_L = 2\Phi M\gamma^{-1}\mathbf{t} \times v_L, \tag{12.62}$$

where $\mathbf{t}$ is the unit vector $\mathbf{t} = \nabla\phi \times \nabla\theta/|\nabla\phi \times \nabla\theta|$ evaluated at the center of the Bloch line. Equation (12.62) is a general result, independent of the specific geometry of Fig. 12.4. This very important relationship is derived in a different way in Section 14 and applied to experiment throughout the remaining sections.

One may equally well consider the gyrovector of an entire domain of

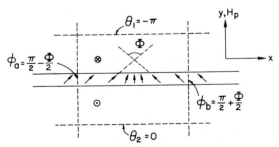

**Fig. 12.4.** Schematic construction for evaluating gyrovector of a vertical Bloch line in an in-plane field, using Eq. (12.60).

arbitrary shape. In this case, contours of constant $\cos\theta$ and $\phi$ in a plane of constant $z$ have the appearance illustrated in Fig. 12.5. Note that $\cos\theta = 0$ at $\infty$ and $\cos\theta = -1$ at the center. In case the winding integer $I$ does not vanish, a "cut" in the field is required, as shown by the line AF in the figure, across which $\phi$ changes discontinuously from 0 to $2\pi I$. The integer $I(z)$ depends on $z$ in stepwise fashion when Bloch points are present.[349] Taking both sides of the cut as boundaries for the region of integration, the integrand of Eq. (12.60) becomes $2 \cdot 2\pi I$, so that Eq. (12.60) reduces to

$$G_z = 4\pi hS, \qquad (12.63)$$

where $S = h^{-1}\int I(z)dz$ is the effective winding number [Eq. (9.18)]. The general result, independent of the particular polarity of Fig. 12.5, is

$$\mathbf{F_g} = 4\pi M\gamma^{-1}hS\mathbf{z_0} \times \mathbf{V}, \qquad (12.64)$$

where $\mathbf{z_0}$ is the direction of the magnetization outside the closed domain, and $\mathbf{V}$ is the steady-state velocity of the domain. When coupled with Eq. (12.57), this result is fundamental in accounting for the skew deflection of bubbles described in Sections 13 and 14.

If the distribution $\mathbf{M(x)}$ is continuous everywhere, then $I(z)$ can only be a constant $I = S$, since a function limited to integer values can only vary discontinuously. However, we have seen that micromagnetic singularities known as Bloch points may be present on Bloch lines in the domain, as

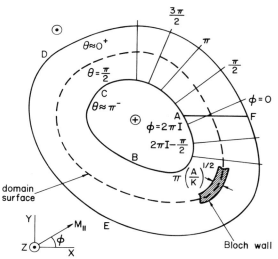

**Fig. 12.5.** Contours of constant $v$ and $\phi$ in the plane $z = $ constant, for a closed domain of general shape with a winding number $I(z)$ (after Slonczewski[349]). AF represents the section of a cut plane across which $\phi$ changes by $2\pi I$.

discussed in Section 9. Consider now the effect of displacing the plane specified by $z$ through a Bloch point. Because a Bloch point represents a gyroflux source of strength $\pm 4\pi$, as shown in Section 9,A, it is clear that

$$\iint_{-\infty}^{\infty} g_z \, dx \, dy$$

changes by $\pm 4\pi$ whenever the plane defined by $z$ passes a Bloch point, implying that $I(z)$ has a step of $\pm 1$ at each $z$ that coincides with a Bloch point. Let us then assume $n$ Bloch points with signatures $N_{hi} = \pm 1$ and $z$-coordinates $z_i$ ($i = 1, 2, \ldots, n$). Thus for the field polarity of Fig. 12.5, we obtain $S = I(h) - h^{-1} \sum_i N_{hi} z_i$, as derived before [Eq. (9.18)]. Applications of this relation to experiment have been discussed in Section 9,B.

# VII

## Low-Velocity Dynamics
## with Vertical Bloch Lines

In the next three chapters we apply the concepts and formalism of Chapter VI (Section 12) to a variety of two- and three-dimensional problems of experimental interest. We break the discussion up according to whether Bloch lines are present in the domain wall to start with or whether they are nucleated or annihilated dynamically. In this chapter we treat the former case, which applies for low-velocity motion in domain walls containing vertical Bloch lines. In Sections 13 and 14, we consider separately the cases of hard walls with many Bloch lines and walls with small numbers of Bloch lines. The comparison of theory to experiment in these sections reveals some of the greatest successes of bubble physics, accounting often in a quantitative way for astonishing phenomena such as bubble skew deflection, dumbbell rotation, hard-bubble generation, and the dependence of dynamic bubble collapse on Bloch-line number.

## 13. Hard-Wall Dynamics

### A. Uniform Drive Fields

As described in Section 8,D, hard walls have vertical Bloch lines packed so closely together that the wall moment orientation angle $\psi(x) = \pm \pi x/s$ varies approximately linearly with distance $x$ along the wall. As in Eq. (8.18), $s$ is the separation between adjacent Bloch lines and the sign corresponds to the sense of Bloch-line twist. First we consider the particularly simple case of a uniform drive field $H$ applied to the wall, as appropriate in treating the dynamic collapse or expansion of hard bubbles.[178] In this case the Bloch lines, illustrated schematically in Fig. 8.7, are unimpeded by boundary conditions because they can chase each other freely around the wall of a hard bubble. More explicitly, precession of all the spins by $d\psi$ corresponds merely to a displacement of the entire Bloch-line array by $dx = \pm s\pi^{-1} d\psi$. Physically, it is therefore clear that there has been no increase in the static internal energy of the wall, and the energy-conserving torques are therefore everywhere zero. Effectively, the usual magnetostatic torques are averaged out as the sign of $4\pi M^2 \Delta \sin 2\psi(x)$ alternates with $x$. Thus the problem reduces to a one-dimensional one.

The only uncompensated torque available to provide wall velocity is the viscous-damping reaction torque $-2M\gamma^{-1}\alpha\Delta\dot\psi$, so that Eq. (12.1) reduces to

$$\dot q = \alpha\Delta\dot\psi, \tag{13.1}$$

where the wall-thickness parameter $\Delta$ is given by Eq. (8.21) for a hard wall. Assuming the only pressure on the wall is due to the drive field $H$, Eq. (12.2) reduces to

$$\dot\psi = \gamma H - \alpha\Delta^{-1}\dot q. \tag{13.2}$$

Solving the last two equations, one finds

$$\dot q = \alpha\gamma\Delta(1 + \alpha^2)^{-1}H, \tag{13.3}$$

$$\dot\psi = \gamma H(1 + \alpha^2)^{-1}. \tag{13.4}$$

By comparison with a normal wall, these results for a hard wall are extraordinary: Equation (13.3) implies that the mobility of a hard wall is reduced by a factor $\alpha^2/(1 + \alpha^2)$ from the normal linear mobility $\alpha^{-1}\gamma\Delta$; this is a huge difference for low damping materials ($\alpha \ll 1$). Furthermore it is noteworthy that the mobility of a hard wall increases with $\alpha$ rather than decreasing as for a normal wall. This result is identical to the free-precession case considered in Section 11,C, and indeed Eq. (13.4) shows that $\psi$ precesses at close to the Larmor frequency provided $\alpha \ll 1$. Using Eq. (8.18), we find that this precession corresponds to a sideways displacement of Bloch lines at a velocity

$$v_{\text{L}} = \pi^{-1}s\gamma H(1 + \alpha^2)^{-1}. \tag{13.5}$$

These remarkable properties of hard walls have been verified in several experiments on hard bubbles[172,178] and stripes.[137] For example Fig. 13.1 shows the vastly reduced mobility of a hard bubble in a EuTmGaYIG film, compared to the much higher initial mobility of normal bubbles. The observed hard-bubble mobility of 1.1 cm/sec Oe compares well with $\alpha\gamma\Delta = 0.96$ cm/sec Oe calculated from the known material parameters. The problems of

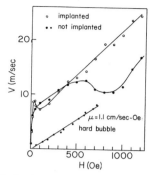

**Fig. 13.1.** Average wall velocity $V$ versus drive field $H$, determined by dynamic bubble-collapse in a EuTmGaYIG film, for normal and hard bubbles in an unimplanted section and for normal bubbles in an implanted section (after Konishi et al.[172]).

determining the much higher linear mobility from such data are discussed in Section 16,A.

Another particularly interesting experiment involved bias-pulsing isolated stripes in a EuTmGaYIG film and observing the transient stripe shape by high-speed photography.[137] When the stripes were pulsed to contract, transient bulges were observed on the stripe walls, indicating sections with a low mobility of 2 cm/sec Oe surrounded by regions of higher mobility. These bulges were interpreted as hard-wall sections of the stripe. The predicted hard-wall mobility from Eq. (13.3) was 0.8 cm/sec Oe. More remarkable was the observation that the bulges displaced sideways along the stripe wall during the pulse, some going one way and some the other, their respective directions of motion reversing when the pulse turned off. Plotted against the bias drive, the instantaneous velocity of bulge displacement along the wall gave a straight line whose slope was a transverse mobility of 110 cm/sec Oe and whose velocity intercept indicated an effective coercivity of 4 Oe. Since the Bloch lines are not squeezed together by geometric constraints in this case but are presumably held together by magnetostatic forces at the equilibrium spacing $\sqrt{2}\pi\Lambda_0$ [see Eq. (8.20)], one can use this spacing to estimate the theoretical transverse mobility of Eq. (13.5) to be 30 cm/sec Oe. The quantitative discrepancies in the forward and transverse mobilities are most likely due to the adjacent normal wall regions accelerating the forward and transverse motions. In any case the bulge displacement effect gives striking qualitative evidence for the precessional effects in the hard-wall dynamics theory. The 4 Oe apparent coercivity described above is considerably larger than the intrinsic coercivity of the sample and implies a "Bloch-line coercivity" different from the usual wall coercivity (see also Section 13,B).

Another aspect of this same experiment is that while the bulges move forward during stripe expansion (pulse on) and back during contraction (pulse off), the displacements do not quite cancel but give a net forward displacement per pulse to the Bloch-line clump. This is one variety of "automotion" effect arising from coercivity, as will be discussed more fully in Section 20. Because of this effect Bloch lines can be expected to move large distances down stripes subjected to repetitive pulsing. This concept may explain the astonishing observation in a variety of garnet compositions that when a dense stripe array is pulsed repetitively, every second wall becomes filled with Bloch lines and moves very little while the other wall is normal and moves much more.[352] The reason may be that to fulfill energy conservation, at least one out of every two walls of a stripe array must move in response to a bias field. Whichever one contains fewer Bloch lines will tend to move more vigorously, thereby pushing its Bloch lines around the stripe end to the side which moves less.

## B.   GRADIENT DEFLECTION OF HARD BUBBLES AND DUMBBELL ROTATION

Probably the most remarkable effect in bubble physics is the deflection effect or skew propagation of hard bubbles.[125,265,353-357] When a gradient field is applied to such a domain it responds by moving at nearly right angles to the force, and different bubbles skew in different directions, some to the right and some to the left. The mobility of these bubbles is also considerably reduced from that of normal bubbles. These effects are illustrated in Fig. 13.2, which shows the results of a gradient propagation of a hard bubble in a EuGaYIG film by nine successive gradient pulses of length 2 $\mu$sec and strength $H_g = |r\nabla H_z| = 4.5$ Oe. The bubble moves at an average angle of $73°$ to the applied gradient direction and has an apparent mobility of only 60 cm/sec Oe as it crosses the center line of the gradient.

To understand the deflection effect, we will balance the applied force against the dynamic reaction force by applying the scheme outlined in Section 12,B. Consider the illustration of Fig. 8.7. In what follows we assume the convention of a downward bubble surrounded by upward magnetization. In a static hard bubble the wall-moment distribution is given by $\psi = S\beta + C$, where $S$ is the winding number discussed in Section 8,C, $\beta$ is the angular position along the circumference of the bubble, and $C$ is a constant. For our model of the dynamic hard bubble we assume that its shape remains circular and that the same $\psi$-distribution applies in the coordinate frame of the bubble moving uniformly with the positive velocity $V$ along the positive $x$ axis. We will calculate the components of the field gradient $\partial H_z/\partial x$ and

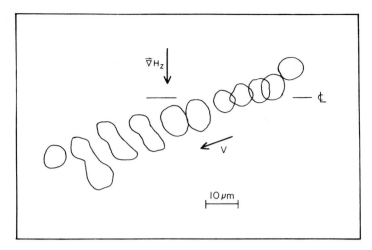

**Fig. 13.2.** Initial and final normal photographs and nine intermediate superimposed high-speed photographs of a hard bubble at the end of each of a sequence of nine gradient pulses of length 2 $\mu$sec and strength $H_g = |r\nabla H_z| = 4.5$ Oe oriented as indicated in a EuGaYIG film. The overall direction of the bubble motion illustrates the skew deflection of hard bubbles and the elliptical transient shape suggests a bunching effect. The horizontal lines indicate the center line of the gradient (after Patterson et al.[357]).

$\partial H_z/\partial y$ required to produce this velocity. This problem is formally equivalent to calculating the velocity components $V_{\parallel}$ and $V_{\perp}$, parallel and perpendicular to a given gradient $\nabla H_z$.

We choose the dynamical variables to be $X$ and $Y$, the coordinates of the bubble center. We have already evaluated the external force $\partial W/\partial X$ on the bubble from the field gradient in Eq. (12.12) and have given the relations of $q$ and $\psi$ to $X$ in Eqs. (12.7) and (12.14), respectively. Substituting these relations into Eq. (12.11) and integrating, we find

$$(\nabla H_z)_{\parallel} = -(\alpha V/\gamma r\Delta)[1 + (S\Delta/r)^2] - (4H_c/\pi r), \qquad (13.6)$$

$$(\nabla H_z)_{\perp} = 2SV/\gamma r^2 . \qquad (13.7)$$

Here we have adopted the notation $(\nabla H_z)_{\parallel} = \partial H_z/\partial x$, $(\nabla H_z)_{\perp} = \partial H_z/\partial y$ to emphasize the fact that these relations give the components of $\nabla H_z$ that are parallel and perpendicular, respectively, to $V$, since the orientation of coordinate axes $X$ and $Y$ was chosen at our convenience.

The fact that $(\nabla H_z)_{\perp}$ is nonzero shows that in general the gradient and the hard bubble velocity are not colinear, as indicated in Fig. 8.7. The angle $\rho$ of deflection of the bubble away from the net gradient-force direction may be defined by

$$\rho \equiv -\tan^{-1}[(\partial H_z/\partial y)/(\partial H_z/\partial x)] = \tan^{-1}(V_\perp/V_\parallel), \qquad (13.8)$$

where $V_\perp$ and $V_\parallel$ are the velocity components relative to the field-gradient direction as shown in Fig. 8.7. Then the deflection angle is given by

$$\rho = \cot^{-1}[\alpha f + (2r\gamma H_c/\pi S V)],$$
$$f = \tfrac{1}{2}(rS^{-1}\Delta^{-1} + S\Delta r^{-1}). \qquad (13.9)$$

Now for very hard bubbles, $\Delta \to r|S|^{-1}$ according to Eqs. (8.21) and (8.28), and so $f \to \pm 1$ depending on the sign of $S$. Thus the large deflection angles of hard bubbles can be understood from Eq. (13.9) if both the coercive term and $\alpha$ are small, as is the case in typical garnet bubble films. Furthermore the two different senses of deflection can be seen to arise from the two possible signs of $S$, corresponding to different senses of Bloch-line winding.

The physical reason for the deflection effect may be seen from Fig. 8.7. It is evident that as the bubble moves along $X$, the local spin precession rates $\dot\psi$ differ in sign at the points $B$ and $D$ fixed in the medium. In order to produce these precessions, the fields $H = \dot\psi\gamma^{-1}$ at these points must also differ in sign; hence a component of $\nabla H_z$ perpendicular to $V$ must be present. It is interesting to note that Eq. (13.7), which gives rise to the deflection effect, has a form similar to the Lorentz force on an electron in a magnetic field, with $S$ playing a role comparable to particle charge. Equation (13.6) shows that the component of gradient parallel to the motion is balanced by damping and coercive terms, as would be expected from energy conservation. If we ignore the correction in the bracket, the damping term reduces to the usual mobility expression. The correction arises from additional dissipation caused by the reduced wall thickness of Eq. (8.21) and by the Bloch line precession losses $\alpha\dot\psi^2$, and it can become significant for very hard bubbles ($|S| > r/\Delta_o$). The net mobility can be found by solving Eqs. (13.6) and (13.7) for $V$ in terms of $|\nabla H_z|$, the net gradient. In the limit of zero coercivity one finds

$$|V| = \tfrac{1}{2}r^2\gamma|S^{-1}\nabla H_z|(1 + \alpha^2 f^2)^{-1/2}, \qquad (13.10)$$

which, for small $\alpha f$, implies a mobility reduced by a factor $r\alpha/2\Delta_o S$ from the normal wall mobility. For a finite coercivity the net mobility cannot be found analytically, but instead Eqs. (13.6) and (13.7) may be combined to give an expression which eliminates explicit appearance of $H_c$:

$$S = \gamma r^2|\nabla H_z|\sin\rho/2V. \qquad (13.11)$$

This expression is the "golden rule" for bubble deflection experiments since it is so useful for extracting the winding state $S$ from experimentally measurable quantities. It also implies that if $V/\sin \rho$ is plotted against the drive field $H_g = r|\nabla H_z|$, one should obtain a straight line extrapolating through the origin.

Experimental tests of the hard-bubble deflection theory[353,355,356] have generally verified the linearity of $V/\sin \rho$ versus $H_g$ predicted by Eq. (13.11), and extracted $S$-values compare reasonably to those from static measurements.[355] However in one study on hard bubbles in a EuGaYIG film, the lines of $V/\sin \rho$ versus $H_g$ did not extrapolate through the origin, as shown in Fig. 13.3.[356] Thus there may be an additional coercive-force component,

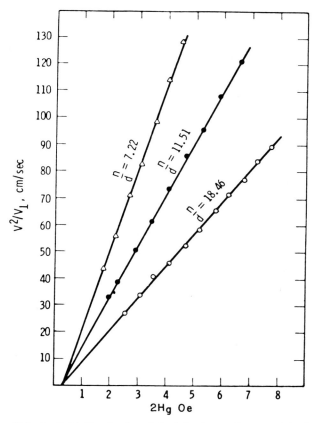

**Fig. 13.3.** Velocity divided by the sine of the deflection angle versus drive field $H_g = |r\nabla H_z|$ in a pulsed gradient propagation experiment on hard bubbles in a EuGaYIG film. $n/d$ represents the deduced Bloch line density in $\mu m^{-1}$, where $n$ is the number of Bloch line pairs and $d$ is the bubble diameter (after Patterson[356]).

presumably arising from the Bloch lines, which is colinear with the gradient, rather than with the velocity as usually assumed. The deflection-angle dependence of Eq. (13.9) was also tested by plotting $\cot \rho$ versus $1/V$, giving a line whose intercept gives $\alpha f$ and whose slope determines $H_c$. Surprisingly the deduced coercivity increased by 1.5 Oe as the Bloch line density $n/d$ increased to 30 $\mu m^{-1}$, implying in another way the existence of a coercivity attributable to Bloch lines (Section 5,A). All the above analysis, however, assumes steady-state motion and must be considered questionable in view of transient effects to be described in the next section.

Another remarkable effect related to hard-bubble deflection is dumbbell rotation.[125,236,358] As discussed in Section 8,D, dumbbells are hard bubbles with so many Bloch lines that they distort into stripes. When such dumbbells are subjected to spatially uniform $z$-field pulses, they are observed to rotate by a given angle per pulse, as indicated schematically in the inset of Fig. 13.4. A series of pulses gives the impression of a whirling propeller. Both senses of rotation have been observed in different domains. As shown in Fig. 13.4, high-speed photography has revealed that in fact the net rotation, which is typically tens of degrees, is the result of a rotation in one sense combined with a stripe length contraction while the pulse is on and a smaller rotation in the opposite sense accompanied by a reexpansion back to the static equilibrium length after the pulse turns off. (Actually there is a small forward rotation in the reexpansion phase at short pulse times, but we ignore this feature in subsequent discussion.) These effects may be understood qualita-

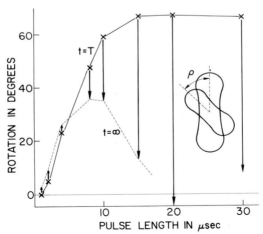

**Fig. 13.4.**  Dumbbell rotation angle $\rho$ at $t = T$ and $t = \infty$ for a fixed bias pulse of strength 6 Oe and duration from $t = 0$ to $t = T$, in a EuGaYIG film. Inset shows schematic dumbbell shape before and during rotation (after Slonczewski *et al.*[125]).

tively as follows: The length contraction due to the bias pulse follows from the static analysis given in Section 8,D, Eq. (8.32). The ends of the dumbbell are typically bulbous (hence the name "dumbbell") and may be likened to two hard bubbles. As these ends move towards each other during the contraction, they deflect just as hard bubbles do in a gradient, and this deflection corresponds to a stripe rotation. The reverse process occurs as the stripe reexpands. The fact that the forward and backward rotations are not the same is presumably due to coercivity. If the pulse has been so short that a new static equilibrium has not been reached by the end of the pulse, then the drive forces reexpanding the dumbbell will be much weaker than the drive forces causing the contraction. Accordingly the reexpansion velocity will be less, the coercive effect greater, and the reverse rotation angle smaller. This is similar to the reduced hard-bubble deflection angle when $V$ is small in Eq. (13.9). For longer pulse times, the figure shows the reverse rotation, indicated by the downward arrows, becoming substantial, until beyond about 20 $\mu$sec, there is on the average no net rotation at all. Net dumbbell rotation occurs only when static equilibrium is not attained during the pulse and coercivity suppresses the reverse motion after the pulse. This general concept also explains a number of other remarkable effects in domain-wall physics, including the bulge displacement effect described in the previous section and the automotion effect discussed in Section 20.

## C. BLOCH-LINE BUNCHING AND WIND-UP

In the model discussed above for hard-bubble deflection, the Bloch-line distribution was assumed to be uniform around the bubble. Yet it is apparent that the gradient drive has a tendency to cause precession in one sense on the front side of the bubble and in the opposite sense on the back side because the actual local drive is down in front, up in back, and zero across the middle (see Fig. 2.5a). Since precession of spins in a Bloch line corresponds to displacement of the Bloch line, there should be a "bunching" of the Bloch lines on one side of the bubble. The bunching effect can be modeled by ignoring coercivity and neglecting all internal wall torques except those arising from exchange energy

$$\sigma_{ex} = 2A\Delta r^{-2}(\partial\psi/\partial\beta)^2. \qquad (13.12)$$

The calculation is similar to that given in Section 13,B except that $\psi(\beta)$ is allowed to vary to minimize $\delta W$ in Eq. (12.10); the reader is referred to the literature for details.[125,355] One finds, considering the geometry of Fig. 8.7,

$$\psi(\beta) = S\beta - (Mr^2/2A\Delta\gamma)V\cos\beta + (\alpha MSr/2A\gamma)V\sin\beta, \qquad (13.13)$$

correct to first order in $\alpha V$, where $\beta$ is defined relative to the velocity direction. The terms in $\cos \beta$ and $\sin \beta$ give rise to the bunching effect. Surprisingly, Eqs. (13.6) and (13.7) are also found, implying that the bubble velocity and deflection angle discussed above are not changed by the bunching in first order. The angle $\beta_{max}$ at which the maximum bunching occurs can be found by calculating the Bloch-line density $\partial\psi/\partial\beta$. One finds

$$\tan \beta_{max} = r/\alpha\Delta S, \qquad (13.14)$$

where we note again that $\beta$ is defined relative to the velocity vector. In the limit $S\Delta/r \to \pm 1$, this angle approaches $\pm\tan^{-1}\alpha^{-1}$ depending on the sign of $S$, which is the same in magnitude as the deflection angle of Eq. (13.9) ($H_c = 0$). Thus we find that if the deflection angle is large, the Bloch lines tend to bunch on the side of the bubble opposite from $-\nabla H_z$, which is intuitively plausible since the Bloch-line reaction force is effectively countering the gradient drive force. Evidence for a bunching effect has been observed in the high-speed photographs of Fig. 13.2, recording the transient position and shape of a hard bubble at the end of each of a series of nine gradient pulses.[357] There is a tendency for runout to occur on one side of the bubble, suggesting that the slower, higher coercivity wall on the upper side of the photographs contains more Bloch lines. Unfortunately a quantitative study of how such transient phenomena affect the analysis of the previous section has not yet been undertaken.

The possibility of bunching implies that when a pulse of strength $H_g = |r\nabla H_z|$ is applied to a static hard bubble, a transient process of rearrangement of the Bloch lines from their uniform static distribution must occur. To estimate how long this rearrangement takes, we ignore the $\alpha$-dependent term in Eq. (13.13) and note that when the pulse is applied, the spins at the front and back of the bubble will initially precess at a rate $\dot\psi = \gamma H_g$ until torques build up and equilibrium is attained. But Eq. (13.13) implicitly gives the velocity sustained by a given $\psi(\beta)$. Differentiating with respect to time, we can find the acceleration $\dot V$ in terms of $\dot\psi = \gamma H_g$. Comparing this result to Eq. (6.1) or (12.29), one finds an effective mass[177]

$$m_{BL} = M^2 r^2/\gamma^2 A\Delta = m_D r^2/\Lambda_0^2, \qquad (13.15)$$

where $m_D$ is the Döring mass ignoring damping corrections [Eq.(11.6)]. This "Bloch-line mass" is generally two orders of magnitude larger than the Döring mass and arises physically from the large exchange energy, corresponding to a kinetic energy, which can be stored in the Bloch lines at a given bubble velocity (see also Section 12,D). From Eq. (6.1), one can deduce the time constant $\tau_{BL} = m_{BL}/b$ of

$$\tau_{BL} = Mr^2/2\alpha\gamma A = (4\pi\alpha\gamma M)^{-1} r^2/\Lambda_0^2, \qquad (13.16)$$

where we have assumed damping losses are dominated by normal Bloch-wall losses, as appropriate if $S\Delta/r < 1$. For typical parameters of low damping 5 $\mu$m garnet films ($\alpha \sim 0.05$), $\tau_{BL}$ is of the order of microseconds. This effect can lead to large ballistic displacements of hard bubbles, as illustrated in Fig. 13.2 by the last two frames on the left showing the displacement of a bubble after the last pulse turned off.

The enormous size of the Bloch-line mass effect raises doubt about the usefulness of the steady-state theory of the previous section in explaining pulsed-gradient experiments. However it will be demonstrated in Section 18 that, provided one ignores coercivity and provided the initial and final spin states of the bubble are the same, the steady-state theory still applies. One must simply use the apparent velocity obtained by taking the initial and final positions of the bubble and dividing by the pulse time, as if one knew nothing about the transient effects. The physical reason for this is that while the bubble may be slow to accelerate, it makes up for its slow start by a large ballistic "overshoot" after the end of the pulse. The net distance is determined simply by the field energy gained from the drive and by the gyrotropic deflection forces, which are conservative and independent of the details of the transient motion with the provisos mentioned above.

Whenever a domain wall is subjected to a nonuniform drive field, as in the case of a bubble in a gradient, the precession rates at different points will be different. We have seen how this effect can cause Bloch-line bunching. If the precession proceeds full cycle, Bloch lines must be generated as well. In Chapter IX we describe in detail the process of Bloch-line generation and find that for low drives magnetostatic as well as exchange forces must be taken into account. However at sufficiently high drives we can ignore any restoring forces. Thus if $H_1 - H_2$ is the difference in drive field between two points, then the number of Bloch lines generated in a time $T$ is

$$n = \pi^{-1}\gamma(H_1 - H_2)T. \tag{13.17}$$

Let us consider the case of the $S = 0$ bubble in Fig. 13.5. Since the sense of precession is opposite on the front and back of a bubble, one finds that Bloch lines of positive sense will pile up on the left-hand flank of the bubble (as one faces the velocity direction) and those of negative sense will pile up on the right-hand flank.[179] The Bloch-line pile-up is analogous to a round rubber band twisted at two opposite points.

The Bloch-line wind-up process gives a clue as to how hard bubbles are formed to start with. Consider for example the geometry of Fig. 13.6, in which stripes run at an angle under conductors that can be pulsed to cut the stripes and form bubbles.[359-361] It was found in experiments on ErEuGaYIG films that hard bubbles were formed when the stripes lay at an angle $\rho \neq 0$ to the conductors, and the hardness increased with $\rho$. A qualita-

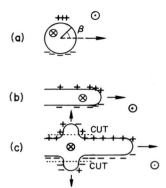

(a)

(b)

(c)

**Fig. 13.5.** Schematic Bloch-line wind-up in bubble motion or stripe extension and buckling. The +'s and −'s indicate twist sense of the Bloch lines (after Voegeli and Calhoun[179]).

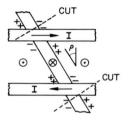

**Fig. 13.6.** Schematic stripe-cutting experiment used for generating hard bubbles.

tive explanation may be given as follows: The bubble is formed when the stripe is cut by local $z$-fields from the striplines. When the striplines are pulsed in opposite senses, the cutting occurs at the dashed lines as shown in the Fig. 13.6. Because the fields are stronger adjacent to the striplines but weaker farther away, there are nonuniform fields acting on the wall causing Bloch-line wind-up with the polarities indicated. After the stripe is cut, it is apparent that for the case shown in the figure, a predominance of Bloch lines of one polarity remains in the central domain. As this domain is contracted to a bubble shape by lifting the bias field, the excess Bloch lines annihilate each other leaving a bubble with a net of negative Bloch lines. Thus, even though equal numbers of Bloch lines of both senses were generated to start with, an asymmetry in the cutting process concentrated the negative Bloch lines in the central domain and the positive Bloch lines in the outer domains. It is apparent by symmetry that when the stripe lies perpendicular to the stripline, this difference can no longer arise, thus explaining why the resulting bubbles are not hard.

A less controlled but easier method for nucleating hard bubbles is to allow rapid demagnetization of a sample from the saturated state. A characteristic highly branched stripe pattern occurs in this case.[267] If the domain pattern is now cut, for instance by a magnetic needle, hard bubbles are found with high probability. To explain this, one may speculate that as the stripe head advances, it experiences large driving fields causing spin precession and creating positive Bloch lines on one side of the advancing stripe head and negative Bloch lines on the other, as shown in Fig. 13.5. The stripe also tends to buckle and form side branches as shown in the figure, and motion of these branches also leads to Bloch-line generation. When the branches are eventually cut off to form bubbles, they typically contain a net of Bloch lines of one polarity.

An additional mechanism must also exist to explain the hard bubble formation method in which a single normal bubble is expanded by a strong bias pulse.[235] High-speed photography has shown that such an expansion process can lead to considerable shape distortion, which gives the nonuniform drive needed for Bloch-line wind-up.[169,170] However since no cutting is assumed to occur, it is hard to understand how a preponderance of Bloch lines of one sign can result. One possibility is a Bloch-line annihilation in regions of high Bloch-line density by means of a generation of Bloch points,[264] leading statistically to a preponderance of Bloch lines of one sign over the other.

## 14. Small Numbers of Vertical Bloch Lines

When the average distance between Bloch lines is greater than $\sqrt{2\pi\Lambda_0}$, they may be considered as localized and their positions to some degree independent of each other and free to move around the bubble wall. It is then useful to localize the dynamic reaction forces associated with them using the Bloch-line approximation (Section 8,A).[225] In Section 14,A, we derive expressions for the dynamic Bloch-line forces as well as the dynamic force of a chiral bubble. These expressions greatly simplify the application of the dynamical-variable formalism of Section 12,B to a given problem. Now the procedure reduces simply to balancing static energy-concerning forces against the dynamic reaction forces derived below. This procedure is applied in Sections 14,B and 14,C to the respective problems of translation and collapse of bubbles containing vertical Bloch lines. It is also used in Section 15 to describe straight-wall motion with horizontal Bloch lines, and in Section 18 to derive Bloch curve shapes in nonlinear bubble translation. It is well to bear in mind that, when this simple treatment is applied to nonsteady motion, one implicitly assumes the quasi-steady approximation (Section 12,D). so that the validity of the results is limited by the criteria of Section 12,E.

The remarkable effects we discuss below, namely the large effect of just a few Bloch lines on the deflection of propagating bubbles or on dynamic bubble collapse, are also of great practical interest, for both have been successfully used for detection of wall states in exploratory bubble devices.[27a,283a,361a,399]

## A. DYNAMIC REACTION FORCES OF ISOLATED BLOCH LINES AND CHIRAL BUBBLES

Consider the Bloch-line configuration of Fig. 8.1, described by

$$\tan(\psi/2) = \exp[-(x - x_i)/\Lambda_0], \qquad \Lambda_0 = (A/2\pi M^2)^{1/2}, \qquad (14.1)$$

where $x_i$ and $y_i$ specify the time-dependent position of the center of the Bloch line. In this context we have

$$\delta q = \delta y_i, \quad \dot{q} = \dot{y}_i, \quad dA = dxdz,$$

$$\dot{\psi} = -\dot{x}_i\Lambda_0^{-1}\sin\psi, \qquad (14.2)$$

$$\delta\psi = -\delta x_i\Lambda_0^{-1}\sin\psi.$$

Treating $x_i$ and $y_i$ as dynamical variables in Eq. (12.11), substituting these expressions and carrying the integral over unit distance in the $z$ direction, we calculate

$$\partial W/\partial x_i = -2M\gamma^{-1}\int_{-a/2}^{a/2}(\dot{y}_i + \alpha\Delta\dot{x}_i\Lambda_0^{-1}\sin\psi)\Lambda_0^{-1}\sin\psi\,dx, \quad (14.3)$$

where $a$ is a constant wall length large compared to $\Lambda_0$. After changing the variable of integration to $\psi$ by means of Eq. (14.1) and approximating $\Delta$ with $\Lambda_0$, one integrates to find[225,231]

$$\partial W/\partial x_i = -2\pi M\gamma^{-1}\dot{y}_i - 4\alpha\gamma^{-1}MQ^{-1/2}\dot{x}_i. \qquad (14.4)$$

This equation expresses the balance of static force $\partial W/\partial x_i$ derived from energy-conserving sources against the dynamic reaction consisting of a gyrotropic term plus a viscous term.

Similarly, the $y$ component of the dynamic reaction is, from Eqs. (12.11) and (14.2),

$$\partial W/\partial y_i = 2M\gamma^{-1}\int_{-a/2}^{a/2}(\dot{x}_i\Lambda_0^{-1}\sin\psi - \alpha\Delta^{-1}\dot{y}_i - \gamma H_c\,\text{sgn}\,\dot{y}_i)dx. \quad (14.5)$$

In this case one must expand $\Delta^{-1}$ to higher order, using Eqs. (8.2) and (14.2) to find

$$\Delta^{-1} = \Delta_0^{-1}(1 + Q^{-1}\sin^2\psi + \cdots). \qquad (14.6)$$

After integration with the help of Eq. (14.1), Eq. (14.5) reduces to

$$\partial W/\partial y_i = 2\pi M\gamma^{-1}\dot{x}_i - 4\alpha M\gamma^{-1}Q^{-1/2}\dot{y}_i - a(2\alpha\Delta_0^{-1}M\gamma^{-1}\dot{y}_i$$

$$+ 2MH_c\,\mathrm{sgn}\dot{y}_i). \quad (14.7)$$

This equation represents the superposition of Bloch-line reaction forces onto a normal Bloch-wall force, the latter proportional to the length of wall, $a$, considered. We explain, parenthetically, that the damping term $4\alpha M\dot{y}_i/\gamma Q^{1/2}$ arises from the increased wall dissipation caused by the "pinching" of the wall in the Bloch-line region, in accordance with Eq. (14.6) as illustrated in Fig. 8.1.

If we omit the normal-wall force term proportional to $a$ from Eq. (14.7), there remains a specific dynamic reaction force per unit length of Bloch line given in vectorial form by

$$\mathbf{F}_{di} = 2\pi M\gamma^{-1}\mathbf{t}_i \times \mathbf{v}_i - 4\alpha M\gamma^{-1}Q^{-1/2}\mathbf{v}_i, \quad (14.8)$$

where $\mathbf{v}_i$ is the velocity of the $i$th Bloch line and $\mathbf{t}_i$ is its unit vector tangent as defined in Eq. (8.10). This is a general result, independent of the specific geometry of Fig. 8.1. The ratio of the magnitudes of the viscous to gyrotropic terms in Eq. (14.8) is $2\alpha/\pi Q^{1/2}$, which is generally much less than one in typical garnet films. Therefore Eq. (14.8) implies that a Bloch line will tend to move normal to a force applied to it. Because the conventions in Eq. (14.8) are difficult to remember, a simple procedure is helpful in determining which direction a Bloch line tends to move when it experiences a drive force. First visualize or draw the Bloch line, as in Fig. 8.1. Considering the direction of the applied field, use the right-hand rule to see which way the spins of the Bloch line turn. Imagine them all rotated by a small angle, say 45°. The direction of motion can then be easily visualized by finding the spin that now points normal to the wall and forms the displaced center of the Bloch line. For example an upward field in Fig. 8.1 moves the Bloch line left.

We also note for future reference that in certain cases such as for an in-plane field $H_p$ normal to the wall, the Bloch-line twist angle may be different from $\pi$, as illustrated in Fig. 8.3 or 12.4. Taking $\Phi$ to be the full angle of Bloch-line twist, defined by $\Phi = 2\cos^{-1}(H_p/8M)$ as in Eq. (8.12), one can derive the dynamic reaction force of the $i$th Bloch line

$$\mathbf{F}_{di} = 2M\gamma^{-1}\Phi\mathbf{t}_i \times \mathbf{v}_i - \tfrac{1}{2}\alpha ME_L(\Phi)A^{-1}\gamma^{-1}\mathbf{v}_i. \quad (14.9)$$

Here $A$ is the exchange stiffness and $E_L(\Phi)$ is the Bloch line energy given by Eqs. (8.13) or (8.14).

In application to bubble motion problems, one superimposes forces of this type for each Bloch line present onto forces associated with a simple chiral bubble. A chiral bubble has a uniform twist with a winding number $S = 1$, because the moment is everywhere tangent to the wall surface, as expressed by the equation $\psi = \beta \pm \pi/2$. Therefore the hard-bubble theory given in Section 13,B may be used to give (ignoring corrections of order $(\Delta/r)^2$)

$$\mathbf{F}_{d\chi} = 4\pi M\gamma^{-1}\mathbf{z}_0 \times \mathbf{V} - 2\pi M\gamma^{-1}\Delta_0^{-1}\alpha r\mathbf{V} - 8MH_c r V^{-1}\mathbf{V} \quad (14.10)$$

for the dynamic reaction force per unit thickness acting on a chiral bubble, where $\mathbf{z}_0$ is the direction of the magnetization outside the closed domain, $\mathbf{V}$ is its velocity, and $r$ its radius.

### B. BUBBLE PROPAGATION AND DEFLECTION WITH SMALL BLOCH-LINE NUMBERS

Now consider the steady motion at velocity $\mathbf{V}$ of a bubble containing $n_+ (> 0)$ $\pi$-lines with positive twist and $n_- (> 0)$ $\pi$-lines with negative twist, as shown in Fig. 14.1. In what follows we assume the bubble has magnetization down inside and up (along $z$) outside. The winding number is $S = 1 + \frac{1}{2}(n_+ - n_-)$, as discussed in Section 8,C. We assume the integers $n_+$ and $n_-$ are constrained to such values that $S$ is an integer of either sign or zero. The forces $\mathbf{F}_{di}$ of Eq. (14.8) dispose themselves in the manner illustrated in Fig. 14.1 for the special case $n_+ = 3$ and $n_- = 1$. The large force $\mathbf{F}_{d\chi}$ of the chiral bubble [Eq.(14.10)] makes a small angle with $-\mathbf{V}$. The smaller Bloch line forces $\mathbf{F}_{di}$ are nearly orthogonal to $\mathbf{V}$ tending to cluster the Bloch lines of like sign on the same flank of the domain. The mobility relation for

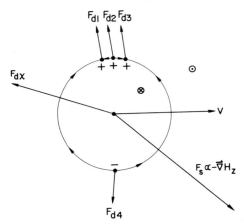

**Fig. 14.1.** Schematic balance of forces in an $S = 2$ bubble with four Bloch lines in a bias-field gradient $\nabla H_z$.

such a domain is obtained by balancing the total dynamic reaction force $F_{d\chi} + \sum_i F_{di}$ against the applied gradient force [Eq. (12.12)]. The resulting vector mobility expression is [225]

$$\nabla H_z = -(\alpha V/\gamma r\Delta_0)\{1 + [2\Delta_0{}^2(n_+ + n_-)/\pi r\Lambda_0]\}$$
$$- (4H_c V/\pi rV) + (2Sz_0 \times V/\gamma r^2). \quad (14.11)$$

The gyrotropic and coercive terms of this equation are identical to those of the corresponding mobility equations for a hard bubble [Eqs.(13.6) and (13.7)]. This result indicates that the gyrotropic force expression is proportional to $S$ for any density of vertical Bloch lines and moreover does not depend on structural details of walls, lines or domain shape. A general proof of this fact is given in Section 18,C. The viscous damping term in Eq. (14.11) has two contributions, one from the $\theta$ damping of the Bloch wall and one from the Bloch-line damping. Clearly if $n_+ + n_-$ is small, the Bloch-line damping term is negligible. In this case, and ignoring coercivity, the deflection angle [Eq.(13.8)] is simply given by

$$\rho = \tan^{-1}(2S\Delta/\alpha r), \quad (14.12)$$

which shows that the angle is inversely proportional to damping and bubble size, provided $S$ and $\Delta$ stay constant. A more general relationship, not ignoring coercivity, is $\sin \rho = 2VS/\gamma rH_g$ [Eq. (13.11)], which is extremely useful in deriving the $S$-state from experiment.

A schematic visualization of the predictions of Eq. (14.11) and the way the deflection would appear in a pulsed gradient-propagation experiment is shown in Fig. 14.2.[175] It is to be noted that bubbles which go straight down a gradient ($S = 0$) have a minimum of two Bloch lines according to Eq. (8.25). By contrast, chiral bubbles with no Bloch lines propagate at an angle. This prediction seems astonishing from the naive point of view that the chiral bubble is the simplest and most symmetric. The deflection arises from the nonnegligible $\psi$-precession on the flanks of the bubble, as discussed in connection with hard bubbles. Next we describe experiments confirming that the chiral bubble propagates not straight but at an angle.

In an early experiment on a EuTbGaYIG film with 5 $\mu$m bubbles, the deflection angles in a conventional pulsed-gradient experiment were determined for different bubbles chosen at random.[125] Three distinct groups were observed at $+13$, 0, and $-13°$, which, when evaluated according to $S = \gamma rH_g \sin \rho/2V$, gave $S$-values of 1, 0, and $-1$, respectively. This experiment gave the first evidence of "quantization" of bubble states and showed that even a single pair of Bloch lines could give a measurable effect on the dynamic properties of bubbles. Figure 14.3 shows the apparent velocity of

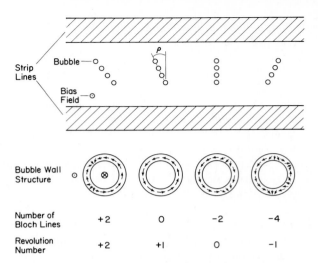

**Fig. 14.2.** Schematic pulsed gradient propagation experiment with four multiple exposures showing bubble displacement and deflection and associated wall states (after Malozemoff[175]).

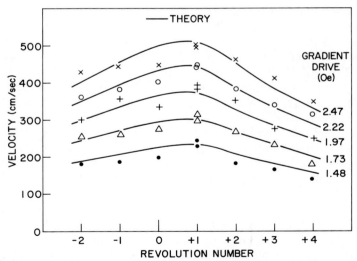

**Fig. 14.3** Net bubble velocity in a pulsed gradient-field experiment versus revolution (winding) number $S$ for different gradient drives $2H_g = 2|r\nabla H_z|$ in a EuTbGaYIG film (after Malozemoff[175]).

different bubble states in this sample as a function of drive and also a comparison to Eq. (14.11).[175] The slight asymmetry in the data, which gives a velocity peak at $S = 1$, indicates that this skew bubble, and not the one going straight ahead, is the one containing no Bloch lines because its mobility is the highest. It should be noted that in fitting the theory to the data in Fig. 14.3, it was necessary to invoke a Bloch-line coercivity proportional to the number of Bloch lines, much as in the hard-bubble experiments discussed in Section 13,B. The coercivity came out to be 0.012 Oe per Bloch line. Good agreement with the deflection-angle formula $S = \gamma r H_g \sin \rho / 2V$ has also been achieved in ion-implanted SmCaGeYIG and SmLuCaGeYIG films.[250,322] In the latter case, the deflection angle of a given state varied from 6 to 22° over a temperature range of $-40°C$ to 75°C. Considering the simultaneous variation in mobility by a factor of four, $S = \gamma r H_g \sin \rho / 2V$ ranged from only 1.3 to 1.1, so that the state may presumably be identified as $S = 1$.

When such experiments were extended to lower damping films, higher angle values were observed for the predominant low-angle skew-deflecting group, in qualitative accord with Eq. (14.12). For example a well-defined group was observed at 45° in a low damping 5 $\mu$m EuGaYIG film[179] and at the very large angle of 75° in a very low damping LaGaYIG film.[213] The prevalence of this angle among randomly generated bubbles suggested that it was an $S = 1$ bubble with no Bloch lines, but quantitatively the experimental $S$ values from Eq. (13.11) came out nonintegral and generally less than 1. Large angles were also observed in a series of films with small bubble sizes qualitatively in accord with the inverse radius dependence in Eq. (14.12), but once again the $S$-values were too small.[301] Similar nonintegral values were also obtained on the EuTbGaYIG film of Fig. 14.3 when gradient drives exceeded the critical drive of $H_g = |r \nabla H_z| = 1.25$ Oe for the onset of nonlinearities in this sample.[260]

One possible reason for these discrepancies was revealed by the observation, using high-speed photography, of angle changes during nonlinear bubble propagation involving overshoot.[135a,357–362] For example, Fig. 14.4 shows the angle $\rho_T$ up to the end of the gradient pulse and the larger angle $\rho_\infty$ the bubble finally achieved.[362] As will be explained in Section 18, theory shows that Eq. (13.11) fails in the nonlinear region if coercivity is present or state changes occur. The success of the early experiments on EuTbGaYIG relied on the fact that the drive was kept below the critical field for nonlinear motion (see Sections 15 and 16).[175] Even in this relatively high-damping sample ($\alpha = 0.12$), that critical field was only 1.25 Oe; so it is unlikely that the linear region could be observed at all in lower damping samples because the critical field will be not much larger than the static coercive field. On the other hand, if $\alpha$ were increased further (e.g., by further Tb addition) to

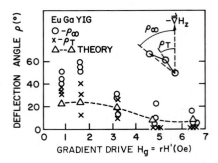

**Fig. 14.4.** Bubble skew-deflection angle $\rho$ at $t = T$ and $t = \infty$ as a function of drive field $H_g = |r\nabla H_z|$, determined by high-speed photography of a pulsed gradient propagation experiment with pulse length $T = 0.5$ $\mu$sec in a EuGaYIG film. Dotted line is theory for $\rho_T$, based on Eq. (13.11) and the fact that the velocity is saturated (after De Luca *et al.*[362]).

increase the critical field, the deflection angle of Eq. (14.12) would become too small to observe easily.

Another possible reason for nonintegral $S$-values is the presence of Bloch points, which can give half-integer values of $S$ in zero in-plane field. The $S$-value and hence deflection angle of such states are expected to decrease with in-plane field,[250,259,260] as discussed more fully in Section 9,B. The cleanest evidence for such states has been obtained in ion-implanted samples, and the theory [Eqs. (9.15), (9.18), and (13.11)] can explain within a few degrees the experimental deflection angles of ($\frac{1}{2}$, 2, 1) and (1, 2, 2) states in a SmCaGeYIG film, shown in Fig. 9.6a.[250] In-plane fields also induce the bubbles to become elliptical with their major axis parallel to the in-plane field, and this ellipticity can change the deflection angle, in a way that depends on the relative orientation of the in-plane field and the direction of motion. Shifts of as much as 8° were observed in the propagation of $S = 1, 0$, and $-1$ bubbles in a EuGaYIG film with an in-plane field of 40 Oe at roughly 45° to the direction of motion. The effect is accounted for by a straightforward extension of the skew angle theory to elliptical bubbles.[187]

The deflection effect also appears to account qualitatively for certain features of stripe expansion or contraction. Often the head of such a stripe leans to one side, presumably because of gyrotropic force.[190] Expansion can therefore give rise to curved stripes or even spirals[236]. A remarkable example emerges from "needle-pricking" experiments on MnBi.[363,364] The coercivity of this uniaxial material is usually so high that after saturating a film with a bias field, the field can be brought to zero without run-in of stripes from the edges. Pricking of the film at one point with a nonmagnetic needle can then cause nucleation and explosive expansion of stripes in all directions. Often the final demagnetized state exhibits a curvature of the stripes in one sense as they emanate from the pricking point. This preferred sense of curvature presumably arises from the gyrotropic effect of a chiral wall without vertical Bloch lines, since even if vertical Bloch lines were being nucleated, one would expect them to appear in equal numbers so that their gyrotropic effect would average out. Under certain conditions needle-pricking gives rise to bubble lattices as well, although the mechanism is not clearly understood.

## C.  BUBBLE COLLAPSE WITH SMALL BLOCH-LINE NUMBERS

When a bubble with just a few Bloch lines collapses or expands radially in response to a uniform pulsed field $H_z$ the motion is not likely to be symmetrical. Indeed the most likely motion is very lopsided because of the static tendency of Bloch lines to cluster as discussed in Section 8,B. Assuming the cluster model shown in the upper left inset to Fig. 14.5, there are four degrees of freedom and hence four dynamical variables, namely the position $(X, Y)$ of the bubble center, the bubble radius, and the angular position $\beta$ of the Bloch line cluster. A complete solution of this problem is given in the literature.[225] Here we derive the salient results semiquantitatively by means of a physical picture. The effect of the pulsed $H_z$ is to produce a pressure $P = 2MH_z$ on the wall. To the extent that this pressure is exerted on the Bloch lines, the resulting velocity $v_L$ is orthogonal, that is tangential to the wall. This is true because, as we have seen in Section 14a, the coefficient of the viscous drag term in Eq. (14.8) is generally smaller than that of the gyrotropic

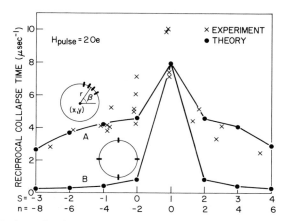

**Fig. 14.5.**  Reciprocal collapse time versus winding number $S$ determined in a dynamic bubble collapse and pulsed gradient propagation experiment in a EuTbGaYIG film. Theory curve $A$ represents the Bloch-line cluster model [Eq. (14.15)] while curve $B$ represents a model with evenly spaced Bloch lines [Eq. (14.16)]. The experimental peak at $S = 1$ confirms that this bubble has the minimum number of Bloch lines (after Slonczewski[225]).

term. The center of the domain, however, responds in an overdamped way, moving with a velocity $\mathbf{V}_{cent}$, for the ratio of the viscous drag term to the gyrotropic term in Eq. (14.10) is $\alpha r/2\Delta_0$, which is not small for many samples. Therefore, as is evident from Fig. 14.5, the Bloch line cluster effectively constitutes a pinning point, drawing the otherwise Bloch-line-free domain into a "sideways" mode of collapse. Assuming the domain remains circular, the appropriate constraint is

$$dX = -\,dr \quad \text{or} \quad \dot{X} = -\,\dot{r}. \tag{14.13}$$

With this constraint, we may substitute the relation $W = 2\pi r^2 MHh$, together with

$$dq = -\,dr(1 - \cos\beta),$$
$$\dot{q} = -\,\dot{r}(1 - \cos\beta), \tag{14.14}$$
$$d\psi = 0, \quad \dot{\psi} = 0, \quad dA = rd\beta dz,$$

into Eq. (12.11). Then since $r$ is the only free variable, we evaluate $\partial W/\partial r$ to find, for $H_c > 0$,

$$\dot{r} = -(2\Delta_0\gamma/3\alpha)(H_z - H_c) \tag{14.15}$$

for the collapse velocity. In this limit, the only effect of the Bloch-line cluster is to reduce the mobility by $\frac{1}{3}$.

Dynamic bubble-collapse results on a EuTbGaYIG film are shown in Fig. 14.5.[225] For each data point, the winding number $S$ was first measured by means of a deflection experiment, using the "golden rule" Eq. (13.11), and then the pulse collapse time $T$ was measured for a fixed drive. The data for $S = 0$ and 2 in this figure exhibit the expected $\frac{1}{3}$ mobility reduction from the chiral case $S = 1$. For $S$ values even further from $S = 1$, the reduction appears to increase progressively. Indeed the detailed calculation shows that increasing the size of the cluster decreases the mobility, slowly at first, but then more rapidly in accordance with curve $A$ of the figure.[225] In the limit where $S$ is so large that the Bloch lines encircle the domain completely, the mobility tends to the low values of the "hard bubble" relation [Eq. (13.3)]. The results of Fig. 14.5 dramatically confirm the identification of the skew-deflecting $S = 1$ state as the bubble with no Bloch lines.

An interesting limiting case is that of a few uniformly spaced Bloch lines at distances greater than $\sqrt{2\pi}\Lambda_0$ [Eq.(8.20)]. The mobility in this case is easily obtained from the above principles and is found to be[225,351]

$$\mu_{unif} = \Delta_0\gamma/\alpha[1 + (\pi^2\Lambda_0/2\alpha^2 a)], \tag{14.16}$$

where $a$ is the distance between neighboring Bloch lines. The collapse behavior, based on this relation and shown by curve $B$ in Fig. 14.5, is much slower than for clustered Bloch lines. In this case the Bloch lines circulate symmetrically around the domain. Since they dissipate energy slowly, the collapse is slow. Depending on whether or not the Bloch lines are clustered, then, one can expect collapse mobilities anywhere between Eqs. (14.15) and (14.16). The fact that agreement with Eq. (14.15) was found in Fig. 14.5 arises from the fact that repetitive pulsing ($\sim 100$ pulses/sec) was used in the collapse experiment, insuring the observation of the *fastest* collapse time. In other bubble-collapse experiments with slower pulsing, the results showed much more scatter, which can be attributed to a lack of consistency in the initial arrangement of Bloch lines.[179] An extension of this effect to ion-implanted films provides a means of discriminating $S = 0$ from $S = 1$ states in an exploratory device concept.[364a]

Many other examples of the interaction of individual Bloch lines with bubble motion could be given. However, further discussion of these cases will be postponed until Chapter IX, where the use of the concept of domain momentum $\bar{\psi}$ simplifies the discussion considerably.

# VIII

## Nonlinear Wall Motion in Two Dimensions

In this chapter we consider a two-dimensional model of wall motion that takes into account variations not only normal to the wall as in the one-dimensional model of Chapter V, but also along the film normal $z$. The $z$-variations arise from any nonuniformity through the thickness of the film, such as from the stray field of the surface magnetic charge distribution, as discussed in Section 8,E, or from material nonuniformities such as capping layers, or from the presence of horizontal Bloch lines. In this chapter we ignore any variation along the $x$ axis, i.e., along the perimeter of the domain wall, such as might arise from vertical Bloch lines or from gradient fields acting on a bubble. Such variations are considered in Chapters VII and IX. If vertical Bloch lines are absent, the two-dimensional model presumably applies to such experiments as bubble collapse or other radial bubble motions, stripe-width expansion or contraction in isolated stripes and stripe arrays, and uniform single-wall motion. We begin with a treatment of the problem in a Bloch-line model that permits easy derivation of the main results.[224] In Section 16, we compare theory to experiment. In Section 17 we describe a more complete theory based on numerical calculations.

### 15. Bloch-Line Model

#### A. Equations of Motion

We start by postulating the existence of a horizontal Bloch line located at some position $z_L$ between the critical points $z = a \equiv h/(1 + e^2)$ and $z = b \equiv he^2/(1 + e^2)$ (where $e = 2.718$) of an isolated plane wall as illustrated

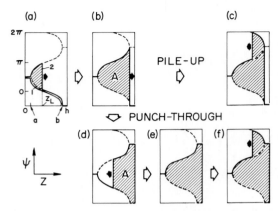

**Fig. 15.1.** Schematic domain-wall spin states $\psi(z)$ during pile-up and punch-through. In (a) a horizontal Bloch line propagates through the film and reaches its instability point in (b), where the wall velocity attains $V_{po}$. If pile-up occurs, the first Bloch line simply stays there while a new Bloch line nucleates and propagates in the reverse direction. If punch-through occurs, the first Bloch line disappears at the surface $z = h$ while a reverse Bloch line nucleates in (d) in such a way that the momentum $A$ is the same as in (b). After this Bloch line denucleates in (e), a new Bloch line starts the next cycle in (f).

in Fig. 15.1a. Such a Bloch line has an energy $E_L(z)$ [Eq. (8.13)] and a rotation angle $\Phi(z)$ [Eq. (8.12)] that are functions of $z$ because of the wall-normal component of stray field $H_y(z)$ [Eq. (8.33)]. As such a Bloch line moves through the thickness of the film, it effectively switches the wall moment from one local magnetostatic equilibrium orientation to another.

We derive here a pair of coupled equations for the simultaneous motion of the wall position $Y$ (velocity $V$) and Bloch line position $z_L$ (velocity $v_L$).[224,226,351] As discussed in Section 14, we simply balance the $y$ and $z$ components of static and dynamic force (per unit distance) on the wall and the Bloch line. Thus we superpose wall pressures due to drive $2M(H - H_c)$ and viscous damping $-2\alpha VM/\gamma\Delta_o$ [Eq. (14.7)], whose balance would give us the conventional mobility relation, onto gyrotropic and viscous-damping Bloch-line forces taken from Eq. (14.9). The $y$-component forces occur in the equation

$$2Mh(H - H_c) - 2\alpha MhV\gamma^{-1}\Delta_o^{-1} - 2\Phi M\gamma^{-1}v_L = 0, \qquad (15.1)$$

where we have neglected the Bloch-line drag in comparison with the wall drag. Of course the fact that the Bloch-line force is concentrated at the Bloch line implies that the wall billows around it as shown in Fig. 12.2 (see section 12,C). The $z$-components of force, which involve only the Bloch line, are

$$-dE_L/dz_L + 2\Phi M\gamma^{-1}V - \tfrac{1}{2}\alpha ME_LA^{-1}\gamma^{-1}v_L = 0, \qquad (15.2)$$

where $-dE_L/dz_L$ is the static Bloch-line force obtainable from Eqs. (8.13), (8.12), and (8.33). This equation represents the torque equation, $dE_L/dz_L$ originating from the Bloch-line torque arising from the stray field. Eqs. (15.1) and (15.2) are coupled equations in $V$ and $v_L$ whose solutions we discuss below.

## B.  BLOCH-LINE NUCLEATION

Suppose that initially, at $t = 0$, the wall has the simple static internal structure represented by curve 1 in Fig. 15.1a. This means Bloch lines are absent. In response to a step increase $H$ of bias field, the wall behaves essentially like a one-dimensional wall, for sufficiently small $H$ or $t$. (However, the mass is substantially increased by stray-field effects discussed in Section 17,B.) But for greater values of $H$, Bloch lines may nucleate by a process of wall breakdown or dynamic conversion. Likely places for Bloch-line nucleation are at the critical points $z_L = a \equiv h/(1 + e^2)$ and $b \equiv he^2/(1 + e^2)$ (see Section 8,E), where the balance of stray field and magnetostatic forces leaves the wall spins most labile. Given the assumed positive polarity of the drive tending to increase $\psi$, and the assumed initial Bloch-wall polarity in Fig. 15.1a, the nucleation is expected to occur at $z_L = a$ [it would occur at $b$ for the other Bloch-wall polarity]. As will be discussed more fully in Section 18d, there is theoretical disagreement about the maximum velocity $V_n$ that can be sustained before a wall breakdown occurs.[224,226,245,351] According to some estimates, this velocity is close to Walker's velocity $V_w = 2\pi\gamma\Delta_0 M$, according to others close to the smaller value $V_{po} = 24\gamma A/hK^{1/2}$ (see below) provided $\Lambda_0 \ll h$ (the Bloch-line limit). The most detailed theory (Section 17,A) predicts $V_n$ is somewhat smaller than $V_{po}$. Complicating the experimental resolution of this point of theoretical disagreement is the possibility that material imperfections could facilitate Bloch-line nucleation and hence lower or smear out the velocity peak in much the way that imperfections facilitate domain nucleation in bulk magnetization reversal.[226] Once the Bloch line has nucleated, the velocity should drop and the Bloch line should begin to move up through the wall as governed by Eqs. (15.1) and 15.2). There are two drive-field regions to consider:

## C.  LOW-DRIVE REGION: THE BLOCH-LINE INSTABILITY VELOCITY

Suppose the drive field is large enough for one Bloch line to nucleate, yet sufficiently small that in time the wall approaches a condition of dynamic equilibrium such that the Bloch line tends to an asymptotic constant position.

Then $v_L \to 0$ and Eq. (15.1) reduces to the conventional mobility expression

$$V = \gamma\Delta_0\alpha^{-1}(H - H_c), \qquad H > H_c, \tag{15.3}$$

while Eq. (15.2) reduces to

$$V = (\gamma/2M\Phi)dE_L/dz_L. \tag{15.4a}$$

The dependence of $V$ on $z_L$ according to this equation is evaluated from Eqs. (8.12) and (8.13) as long as $z_L$ lies in the range $a < z_L < b$ (curve 2 in Fig. 15.1a). The function $V(z_L)$ has a symmetric form in this range:

$$V = \gamma A(2\pi/K)^{1/2}[z_L^{-1} + (h - z_L)^{-1}], \tag{15.4b}$$

with a minimum at the midplane of the film, as illustrated in Fig. 15.2. A more exact theory (Section 17,A) substantially agrees with this model for sufficient thickness $h$, except near $z_L = a$ where the Bloch-line model fails because the Bloch-line width tends to infinity [see Eq. (8.15)]. The dashed curve in Fig. 15.2 indicates schematically the behavior given by the more exact theory. Wherever valid, the formula for $V(z_L)$ can be solved to give the steady-state value of $z_L$.

For $z_L$ in the range $b < z_L < h$, where the Bloch-line angle is $\Phi = 2\pi$, the energy expression (8.14) must be used. Applying this, one finds the expression

$$V = [2\sqrt{2}\gamma Ah/(\pi K)^{1/2}z(h \qquad z)]\arcsin[\tfrac{1}{2}\ln(z/h - z)]^{-1/2}. \tag{15.4c}$$

According to this equation, $V$ is continuous at $z_L = b$ and, after a small dip, increases with $z_L$ as shown in Fig. 15.2. Its maximum value in this region is

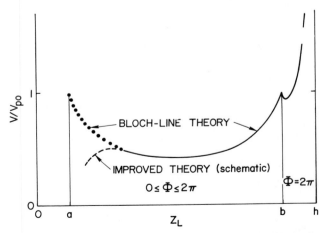

**Fig. 15.2.** Reduced velocity of a plane wall versus position of horizontal Bloch line $z_L$. The critical positions $z_L = a = h/(1 + e^2)$ and $b = he^2/(1 + e^2)$ are the same as those in Fig. 15.1.

estimated from a more accurate theory, which eliminates the singularity at $z_L = h$, to be $\approx 1.3|\gamma|A^{1/2}$, with little dependence on film parameters.[226] Noteworthy is the fact that realistically the distance between $z_L = b$ and $z_L = h$ is too small compared to the Bloch-line width for calculated $V(z_L)$ relations in this region to be taken very seriously.

However, the following application of the Bloch-line model predicts that $z_L$ cannot exceed $b$ in a state of constant $V$: Suppose the wall contains a Bloch line in this region. Consider the nucleation of a second Bloch line in the manner shown in Fig. 15.1c. According to the Lagrangian principle (12.23a), the wall energy $\bar{\sigma}$ is stationary with respect to the momentum $\bar{\psi}$. But note that $V(z_L)$ is greater in the $z_L > b$ region than for a freshly nucleated Bloch line in the $z_L < b$ region. This implies, according to Eq. (12.23b), that when two Bloch lines are present a virtual energy change $\delta\bar{\sigma} = 2M\gamma^{-1}V\delta\bar{\psi}$ due to a virtual Bloch-line displacement in the region $z_L > b$ is greater than one due to a displacement in the other region. Therefore it follows that $\bar{\sigma}$ decreases at constant $\bar{\psi}$ (represented by the shaded area in Fig. 15.1c) by simultaneous displacement of both Bloch lines toward the film center, since the energy decrease in region $z_L > b$ more than compensates for the energy increase in region $z_L < b$. Thus, according to the Bloch-line model the peak velocity for steady-state motion is given by $V_p = V(b)$, and the corresponding critical drive field $H = H_{max}$ is given by

$$H_{max} = H_c + \mu^{-1}V_p. \tag{15.5}$$

For the particular case of an isolated plane wall, using the stray field of Eq. (8.33), one finds that the maximum velocity occurs when the Bloch line reaches $z_L = b \equiv he^2/(1 + e^2)$ (see Fig. 15.1b), where $V_p = V_{po}$,[224]

$$V_{po} = 4\gamma Ah^{-1}(2\pi/K)^{1/2}\cosh^2 1 \equiv 23.8\gamma A/hK^{1/2}. \tag{15.6}$$

This important relation represents the critical velocity for a Bloch-line instability, whose nature will be discussed more fully below and in Section 17. $V_{po}$ is smaller than the Walker velocity by the factor

$$V_{po}/V_w = 3.8A^{1/2}/hM = 9.5\Lambda_0/h. \tag{15.7}$$

What this means is that in two dimensions the formation of a Bloch line eases the wall structure around the energy barrier provided by local wall demagnetization. Therefore the wall structure becomes unstable at a lower torque, or equivalently at a lower velocity. The reduction is in proportion to the ratio of $\Lambda_0/h$ because in the Bloch-line limit, the energy is concentrated in that portion of the film thickness lying within the width of the Bloch line. A recent calculation, which includes corrections of order $1/Q$ as compared to the foregoing theory, leads to the multiplicative correction factor $(1 - 1.12/Q)$ in Eq. (15.6).[364b]

## D.  HIGH-DRIVE REGION

Steady-state wall motion is not possible for $H > H_{max}$ of Eq. (15.5). We expect that $V(t)$ and $v_L(t)$ obey the transient equations (15.1) and (15.2) only until the Bloch line reaches the point $z = b \equiv he^2/(1 + e^2)$. Because of the differing wall velocities required by the two Bloch lines discussed above, the Bloch-line model cannot unambiguously describe the wall behavior beyond this point. No detailed theory of the ensuing unstable motion exists, but there are at least two possibilities.[224,226,327] One is that the Bloch lines "pile up" or "stack up" near both critical points by repeated Bloch-line nucleation in the manner of Fig. 15.1c. Bloch-line nucleation and pile-up can continue until the piled-up Bloch lines fill up the entire thickness of the film, at which point the outermost ones become repelled to the film surfaces and unwind. Another possibility, called "punch-through," is that the initial Bloch line simply continues to the surface from inertia and unwinds there, its stored energy being dissipated in lattice vibrations and bulk spin waves. Cycles of such Bloch-line nucleation and punch-through can continue ad infinitum if the driving field is maintained. There is experimental evidence for both processes in different films (see Sections 16 and 19). Pile-up usually occurs whenever punch-through is resisted by exchange coupling of the wall magnetization to an in-plane layer at the film surface.

Estimates have been made of the average or "saturation" velocity of the cyclic process, based on rather arbitrary assumptions about the punch-through.[224,226,327] In one approach it has been noted that the area swept out by the $\psi(z)$ contour corresponds to the average momentum of the wall [Eq. (12.21)]. If the horizontal Bloch line of Fig. 15.1b punches through to the surface, it sweeps out an area $2\pi h(1 + e^2)^{-1}$ (assuming the stray field of an isolated wall, see Section 8,E). Suppose that during the ensuing phase of unstable motion the wall settles into a new quasi-steady state in a time $T_1$, which is short compared to the half Larmor period $\pi/\gamma \bar{H}$ of the cyclic wall motion. Then this swept area must, in effect, be instantly compensated by the nucleation of a new Bloch line in such a position that the total area $A$ (proportional to $\bar{\psi}$) remains constant, as indicated by the transition from frame (b) to frame (d) of Fig. 15.1. Thus, as the first Bloch line punches through, a second Bloch line quickly nucleates to take up the momentum. Since this Bloch line has the reverse twist from the Bloch line in Fig. 15.1a, the velocity during the ensuing Bloch-line motion must be negative and can also be calculated from Eqs. (15.1) and (15.2). The Bloch line of Fig. 15.1d moves to the left and soon disappears (Fig. 15.1e). Then a new cycle begins with Bloch-line nucleation near the other surface (Fig. 15.1f), and the velocity becomes positive once again. Using Eqs. (15.1) and (15.2), and ignoring damping, one can calculate the average velocity over one cycle of such a process to be[224]

$$V_{s01} = 0.29V_{po}. \tag{15.8}$$

Obviously the instantaneous velocity in this model has strong swings from positive to negative values.

In an alternative model for punch-through, the spin state of the wall is assumed to jump instantaneously from that of Fig. 15.1b to that of Fig. 15.1e, without the intervening state of Fig. 15.1d, even though momentum is not conserved.[226] In this model negative wall motion is absent and the averaged velocity is

$$V_{s02} = 0.55 V_{po}. \tag{15.9}$$

This average should also apply to the initial cycles of pile-up assuming the pile-up occurs at *both* surfaces, as might well be the case when in-plane capping layers are provided on both sides of the film.[327]

What becomes of the energy liberated during the Bloch-line annihilation leading to either Eq. (15.8) or (15.9) remains to be clarified. Energy-conservation would require this energy to reside initially in flexural wall oscillations (Section 22). It may then decay and thermalize through unspecified spin–spin or spin–lattice interactions, as discussed in more detail in Section 17,C.

In the case of $\alpha \neq 0$, the moving Bloch line will contribute a damping that combines with the wall damping to give a high-drive mobility or slope to the saturation velocity versus drive.[225,365] To obtain a simple estimate of this mobility, we use Eqs. (15.1) and (15.2), substituting the values of $E_L$, $dE_L/dz_L$, and $\Phi = \pi$ appropriate to the Bloch line at the center of the film. The result is of the form $V_s = V_{so} + \mu_h H$, where $\mu_h = \mu_1$,

$$\mu_1 = \gamma \Delta_0 [\alpha + (\pi^2 \Delta_0 Q^{1/2}/2h\alpha)]^{-1}. \tag{15.10}$$

It is interesting to note that the Bloch-line damping contribution often exceeds the wall contribution, essentially because the wall is moving slowly while the Bloch line is moving fast. In typical 5 $\mu$m garnet films with linear initial mobilities $\gamma \Delta_0 \alpha^{-1} > 1000$ cm/sec Oe, one finds a high-drive mobility of only 10–100 cm/sec Oe. This may also be compared to the free precession case, which would be more appropriate for even higher drives [exceeding $H_{qs}$ of Eq. (12.39)], giving the mobility $\mu_h = \mu_2$,

$$\mu_2 = \gamma \Delta_0 (\alpha + \alpha^{-1})^{-1}. \tag{15.11}$$

This is often in the range of 1–10 cm/secOe for typical 5 $\mu$m garnet films.

## E. EFFECT OF STRAY AND APPLIED IN-PLANE FIELDS AND IN-PLANE ANISOTROPY

The above results apply to the case of an isolated plane wall. Analytic expressions also exist for walls in an infinite stripe array with stripe width $w$. For a general but constant $z_L$, the steady wall velocity is[242]

$$V = 4\gamma A h^{-1} (2\pi/K)^{1/2} \frac{\pi\rho \sinh(\pi\rho)\cosh(\pi z_L'/w)}{\sinh^2(\pi\rho) - \sinh^2(\pi z_L'/w)}, \tag{15.12}$$

where $\rho = h/2w$ and the Bloch-line coordinate $z_L' = z_L - (h/2)$ in this formula is measured from the midplane of the film. The maximum $V_p$ and minimum $V_{min}$ are given by

$$V_{min} = 4\gamma Ah^{-1} (2\pi/K)^{1/2}\pi\rho/\sinh(\pi\rho), \tag{15.13}$$

$$V_p = V_{min} \cosh^2(1) [1 + \sinh^2(\pi\rho)\tanh^2(1)]^{1/2}. \tag{15.14}$$

In the case of cylinders of diameter $d$, an analytic expression corresponding to $H_y(z)$ is not available. Resort to numerical integration resulted in the diameter dependence of $V_p$ shown in Fig. 15.3.[224] The maximum velocity in both these cases becomes large for either $w$ or $d$ small because the surface charges become more closely spaced, thus producing stronger gradients of $H_y(z)$ for $z$ close to the film surface. The effect of these different stray fields on the averaged or saturation velocity is weak, increasing it by only 15% over Eq. (15.8) for the case of a bubble with $d = h$.

In Section 11,C, we discussed the beneficial effect of in-plane anisotropy $K_p$ on the maximum wall velocity in the one-dimensional model. Incorporating $K_p$ into the Bloch-line model, one finds for the maximum velocity of an isolated plane wall[337]

$$V_p = [V_{po} \cosh^2(1 \pm \eta)]/(1 \pm \eta)^{1/2}, \tag{15.15}$$

$$\eta = K_p/2\pi M^2 . $$

The parameter $\eta$ lies in the range $1 \ll \eta \ll Q$, and $\pm$ corresponds to the in-plane easy axis parallel or perpendicular to the wall, respectively.

The effect of an applied in-plane field $H_p = H_x$ parallel to the wall plane and in the same direction as the average wall magnetization is to increase the critical velocity for Bloch-line nucleation $V_n(H_x)$ according to

$$V_n(H_x) = V_{no} + (\pi/2)\gamma\Delta_0 H_x, \tag{15.16}$$

where $V_{no}$ is the nucleation velocity in the absence of in-plane field. A proof of this result will be described in Section 17,A. The effect of the in-plane field $H_x$ on the Bloch-line instability velocity $V_p$ is to increase it for the stable case (mean wall moment parallel to $H_x$ and decrease it for the metastable case (mean wall moment antiparallel to $H_x$).[294b,365a] If the applied in-plane field $H_p = H_y$ is perpendicular to the wall it adds to the stray field and one can visualize a sideways

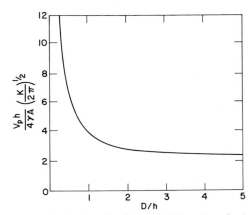

**Fig. 15.3.** Normalized peak velocity of uniform radial motion of a bubble versus ratio of bubble diameter to film thickness (after Slonczewski[224]).

shift (along $z$) of the minimum energy contour in Fig. 15.1. The critical velocity for Bloch line instability is[294b]

$$V_p = 4(2\pi)^{1/2}\gamma A h^{-1} K^{-1/2} \cosh^2[1 - (H_y/16M)], \qquad (15.17)$$

assuming a Bloch line with the polarity of Fig. 15.1a and the stray field of an isolated wall. A positive applied field $H_y$ reduces the critical velocity because it pushes the critical point away from the surface of the film, into a position where $|dH_y/dz|$ is diminished. For a Bloch line with the opposite polarity, a plus sign replaces the minus in Eq. (15.17), and the critical velocity is increased. Clearly, an in-plane field makes the upward and downward cycles of horizontal Bloch-line propagation inequivalent.

## 16.  Comparison to Experiment

Comparison of the theory of nonlinear wall motion to experiment, both here and in Section 19,A, will reveal an important discrepancy, namely in the thickness dependence of the saturation velocity. One may wonder why we devote so much attention to this theory in the present and the next chapters, considering this glaring failure. On the other hand, we shall also describe some remarkable successes of the theory in explaining experiment, particularly in the case of chiral switching and other dynamic state changes. Furthermore, it is good to bear in mind, as we go through the litany of detailed experimental comparisons, that the theory is based on a consistent set of relevant assumptions, and therefore it should form part of the framework for any new and better theory that might emerge in the future.

## A.   Peak Velocity

In the previous section we have seen that dynamic equilibrium motion with the conventional mobility $\mu$ is expected in a two-dimensional wall up to one peak velocity ($V_n$) for Bloch-line nucleation or another peak velocity ($V_p$) for Bloch-line instability. Below these velocities and their corresponding drive fields [Eq. (15.5)], the dynamic equilibrium motion obeys the same relations as in the one-dimensional theory of Chapter 5, even though Bloch lines may be present. Because the critical drive fields are often no more than an oersted for typical high-mobility bubble materials, this "linear" region is hard to observe in any experiment. It becomes easier to observe whenever the critical drive fields are large, namely in high-$g$ materials (because of their large $\alpha$, see Section 3,C), or in the presence of large in-plane fields [Eq. (15.17)] or in-plane anisotropy [Eq. (15.15)]. These considerations explain our applications of the one-dimensional theory in Section 11. However, even under these conditions (i.e., high $g$ or large $H_p$ or $K_p$), the effective mass and critical velocity can be drastically altered from the predictions of the one-dimensional theory if Bloch lines are present.

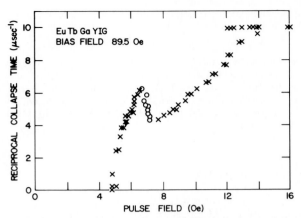

**Fig. 16.1.** Reciprocal collapse time $T^{-1}$ versus drive field $H$ in a dynamic collapse experiment on $S = 1$ bubbles in a EuTbGaYIG film (after Malozemoff[175]).

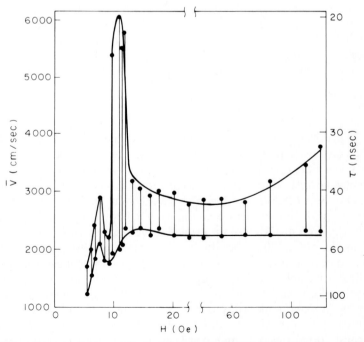

**Fig. 16.2.** Average apparent wall velocity during dynamic bubble collapse versus bias pulse field $H$ for a bubble in a LuGdAlIG film biased 3 Oe from collapse (after Vella-Coleiro[174]).

We consider first the observation of velocity peaks in films without the special conditions described above, in an effort to test the simple Eq. (15.6) or its variants in Eq. (15.14) and Fig. 15.3. In this connection the rough criteria [Eqs. (12.35), (12.39), and (12.40)] for the quasi-steady approximation should be borne in mind. For typical 5 $\mu$m materials they limit effective drives to < 10 Oe and observation times to more than 25 nsec. Many experiments do not satisfy these conditions and often reveal very high velocities during the initial phase of the motion. These interesting results have been discussed in Section 12,E. In what follows we limit ourselves to discussing those experiments for which the theory of the previous section should apply.

Even under these conditions, velocity peaks have usually not been observed in bubble-collapse experiments, and the velocity simply rises monotonically to saturation as a function of drive.[163,171,328,366] One exception is early data on a hexaferrite platelet of composition $Sr_{0.5}$-$Ca_{0.5}Al_4Fe_8O_{19}$ that shows a peak at 250 cm/sec at a pulse field of 60 Oe.[171] These values are roughly a hundred times larger than the theoretical prediction, but too little has been done on hexaferrites to draw any general conclusions.[224] Another exception is a bubble collapse experiment on a EuTbGaYIG film in which the restoring force field was kept smaller than the critical field by biasing the bubble within 1.5 Oe of its collapse field.[175] A velocity peak of ~800 cm/sec was observed as shown for a different biasing condition in Fig. 16.1. This is close to the theoretical prediction for $V_p$ of 750 cm/sec (including stray field corrections) and six times smaller than Walker's velocity for this material. A third case is a similar experiment on a LuGdAlIG film, shown in Fig. 16.2, in which a peak of 3000 cm/sec was observed at a drive of 8 Oe.[174] The velocity peak is about two times larger than the theoretical prediction for $V_{po}$. An additional higher velocity peak approaching Walker's velocity is also observed in the figure and is possibly attributable to the breakdown of the quasi-steady approximation (Section 12,E). A fourth case, illustrated in Fig. 13.1, is data on a EuTmGaYIG film, part implanted and part nonimplanted.[172] Neither of the observed peaks, at 40 Oe for the implanted and 600 Oe for the nonimplanted samples, is likely to be the velocity peak of interest because the drives are much higher than the expected 0.5 Oe critical drive, and the observed low-drive mobility of 30–40 cm/sec Oe is much lower than the expected mobility of 1400 cm/sec Oe for this composition.

In addition to bubble-collapse measurements, low-drive velocity peaks have been observed in photometric measurements.[113,336] In experiments on a stripe array in a EuGaYIG film, for example, the peak occurred at 1300 cm/sec with 3 Oe drive. By contrast several other photometric experiments have failed to show the peak,[95,109,367] for example, in a study of an isolated bubble in a similar EuGaYIG film, as illustrated in Fig. 4.2.[109] One possible explanation for the lack of a peak is that the critical drive for the velocity peak is lower than the static coercive field so that it simply cannot be observed. Another possible explanation is that imperfections smear out the velocity peak. However the discrepancies between these different experiments have not been fully resolved.

## B. SATURATION VELOCITY AND CHIRAL SWITCHING

The saturation velocity is one of the most frequently measured quantities in bubble dynamics, particularly since it is of such importance in determining the maximum data rate of devices. Nevertheless, comparison of this quantity to the theory we have described in preceding sections has proven extraordinarily complicated, and it must be admitted at the outset that no satisfactory picture has yet emerged to unite theory with experiment. The complexity comes from the fact that there are many different experimental configura-ations—straight-wall motion, bubble collapse or expansion, bubble transla-tion, stripe expansion, stripe run-out or run-in—and for each of these, theory predicts somewhat different velocities. Even for a single one of these experiments, there may be two different nonlinear regions as a function of drive field, each characterized by its own saturation velocity. For this reason we have chosen in this review to break up the discussion into sections so as to treat each theoretical prediction separately. In the present section we treat only experiments involving two-dimensional wall motion, that is, bubble collapse[163,171–172,175,328,366] and expansion,[103,365,368] stripe ex-pansion,[168] and straight wall motion.[94,153,327,335,367,369] We defer until Section 19,A the discussion of saturation velocities in experiments involving bubble or stripe-head translation. Furthermore, in each of the two sections we treat separately the cases of low or high drive. We also treat effects of surface layers and in-plane fields separately. Finally, in Section 21, we con-sider the effect of saturation velocities on device operation.

The danger of such an analytical approach is that one may lose the forest for the trees. To give some perspective we describe briefly the conclusions of a recent overview of the subject from an experimental view-point by de Leeuw.[369a] He suggests that if one ignores an initial nonlinear region, one generally finds a well-defined saturation velocity in most materials, and for any given material, this velocity is usually the same, within about 20%, irrespective of the experimental technique. Furthermore this velocity is approximately described by the empirical relation

$$V_s = 0.4\pi\gamma\Delta_o M. \tag{16.1}$$

Such a formula is difficult to justify theoretically under typical experimental conditions, although it is close to a formula for the saturation velocity in an infinite medium [see Eq. (17.20)].

We turn now to the arduous task of examining the detailed experimental results and seeing in what ways the theory does or does not work. We start by noting that many different kinds of velocity vs. drive behavior have been observed in the nonlinear region. In some experiments,[153,163,367] ve-locity was found to increase as $(H - H_{crit})^{1/2}$, where $H$ is the applied drive field and $H_{crit}$ the critical field for onset of nonlinearity. An example is shown in Fig. 5.5. In many other experiments, the velocity increased linearly with field as $V_{so} + \mu_h H$, where $V_{so}$ is the saturation velocity extrapolated to zero drive field and $\mu_h$ is a high drive mobility. An example is shown in Fig. 11.6 (in this case $\mu_h \approx 0$). In a few experiments more complex behavior is observed. For example, the nonimplanted sample of Fig. 13.1 has two velocity vs. drive regions, showing that a break may arise at very large drive fields (comparable to the anisotropy field). In Fig. 16.1 there also appear two velocity vs. drive regions, in addition to the linear region below the velocity peak. In this case the crossover between the two nonlinear regions occurs at a drive of about 12 Oe. It is unclear whether or not such a crossover is characteristic of the nonlinear behavior of most materials, for few experiments have been conducted with drive fields spanning the field range of Fig. 16.1. Most "two-dimensional" wall-motion experiments (bubble collapse, straight-wall motion, etc.) have used drives that are too high, while most bubble-translation-type experiments have used drives that are too low.

According to theory (Section 12,E), such a crossover is to be expected as a transition from "quasi-steady" to "unsteady" wall motion, that is, as a transition between drive field regions in which the quasi-steady approximation either does or does not hold. In the case of Fig. 16.1 the observed crossover field of 12 Oe compares well to the theoretically predicted $H_{qs} = 11$ Oe [see Eq. (12.39)]. A predicted critical field $H_{qs}$ of order 10 Oe is in fact typical of many samples. Clearly it is important to determine whether the drive fields of a given experiment lie below or above $H_{qs}$ before attempting comparison with theory.

If the drive fields lie below $H_{qs}$, then the saturation velocity $V_{so}$ should be given by Eq. (15.8) or (15.9) and the high drive mobility $\mu_h$ by Eq. (15.10). In the case of Fig. 16.1, the experimentally observed $V_{so} = 100$ cm/sec compares to the theoretically predicted $V_{so1} = 200$ cm/sec [Eq. (15.8)], and the observed $\mu_h = 120$ cm/sec Oe compares to the predicted $\mu_1 = 35$ cm/sec Oe. Such factor-of-two agreement in $V_{so}$ and factor-of-four agreement in $\mu_h$ characterizes several other similar comparisons of theory and experiment in this regime.[94,172] But a systematic test of these formulas has not yet been carried out by "two-dimensional" wall-motion experiments in the quasi-steady regime. For example, there have been no studies of the effect of film thickness on the low-drive saturation velocity, except for one study of LaGaYIG films,[327] in which the velocity was found to decrease with decreasing film thickness. This result appears to contradict the inverse film-thickness dependence predicted by Eq. (15.8). However, the magnetization of these films was so small that the Bloch-line width $\pi\Lambda_0 \sim 1$ $\mu$m was not small enough compared to film thickness (from 1 to 4 $\mu$m) for the Bloch-line theory to apply. Better tests of the low-drive theory have been carried out by bubble propagation experiments, and, as will be described in Section 19,A the predicted thickness dependence is *not* observed.

For drive fields larger than $H_{qs}$, the two-dimensional wall-motion experiments yield saturation velocities well described by the empirical relation Eq. (16.1).[369a] The high-drive mobilities generally lie somewhere between the value predicted for the quasi-steady model [Eq. (15.10)] and the value predicted for free precession [Eq. (15.11)].[365] This suggests that in this "unsteady" regime of wall motion, the correct model should lie somewhere between the quasi-steady and the free-precession models. However, as mentioned earlier, no such model has yet been developed.

We mention in passing that after a wall has been propagated at its saturation velocity, its properties—in particular its coercivity—are often altered.[370] The increased coercivity is difficult to explain unless one invokes Bloch lines. Therefore the observation of such increased coercivity gives support to a basic feature of the theory, namely, that motion at the saturation velocity involves nucleation of Bloch lines.

Another aspect of the Bloch-line saturation-velocity theory is that it depends on a cyclic process of Bloch-line nucleation and propagation. Strong velocity variations, $3:1$ in one picture,[226] negative velocity motions in another,[224] are predicted (Section 15,D) with a time periodicity of roughly

$$\tau = \pi/\gamma(H - H_c - \mu^{-1}V_s), \qquad (16.2)$$

where $V_s$ is the observed average saturation velocity. This periodicity is generally in the range of $10–100$ nsec for drives from 20 to 2 Oe, respectively. While these times fall in an experimentally accessible range, such velocity variations have in many cases not been observed in experiment. Instead the velocities appear uniform, i.e., truly saturated, as illustrated in the photometric results of Fig. 11.5, where the period should have been $\sim 25$ nsec.[335]

More recently, however, the cyclic nature of wall motion has found support in chiral-switching experiments,[229] which we discuss next, and in bubble collapse[371,372] or radial expansion experiments,[373] which we discuss in Section 17,B. The Bloch-line model of Section 15, and indeed the general dynamics theory of Section 12, imply that bias pulses can cause changes in the polarity of the Bloch walls of bubbles via the nucleation, propagation, and punch-through of Bloch-line rings. In terms of the bubble states described in Section 8,C, this means that the bubble can be dynamically switched between the $(1, 0)^+$ or $\chi^+$ and $(1, 0)^-$ or $\chi^-$ states, which correspond to the two possible Bloch-wall polarities (see Figs. 8.5b and 8.5c).[229] It is of fundamental importance to test this chiral switching because the precession is the cornerstone of the dynamics theory. It is also of practical interest for control of information stored in the bubble wall state. The dynamic experimental methods used to identify $\chi^+$ and $\chi^-$ will be described in Section 19,D.

Chiral switching has been demonstrated in bias-pulse experiments on bubbles in a GdTmGaYIG film.[229] The bias pulse has a short rise time $\tau$ (i.e., $H_a = H_s t/\tau$, where $H_s$ is the maximum field value) and a much longer fall time. The drive field is partly compensated by the bubble restoring-force field, which, if the wall is assumed to move at a saturation velocity $V_s$, is given by $kV_s t/2M$. The difference between these fields impels the Bloch line according to Eq. (15.1). If we assume that the single Bloch line present at all

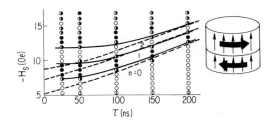

**Fig. 16.3.** Operating margins for "chiral switching" of $\chi$ bubbles by a pulsed bias field of maximum strength $H_s$ with rise time $\tau$. Chiral states were identified by means of conversion to automoting $\sigma$ states. Solid lines represent approximate experimental boundaries for $n = 0$, 1, and 2 switches, while dashed lines give the theory of Eq. (16.5) (after Dekker and Slonczewski[229]). Empty circles signify consistent observation of one chirality, solid circles the other, and half-solid circles of a mixture.

times during the motion is a $\pi$-line, and if we ignore damping and coercivity, Eq. (15.1) becomes

$$(2MH_s\tau^{-1} - kV_s)t = 2\pi M\gamma^{-1}h^{-1}\dot{z}_L \qquad (0 < t < \tau), \qquad (16.3)$$

during the pulse rise and

$$2MH_s - kV_s(t - \tau) = 2\pi M\gamma^{-1}h^{-1}\dot{z}_L \qquad (\tau < t < 2MH_s/kV_s), \quad (16.4)$$

afterwards. Here $z_L$ is the Bloch-line position, or accumulated Bloch-line displacement if more than one nucleates. We wish to calculate the conditions under which Bloch lines have traversed an integral number $n$ of cycles up and down the wall during the rise time of the pulse and until wall motion has ceased. Integrating from $t = 0$ to $t = 2MH_s/kV_s$, one finds, after setting the final $z_L = nh$,[229]

$$\tau = 2MH_sk^{-1}V_s^{-1} - 2\pi n\gamma^{-1}H_s^{-1}. \qquad (16.5)$$

This equation gives the thresholds for $n$ switches back and forth between chiral states as a function of pulse rise time $\tau$ and strength $H_s$. Comparison with experiment is shown in Fig. 16.3.[229] Bands representing a single chiral switch and a double chiral switch are clearly distinguished. There is apparently too much scatter to permit high-order switching to be observed clearly.

## C. EFFECTS OF IN-PLANE FIELD AND IN-PLANE ANISOTROPY

The effect of an applied in-plane field $H_p$ on straight-wall velocities is dramatic. While velocity peaks are rarely observed without $H_p$, a sharp peak usually develops with $H_p$, as illustrated in Fig. 11.6.[94] If the in-plane field is sufficiently large ($H_p \gg 8M$), the experimental results on peak velocities can be interpreted by means of the one-dimensional theory (Section 11,C) because the applied field overwhelms the stray fields that are responsible for horizontal Bloch-line nucleation and the wall spin structure becomes essentially uniform. As far as saturation velocities are concerned, they are generally observed to increase first in a quadratic fashion with in-plane field and then linearly at sufficiently high in-plane fields.[94,95,327,335,374] The linear dependence will be discussed further in Section 17,C.

The effect of small in-plane anisotropy on wall saturation velocity has been studied by high-speed photography of bubble expansion, as illustrated for a EuGaYIG film in Fig. 5.4.[103,365] In this film a 2.2° tilt of the crystallographic [111] axis away from the surface normal gives rise to a substantial in-plane anisotropy of over 2000 erg/cm$^3$.[48] As a result the static bubble is elliptical as indicated in the figure. Theoretically the wall velocity should be reduced along the easy axis, increased along the hard axis [see Eq. (15.15)]. Thus in bubble-expansion experiments the bubble expands more rapidly along its original minor axis, and the major–minor axes invert in Fig. 5.4 after about 0.2 μsec. The velocity anisotropy is about 2:1. This effect is responsible for the topological-switching phenomenon described in Section 5,E. As the size of the crystallographic tilt decreases in different samples, the velocity anisotropy is also observed to decrease.[365]

Similar ellipticities have been observed in bubble expansion experiments with in-plane field instead of in-plane anisotropy in a EuGaYIG film, as shown in Fig. 16.4.[375] Wall velocities are larger perpendicular to the field than parallel to it. However in this case the slow sides of the bubble are observed to become diffuse in the high-speed single-shot photograph so that the wall has an apparent width that increases with drive field and with in-plane field. Widths of 25 μm were attained with a drive of 240 Oe and an in-plane field of 100 Oe. This phenomenon is not understood. However it can be remarked that the applied in-plane field perpendicular to the wall reinforces the stray field at one surface and bucks it at the other, thus accentuating the net in-plane field difference from the top to the bottom of the wall. In such a case the transient velocities at the two surfaces will also be different, encouraging wall "bowing." The observation of a diffuse wall in this special case serves to emphasize that most other high-speed photography observa-

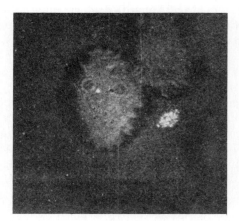

**Fig. 16.4.**  High-speed photograph of bubbles in a EuGaYIG film expanding under the influence of a 240 Oe pulsed bias field and a 100 Oe dc in-plane field directed horizontally. The polarizer and analyzer are almost crossed (wall-contrast mode). The scale is roughly 85 μm for the vertical dimension of the largest bubble. Under these conditions the bubble wall appears "diffuse" (after Zimmer et al.[375]).

tions of walls moving at a saturation velocity have shown wall contrast as sharp as static wall contrast (e.g., see Fig. 4.1). This fact is important for the plausibility of the quasi-steady theory in usual cases.

In summary, although some qualitative predictions of the theory for in-plane field and anisotropy are followed, no quantitative test of Eq. (15.15) or (15.17) has yet been reported in a plane-wall environment, and there are still a number of unexplained phenomena. Further tests are afforded by bubble-propagation and stripe-runout experiments, and these are discussed in Section 19,D.

## D. CAPPING LAYERS AND THICKNESS INHOMOGENEITIES

Qualitatively, one effect of a capping layer is to modify the stray fields near the capped surface, which should lead to modified critical and saturation velocities. The experimental results are somewhat contradictory, in some cases showing small,[172,365,368] in other cases large,[95,327] differences between capped and as-grown samples. In a pair of EuTmGaYIG samples, for example, the high-drive bubble-collapse saturation velocities, back-extrapolated to zero field, were 500 cm/sec and 660 cm/sec for the implanted and nonimplanted samples, respectively, as shown in Fig. 13.1.[172] However this figure also shows a remarkable difference between the two cases at extremely high-drive fields ($> 600$ Oe). The case of free-precession, giving the mobility $\mu_2$ of Eq. (15.11), is represented by the data for the hard bubble. It appears as if the nonimplanted sample falls back to the free-precession line above 600 Oe, while the implanted sample does not, suggesting that the capping layer somehow impedes the onset of free precession through its exchange coupling to the film.[172]

Low-drive experiments on LaGaYIG films with nonuniformities through their thickness show some remarkable asymmetries with respect to the in-plane field.[95,327] For example, in a field of 130 Oe perpendicular to the wall, the peak velocity was found to be 18500 or 26000 cm/sec depending on the sign of the field.[95] Similarly the saturation velocities increase quadratically with in-plane field, but with the minimum centered around $-45$ Oe, implying that the sample somehow has a built-in effective in-plane field of $+45$ Oe. The offset field is also seen as an asymmetry in the oscillation frequency data of Fig. 11.2.[95] The effective field presumably represents some average over the stray field that is nonzero because the stray field is different at the top and bottom of the film.

Since the saturation velocity of a capped film can be less in the presence of an in-plane field $H_p$ than without it, the differences between capped and uniform films are often best revealed by measurements as a function of $H_p$. The *minimum* saturation velocity of a capped film is then generally found to be much larger than the minimum for a uniform film, even though the saturation velocity at $H_p = 0$ may not be very different. For example, in a uniform film of thickness 2.5 $\mu$m, the minimum saturation velocity was only 500 cm/sec, compared to a predicted 1100 cm/sec from Eq. (15.8). In a triple-layer film with a similar thickness but with negative-anisotropy capping layers on both surfaces, the minimum saturation velocity was 2200 cm/sec, which may be compared to the predicted velocity for Bloch-line pile-up Eq. (15.9), giving 2000 cm/sec.[327] Although such excellent agreement with theory has been reported for several multilayer films, the less good agreement in the uniform film remains puzzling.

Capping layers offer a new way of obtaining higher-velocity materials for devices without using large applied in-plane fields. The method may be an attractive alternative to the use of high-$g$ materials or ones with large in-plane anisotropy. The in-plane surface layers may be prepared by means of growth transients using a single melt composition for the liquid phase epitaxy.[327] Triple-layer films have also been prepared using a low anisotropy GdYIG composition for the surfaces, while the middle layer is a high-anisotropy LaSmGdTmGaYIG or

EuGaYIG composition.[376] Bubble-collapse measurements on these films have indicated apparent velocities of over 8000 cm/sec compared to saturation velocities of 2000 cm/sec in comparable compositions that had no triple layers but were simply implanted. As will be discussed further in Section 19,D, it is likely that the high apparent velocities are at least partially due to a ballistic overshoot effect arising from the fact that Bloch-line punch-through is suppressed by the GdYIG layers.

Another triple-layer geometry has been proposed, in which two halves of a normal bubble film would be separated by a thin layer of nonmagnetic material like GGG.[226] This layer has the effect of breaking the exchange coupling through the thickness of the film and thereby preventing Bloch-line propagation. Critical velocities close to Walker's velocity are expected. Such a structure has not yet been experimentally realized.

The effect of an ion-implanted capping layer in preventing Bloch-line punch-through has been used to distinguish between $\chi^+$ and $\chi^-$ bubbles in a bubble-collapse experiment.[376a] Since a pulsed-bias field causes Bloch-line nucleation and propagation toward the capped surface in one case and towards the "uncapped" interface in the other, the wall velocities and hence collapse times are different for the two bubbles. Moreover, application of single pulses of controlled shape and sign were found to *generate* $\chi^+$ or $\chi^-$ at will without regard to initial state. These results provide further support for the chiral-switching model.

## 17. Advanced Topics in Two Dimensions

### A. STEADY-STATE STRUCTURES: NUMERICAL CALCULATIONS

The Bloch-line model relied upon in Section 15 to describe wall motion in a film is valid only if the Bloch-line width is small compared to the film thickness. However, in the instant of Bloch-line nucleation at a critical point, its thickness is infinite in the Bloch-line model [Eq. (8.15)]. Clearly a better approximation is needed to describe the nucleation process in general and to understand the critical velocity corresponding to instability of the Bloch line at the second critical point, particularly at smaller film thicknesses. The calculations described here provide answers to these questions and also a better basis for understanding phenomena beyond the critical velocity.

We seek then to calculate more accurately the steady-state spin structures of a planar wall in a film of thickness $h$. According to Eqs. (12.5) and (12.19) and the discussion of Sections 12,A and 12,D, the problem reduces to solving the differential equation for $\psi(z)$:

$$V/\gamma\Delta_0 = -(2A/M)(\partial^2\psi/dz^2) + 2\pi M \sin 2\psi - (\pi/2)H_y(z)\cos\psi, \qquad (17.1)$$

with the "unpinned" boundary conditions

$$\partial\psi/\partial z(z = 0, h) = 0, \qquad (17.2)$$

which hold in the absence of surface anisotropy or other special surface conditions. Equation (17.1) shows that torques due to exchange as well as demagnetizing and stray fields contribute to the wall velocity. This equation is the simplest one possible that can improve on the Bloch-line model of Section 15, for if any term were omitted, even the Bloch-line limit could not be recovered.

Equation (17.1) may be integrated by the Runge–Kutta method directly with the computer. However, satisfying the simultaneous boundary conditions $\partial\psi/\partial z = 0$ at $z = 0, h$ presents numerical difficulties at large $h$.[377] We discuss below a more successful variational computation by Hubert[245] using the equivalent Lagrangian formulation of the problem [see Eq. (12.20)]. This computation included several refinements of the model presented above. Although they are not obviously necessary for understanding the phenomenon, they improve quantitative accuracy. The principal refinement was to allow the wall profile $\theta(y - q, z)$, $\phi(y - q, z)$ to depend on $z$ owing to the fact that the bounding values of $\theta$ and $\phi$ in the domains (at large $|y - q|$) are altered by the tilting effect of $H_y(z)$. Computation proceeded in two successive steps: First, a particular trial function $\psi(z)$ was written with nine adjustable parameters (Ritz method). Variation of these parameters minimized the refined expression for the Lagrangian $L$. Second, a finite difference procedure reduced $L$ further, using the Ritz solution as a starting point, until the refined version of Eq. (17.1) was satisfied to numerical accuracy. By starting with different trial functions, Hubert attempted to determine all possible stable solutions $\psi(z)$ and their corresponding values of

$$\bar{\psi} = h^{-1} \int_0^h \psi(z)\, dz \tag{17.3}$$

as a function of $V$. The stray field $H_y(z)$ was given by the analytic expression Eq. (8.34) for the case of a periodic stripe array with film thickness $h$ and stripe width $w$.

Figure 17.1 shows the computed relationship between velocity and reduced momentum $\bar{\psi}$, for one set of values of $Q$, $w/h$, and $h/l$.[245] The solution has a finite number of branches denoted by the index $k$. They are understood to be repeated with period $\pi$ in $\bar{\psi}$ (not shown). The branch

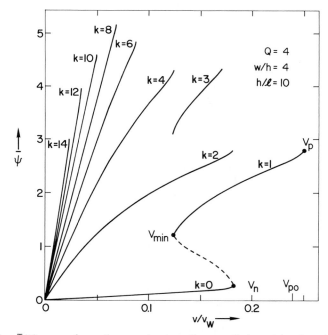

**Fig. 17.1.** $\bar{\psi}(V)$ curves for ordinary and extraordinary walls in a stripe domain array (see Fig. 8.9), from a numerical calculation for a two-dimensional wall model. Dashed line indicates the schematic location of the unstable regions predicted by the Bloch-line model. $V$ is normalized to the Walker velocity $V_w = 2\pi\gamma\Delta_0 M$ (after Hubert[245]).

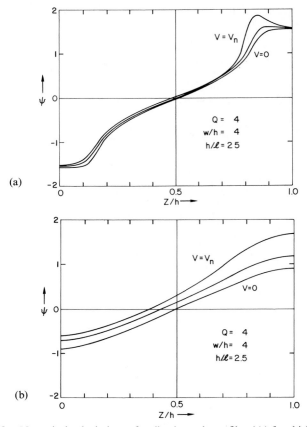

**Fig. 17.2.** Numerical calculations of wall spin angle profiles $\psi(z)$ for thick (a) and thin (b) films, showing ordinary walls which are static ($V = 0$), and which propagate at an intermediate and peak velocity ($V_n$). In (a) a Bloch line nucleates at the critical velocity, while in (b) there is a surface spin instability (after Hubert[245]).

$k = 0$ has the lowest energy and corresponds to the "ordinary" static wall discussed in Section 8,E.

Figure 17.2 shows how the profile $\psi(z)$ varies with velocity $V$ for this branch.[245] These profiles are to be compared to those of Fig. 8.10, except that the opposite starting Bloch-line polarity is assumed. The profiles of Fig. 17.2 are simply exchange-smoothed versions of those shown earlier in the Bloch-line approximation. Consider first the thick film $h/l = 25$, (Fig. 17.2a). When $V$ approaches the critical value $V_n$, the profile becomes unstable in the region $z/h = 0.8$ to $0.9$ where a Bloch line can nucleate. This compares well to the critical point $z_c/h = e^2/(1 + e^2) = 0.88$ of the Bloch-line model (see Section 15). Within the film-thickness range of interest, the computed $V_n$ is somewhat smaller than $V_{po}$, the characteristic critical velocity of the Bloch-line model [Eq. (15.6)]. (As mentioned in Section 15,C, $V_n$ cannot be calculated in the Bloch-line model because the Bloch-line width tends to infinity at the nucleation point.)

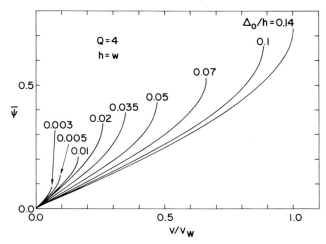

**Fig. 17.3.**  Numerical calculations of momentum $\bar{\psi}$ versus normalized velocity $V$ for an ordinary wall in films of different thicknesses $h$ (after Hubert[245]).

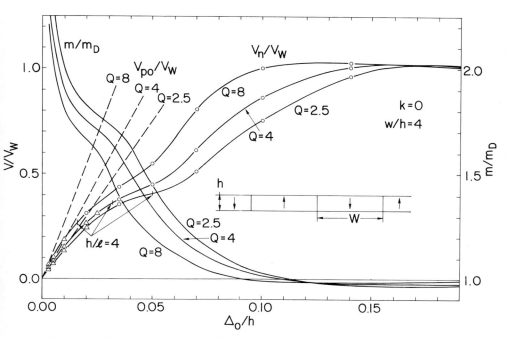

**Fig. 17.4.**  Numerically calculated critical nucleation velocities $V_n$ (normalized to the Walker velocity $V_w = 2\pi\gamma\Delta_0 M$) and masses $m_0$ (normalized to the Döring mass $m_D = (2\pi\gamma^2\Delta_0)^{-1}$) for ordinary walls as a function of film thickness $h$. Dotted lines give predicted critical velocities $V_{po}$ according to the Bloch-line approximation (after Hubert[245]).

When $w$ and $h$ are both ten times smaller, the twist is reduced in total (Fig. 17.2b; compare to Fig. 8.10a). In this case a large change of $\psi$ near the surface $z = h$ is noted, in connection with the instability occurring at $V = V_n$, contrary to the idea of Bloch-line nucleation. Figure 17.3 shows how $V(\bar{\psi})$ depends on $h$ and $w$.[245] When $h(= w)$ is small relative to $\Delta_0$, $V(\bar{\psi})$ approaches the one-dimensional Walker solution $V = V_w \sin 2\bar{\psi}$. At greater $h(= w)$, $V$ is much reduced by the twisting tendency of the stray fields. Paradoxically, the limit $h \to 0$ rather than $h \to \infty$ approaches the behavior of the one-dimensional theory. The reason is that perturbing torques due to the magnetic charges on the two surfaces tend in opposite directions. When $h$ is small, the exchange stiffness is better able to resist the twisting effect and balance out these torques to render them ineffective. The dependence of $V_n$ on $h$ for various $Q$ is shown in Fig. 17.4, together with $V_p$ and the initial mass $m$ discussed in the following section.[245]

Let us return to Fig. 17.1 and consider the branch $k = 1$. Every solution on this branch has one horizontal Bloch line, represented by the steep portions of the $\psi(z)$ curves in Fig. 17.5 (compare to curve 2 of Fig. 15.1a). Indeed the Bloch-line model would have the first two branches connected as indicated schematically in Fig. 17.1 by the dashed curve between the points $V_n$ and $V_{\min}$. Numerical instability presumably prevents the connection from appearing in Hubert's computations. The relationship $V(\bar{\psi})$ for $k = 1$ is consistent with expectations based on the Bloch-line model for films in the range of interest such as this one with $h/l = 10$, because $h$ is considerably greater than the Bloch-line width $\pi\Lambda_0 = \pi Q^{1/2}\Delta_0$ in this range. This is not so, however, for much thinner films in which the maximum $V_n$ disappears and the curves $k = 0$ and $k = 1$ merge. The critical velocity $V_p$ is only slightly larger than $V_{po}$, as indicated in Fig. 17.1, for the case $h/l = 10$. The critical velocity $V = V_p$ corresponds to an instability localized near the surface $z = 0$, indicated by an increasingly rapid increase in $\psi$ with $V$ near this point in Fig. 17.5.

Surprisingly, these curves provide little support for the possibility that a second Bloch line could nucleate at the second unstable critical point (e.g., see Fig. 15.1c).[224] If an instability of this nature existed one would expect it to be foreshadowed by a corresponding protrusion at the peak of the curve labeled $V = V_p$ in Fig. 17.5. This is a physically significant question since periodic repetition of such a nucleation in a long drive pulse would lead to "stacking" of Bloch lines of opposite signature against the opposing critical points.[226] The related question of state conversion is considered at length in Section 18,D.

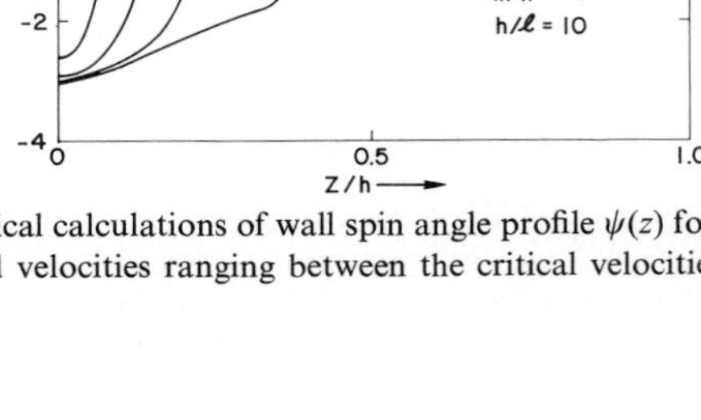

**Fig. 17.5.** Numerical calculations of wall spin angle profile $\psi(z)$ for walls with a single horizontal Bloch line and velocities ranging between the critical velocities $V_{\min}$ and $V_p$ (after Hubert[245]).

Returning once again to Fig. 17.1, the curves with $k \geq 2$ represent so-called "anomalous" or "heavy" or "extraordinary" walls with one or more $2\pi$-Bloch lines analytically continued from the static states described in Section 8,E.[247] Each odd $k$ has one more Bloch line than $k - 1$, and this additional Bloch-line is dynamically nucleated. Note that no solutions corresponding to odd $k > 3$ or even $k > 14$ were found. Although additional ones might exist, the total number must be finite because the total twist increases steadily and its equilibrium is limited by the finite "pinning strength" of the stray field near the film surfaces.

Finally we consider the effect of an in-plane $H_x$ parallel to the domain wall on the critical velocity $V_n$ for Bloch-line nucleation.[245,378] This introduces a term $(\pi/2)H_x \sin \psi$ on the right-hand side of Eq. (17.1). If $\Lambda_0 \ll h$, Bloch-line nucleation occurs near the critical points where $\psi \to \pm \pi/2$, $\sin \psi \to \pm 1$. In this region the term in $H_x$ may be treated as an additive constant and therefore may simply be added to the numerically calculated critical velocity. This procedure yields Eq. (15.16) for the case of $H_x$ pointing along the average Bloch-wall magnetization direction (rather than opposed to it). This dependence on in-plane field has also been borne out by detailed numerical calculation.[378]

## B. Effective Mass

In Section 12,D we showed how the concept of effective mass, found useful in the one-dimensional theory (Section 11,B), can be extended to two or more dimensions by using the quasi-steady approximation. In the case of the quasi-planar wall, Eq. (12.33) applies and thus the effective mass is proportional to the slope of the curve $\bar{\psi}(V)$. For instance the value $\bar{m}_0$ relevant to small signal oscillation experiments is proportional to the slope of the $k = 0$ $\bar{\psi}(V)$ curve of Fig. 17.1.[245] An analytic estimate based on the Bloch line approximation is unavailable because of the problem of the Bloch-line width divergence at $\bar{\psi} = 0$. Figure 17.4 shows the results for $\bar{m}_0$ based on Hubert's numerical computations for $\bar{\psi}(V)$ for $k = 0$.[245] Notice that $\bar{m}_0$ is less sensitive to the effect of stray fields than $V_n$ (considering the displaced origin of the $\bar{m}_0/m_D$ scale). This follows from the fact the $V_n$ is governed by a narrow region of the wall that is greatly "softened" by the stray field, whereas $\bar{m}_0$ is an average over the entire wall including much area that is subject to little stray field. The fact that $\bar{m}_0$ is somewhat greater than the Döring one-dimensional mass $m_D = (2\pi\Delta_0\gamma^2)^{-1}$ of Eq. (11.6) is understood from the dependence of $m$ on the stray field $H_y$, ignoring exchange stiffness [see Eq. (11.11) and Fig. 11.1]. Figure 11.1 indicates that $m > m_D$ over most of the film thickness, and it is infinite at the critical points where $H_y = \pm 8M$. These singularities are nonintegrable and consideration of the exchange stiffness, which is indispensible near these singularities, gives a finite result for the mean $\bar{m}_0$, as shown by the numerical results [245,246,345] and analytical arguments.

Small-signal oscillation experiments may be compared to these predictions. Best agreement has been achieved with the photometrically detected resonances of stripe arrays in a series of GdGaYIG films, where masses ranging from 1.5 to 3.5 times the Döring mass are in qualitative agreement with the theory.[306] However in another study of EuGaYIG films, resonance frequencies are down by a factor of 10 from the Döring prediction, indicating mass discrepancies of order 100.[112,163] Furthermore the oscillations often take on a triangular appearance, as illustrated in Fig. 4.2. for bubble oscillations in a EuGaYIG film, and the period increases rapidly with drive.[95,109] We will return to these problems below.

Initial masses of anomalous walls with multiple horizontal Bloch lines (see previous subsection or Fig. 8.10c) are validly treated in the small-number Bloch-line limit as long as the total width of the cluster of $n$ $2\pi$-Bloch lines is small compared to the film thickness. Combining Eqs. (8.12), (8.13), (8.34), (12.33), and (17.3), one finds for this case

$$\bar{m}_n/m_D = (\pi/2)^2 nh\Lambda_0^{-1}[\sinh(\pi h/2w)/(\pi h/2w)]^2, \qquad (17.4)$$

where $h$ is film thickness and $w$ is the stripe period, as illustrated in Fig. 17.4. This mass is indeed large for film thicknesses in the device range. Its magnitude, according to Eq. (12.33), stems from the fact that, once a Bloch line is present, displacement of its position involves little change in average wall energy $\sigma$ for a given change in $\psi$. Since such anomalous walls have not been experimentally identified, there are as yet no tests of Eq. (17.4).

A simple expression for $m$ is also available in the case that simultaneously (1) the stray field $H_y(z)$ is neglected except as an agent to "pin" $\psi$ at the film surface, and (2) the distance between neighboring Bloch lines falls between $\pi\Delta_0$ and $\sqrt{2\pi\Lambda_0}$. Under these conditions Eq. (17.1) reduces to

$$V/\gamma\Delta_0 = -(2A/M)d^2\psi/dz^2,\tag{17.5}$$

with the solution

$$\psi = -(MV/4A\gamma\Delta_0)z(z-h) + c_1 + c_2z,\tag{17.6}$$

in which the constants of integration $c_1$ and $c_2$ depend on the "pinned" values of $\psi$ at the surfaces $z = 0, h$ but not on $V$. Evaluating $\bar{\psi}$ from Eqs. (17.3) and (17.6) and $\bar{m}$ from Eq. (12.33), one readily finds

$$m_{\text{ph}}/m_{\text{D}} = \pi h^2 M^2/12A = h^2/24\Lambda_0{}^2\tag{17.7}$$

for the mass of a wall with pinned horizontal Bloch lines, relative to the Döring mass $m_{\text{D}}$ [Eq. (11.6)]. This result is closely analogous to the hard-bubble mass of Eq. (13.15).[177] The cause of its large magnitude is the inherent weakness of exchange stiffness, requiring a large amount of twist to build up a significant amount of kinetic energy $\bar{\sigma} - \sigma_0$. Equation (17.7) may be applied to photometric straight-wall experiments in nonuniform LaGaYIG films where $\psi$ appears to be effectively pinned at one rather than both surfaces, so that the right-hand side of Eq. (17.7) should be multiplied by 4.[95,327] Large numbers of Bloch lines are presumably generated during the large-amplitude oscillations. The predicted oscillation frequency $(2\pi)^{-1}(k/m)^{1/2}$ is 2 MHz, as compared to the observed 1 MHz.[95,115]

So far the masses we have considered are "linear" in the sense that they describe a linear portion of the $V(\bar{\psi})$ curve around the static value of $\bar{\psi} = 0$. However the range of $\bar{\psi}$ for which such a linear mass is valid is very small in some cases. For example, in Fig. 17.1, the linear range for the $k = 0$ curve goes out to $\bar{\psi}$ of order only 0.1. For more realistic experimental conditions, Eqs. (12.26) and (12.27) would have to be integrated numerically using calculated $V(\bar{\psi})$ curves such as those of Fig. 17.1. A simple approximation that has permitted analytical progress is to take $V(\bar{\psi})$ to be a step function rising from 0 to a "saturation velocity" $V_s$ at $\bar{\psi} = 0$.[115,379] This approximation simulates in a crude way the $k = 0$ and $k = 1$ $V(\bar{\psi})$ curves of Fig. 17.1. Physically this means that a horizontal Bloch line nucleates as soon as $\bar{\psi} > 0$, and as long as the Bloch line is present in the wall, the wall moves at a velocity $V_s$. This approximation will be discussed in more detail in Section 18,E.

Now let us consider a restoring force described by an effective gradient $H'$, acting on a wall whose displacement is described by $x$ and whose initial static position is $x = 0$. When a step drive field of strength $H_a$ is applied, the equation of motion for $\bar{\psi}$ [Eq. (12.26)] reduces to[115]

$$\dot{\bar{\psi}} = \gamma(H_a - H'x - H_c - \mu^{-1}V_s).\tag{17.8}$$

As long as $\bar{\psi}$ remains greater than zero, i.e., as long as the Bloch line is present in the wall, the wall continues to move forward at a velocity $V_s$. Integrating Eq. (17.8) shows that $\bar{\psi}$ climbs to a peak

$$\bar{\psi}_{\text{max}} = \gamma(H_a - H_c - \mu^{-1}V_s)^2/2H'V_s\tag{17.9}$$

and then drops back to zero, while the wall overshoots its equilibrium position $x_{eq} = H_a/H'$ to the peak displacement

$$x_{max} = 2(H_a - H_c - \mu^{-1}V_s)/H' \tag{17.10}$$

at the moment $\bar{\psi}$ returns to zero. The physical reason for the ballistic overshoot is the momentum stored in the Bloch line. Next a reverse cycle begins with the wall oscillating back towards $x_{eq}$ and overshooting in the reverse direction. The net result is a series of triangular peaks, as illustrated by the dashed lines in Fig. 4.2, with a time of[115]

$$\tau_i = 2[H_a - (2i - 1)(H_c + \mu^{-1}V_s)]/V_sH' \tag{17.11}$$

between the $i$th and the $(i + 1)$st peak. These oscillations are highly nonlinear in nature, with the period increasing linearly with drive as shown by Eq. (17.11) and with the amplitudes of each successive peak dying away faster than linearly.

These characteristics are often observed in experiment,[95,109,115,327,373] and comparison with radial bubble oscillations in a EuGaYIG film is shown in Fig. 4.2.[115] If the drive field is sufficiently large, $\bar{\psi}_{max}$ of Eq. (17.9) will exceed $\pi$. In this case the Bloch line responsible for the overshoot may annihilate at a surface of the film and the momentum is lost. Thus, as a function of drive one may expect oscillatory motion interspersed with nonoscillatory motion. Such "wall structure changes" have been observed in radial oscillation experiments in EuGa and LaGa films.[327,373] In spite of its crudeness, the ballistic-overshoot model is considerably more successful in accounting for wall-oscillation experiments than the linear masses considered earlier.

In bubble-collapse experiments, the ballistic overshoot can lead to two interesting effects. In short-pulse-time experiments, the apparent velocity, calculated by dividing the pulse time into the dynamic bubble-collapse distance (see Section 5,E), may be appreciably larger than the true velocity because some of the wall motion could occur in the overshoot phase of the motion.[115,172] It has been suggested that such an effect could explain the anomalously high velocities observed in short-pulse-time experiments,[172] but so far these velocities have been found to be real in those cases where direct stroboscopic measurements have been performed (e.g., see Fig. 4.3).[129] Another effect, called "dynamic barrier penetration," appears in examining the minimum drive for bubble collapse at infinitely long pulse length.[115] For example, in a EuTbGa-YIG film, when the bias field was raised quasistatically from 85.5 Oe, the bubble was found to collapse at 95.0 Oe.[175] However when the field was applied as a step pulse with a 15 nsec rise-time, the bubble collapsed at 93.0 Oe. The domain wall had thus apparently "tunneled" through the magnetostatic field barrier. Ballistic overshoot can explain such a tunneling and account for the 2 Oe discrepancy.[115]

## C. General Theory of Nonlinear Wall Motion

The Bloch-line model of nonlinear motion given in Section 15 is the simplest known in two dimensions. We treat nonlinear motion here in a way that does not rely on the Bloch-line model.[224] In addition to indicating the extent to which the existence of velocity saturation is independent of the Bloch-line approximation, it provides, in principle, a means of improving on estimates of the saturation velocity. The theory does, however, rest on the quasi-steady state approximation and therefore is subject to the limitations of Section 12,E.

Turning back to the general steady-state case described in Section 12,D, we recall that in general there will be not one but $n$ locally stable minima of $\bar{\sigma}(\bar{\psi})$ for a given $\bar{\psi}$ corresponding to $n$ different steady state solutions $\psi(z)$. As before we describe these different solutions by the subscript $i (=1, \ldots, n)$ and recall that to each solution corresponds a $\bar{\sigma}_i(\bar{\psi})$ and a $V_i = (\gamma/2M)d\bar{\sigma}_i/d\bar{\psi}$ [Eq. (12.23b)]. In general the stable structure $\psi_i(z)$ will depend continuously on $\bar{\psi}$ but may terminate at a critical value of $\bar{\psi}$ where the solution becomes unstable and where $V(\bar{\psi})$ has a peak $V_p$. Examples are shown in the different $k$-branches of the computed $V(\bar{\psi})$ plots of Fig. 17.1 and also in Fig. 17.6, which shows a schematic periodic series of branches. We have previously discussed the qualitative behavior of the wall motion for the case that the average instantaneous net drive field $H$ is less than $\mu^{-1} V_p$. In this case the wall motion responds to a step in drive field by approaching exponentially a dynamic equilibrium condition as illustrated in Fig. 12.3. Here we we consider in more detail the nonsteady-state case for which $H > \mu^{-1} V_p$ and address the question of what happens beyond the instability point.

Fairly concrete, yet simple, conclusions are drawn when both $H$ and $\alpha$ (or $\mu^{-1} V_p$) are considered first-order infinitesimals. In any interval of $\bar{\psi}$ not including a critical point we can use the quasi-steady-state equations (12.26) and (12.27). In the absence of coercivity they have the form

$$\dot{\bar{\psi}} = \gamma[H - \mu^{-1} V_i(\bar{\psi})], \tag{17.12}$$

$$\dot{q} = V_i(\bar{\psi}). \tag{17.13}$$

Let us assume we start on the branch $i = 1$, with the initial values $\bar{\psi} = 0$, $V_1 = 0$ at $t = 0$, as shown in Fig. 17.6. A step drive field causes $\bar{\psi}$ and $\dot{q}$ to increase with $t$ to $\bar{\psi}_1$ and $V_p$, respectively, at the first instability point. Immediately afterward, the now unsteady motion obeys neither of these equations and a description of this unstable episode would require, at the very least, the full partial differential equations (12.1) and (12.2). These equations describe rapid fluctuations of $q(x, z, t)$ and $\psi(x, z, t)$ owing to internal forces not vanishing with $H$ and $\alpha$. We assume that with the passage of time nonlinearities in the equations cause these fluctuations to increase rapidly in spatial and temporal frequency so that, after some characteristic time $\tau$ that is finite in the limit $\alpha = H = 0$, the fluctuations may be considered thermal and therefore negligible. Since $\bar{\sigma}$ is considered to exclude thermal energy, we adjust $\bar{\sigma}$ by substracting the amount of thermalized

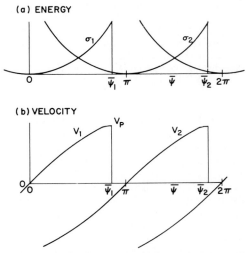

**Fig. 17.6.** Schematic energy $\sigma_i$ and velocity $V_i$ versus momentum $\bar{\psi}$ for stable kinetic domain-wall structures (after Slonczewski[224]).

energy. We note that the attendant temperature rise is, in practical cases, a very small fraction of 1 K, so that the material parameters are hardly changed. Moreover, given sufficient time, the thermal energy diffuses away from the wall via lattice vibrations and bulk spin waves. At room temperature, then, the "renormalized" equations of motion (12.1) and (12.2), neglecting thermal fluctuations, are practically the same as before. It follows that the system may tend, after the time $\tau$, to some new stable branch, say $i = 2$ with diminished $\bar{\sigma} = \bar{\sigma}_2(\bar{\psi})$ and a new velocity $V_2(\bar{\psi})$ (see Fig. 17.6). If the drive $H$ remains undiminished, the velocity $\dot{q}$ can in many cases be periodic in time, with reversals in sign possible, as illustrated in Fig. 12.3 or 17.6.

The important point to grasp is that even though Eqs. (17.12) and (17.13) are invalid during the unstable episode of duration $\tau$, the general momentum-conservation equation (12.26) *is* valid under general assumptions. Indeed the roots of momentum conservation lie more deeply within the Landau–Lifshitz equations than the domain-wall model itself (Sections 18,B and 18,C). The principal doubt one may have about this argument concerns the mechanism by which the thermal energy evolved during the unstable motion is conveyed to bulk spin-wave modes or to lattice vibrations, for some wall momentum may thereby also be lost (or gained). Coupling of wall motion to spin waves has been discussed in other contexts[343] (see also Sections 22 and 23).

A detailed argument shows that the wall displacement occurring during one of the unstable episodes of the motion when $\dot{q}$ is not given correctly by $V_i(\dot{q})$ tends to a finite limit as both $H$ and $\alpha$ are proportionately decreased.[224] In essence, if $H$ and $\alpha$ are both reduced in constant proportion, the period of oscillation (of order $\pi/\gamma H$) increases in inverse proportion. Since the number of unstable episodes of motion is fixed (say one) in a given period, they contribute proportionately less to the time-average velocity

$$\langle \dot{q} \rangle = T^{-1} \int_0^T \dot{q}\,dt \tag{17.14}$$

as $H$ and $\alpha$ are decreased. It follows that to first order in $H$ and $\alpha$, the mean velocity

$$\overline{V}(H) = \lim \langle \dot{q} \rangle = T^{-1} \int_0^T V_i[\bar{\psi}(t)]\,dt \tag{17.15}$$

in the limit of $\alpha \to 0$ with $H/\alpha$ constant is given correctly by integrating the discontinuous Eq. (17.12) while changing $i$ appropriately at critical points of instability.

In the special case of a plane wall in the Bloch-line limit, Eqs. (17.12) and (17.13) are found to reduce to Eqs. (15.1) and (15.2), provided the term proportional to $\alpha$ in Eq. (15.2) is neglected. The computed result of Eq. (17.15) for a flat wall in the Bloch-line limit is shown in Fig. 17.7, which includes the $H < H_{max} \equiv \mu^{-1} V_p$ behavior $V = \mu H$ as well.[224] The general characteristic is a precipitous drop at $H = H_{max}$, followed by an approach to a saturation value $V_{s0}$.

A general expression for the saturation velocity $V_{s0}$ is easily obtained by taking the limit

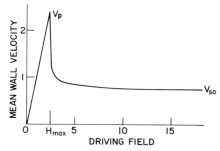

**Fig. 17.7.** Normalized mean wall velocity $V$ versus normalized drive field $Hh/4\alpha(2\pi A)^{1/2}$ for a plane wall in the Bloch-line approximation (after Slonczewski[224]).

$\alpha \to 0$ at constant but infinitesimal $H$, for then $H_{max} = \mu^{-1}V_p \to 0$. In this limit it is clear that the mean input power $2MV_{s0}H_{max}$ is converted into thermal energy during the unstable episodes. But thermal energy evolves at a rate given by summing the energy steps $\bar{\sigma}_i(\bar{\psi}_i) - \bar{\sigma}_{i+1}(\bar{\psi}_i)$ in one period of motion and dividing by the time $2\pi N/\gamma H$ required for the $N$ Larmor cycles in the period, according to Eq. (17.12). Equating the input power $2MHV_{s0}$ to the dissipation, one finds[224]

$$V_{s0} = (\gamma/4\pi MN) \sum_i [\bar{\sigma}_i(\bar{\psi}_i) - \bar{\sigma}_{i+1}(\bar{\psi}_i)]. \tag{17.16}$$

A most simple application of Eq. (17.16) can be made for the case of a wall subject to a constant drive $H_z$ and a large constant in-plane field $H_x$ ($\gg 4\pi M$).[94] Then, if $\psi$ is initially independent of $x$ and $z$, the leading term in the velocity is, from Eqs. (12.1) and (12.5),

$$\dot{q} = +(\pi\Delta_0\gamma/2)H_x \sin\psi. \tag{17.17}$$

In this case, assuming $\alpha \doteq H_c = 0$, we have $\psi = \gamma H_z t$ so that $\dot{q}$ varies sinusoidally according to Eq. (17.17). This oscillatory motion is unstable whenever the mass is negative, for a negative mass can easily be seen to imply that any infinitesimal initial wavelike perturbation of the wall would grow exponentially. Evaluating the mass from Eq. (12.33), we find it is proportional to $\cos\psi$ and hence becomes negative at $\psi = (\pi/2) + 2\pi n$. A simple assumption is that uniform precession occurs until the time when $\psi$ reaches these values, where the critical velocity is

$$V_p = (\pi\Delta_0\gamma/2)H_x. \tag{17.18}$$

Then one can imagine that the wall structure quickly reforms at constant $\bar{\psi}$ into large regions with $\frac{3}{4}$ of the area satisfying $\psi = 2\pi n$, and $\frac{1}{4}$ satisfying $\psi = 2\pi(n + 1)$, separated by (perhaps curved) $2\pi$-Bloch lines. The absence of static forces on these Bloch lines implies $\dot{q} = 0$. In order to satisfy the relation $\bar{\psi} = \gamma H_z$, the regions with $\psi = 2\pi(n + 1)$ grow in size at the expense of those with $\psi = 2\pi n$. At the instant when $\bar{\psi} = 2\pi(n + 1)$ the Bloch lines all disappear, $\psi = 2\pi(n + 1)$ is satisfied everywhere, and uniform precession begins anew. The time-average $V_0$, calculated from Eqs. (17.12) and (17.13) (neglecting Bloch-line energy) or directly from Eq. (17.16) using $\bar{\psi} = \gamma H_z t$ is[94]

$$V_{s0} = \frac{1}{4}\gamma\Delta_0 H_x. \tag{17.19}$$

Equation (17.19) shows the saturation velocity is reduced by a factor $1/2\pi$ from the peak velocity of Eq. (17.18). Equation (17.19) has been tested successfully in photometric experiments on single plane walls stabilized in an external field gradient in a series of LaGaYIG films.[94,327] A typical plot of wall velocity versus drive is shown in Fig. 11.6. Sufficiently far beyond the peak, the velocity is saturated at 110 m/sec in good agreement with 88 m/sec predicted from Eq. (17.19). However at still larger drives ($> 20$ Oe) the velocity begins to rise strongly again, and at any given drive the in-plane field dependence is closer to Eq. (17.18) than (17.19).[135] This latter effect is not understood.

A derivation similar to that just given for Eq. (17.19) can also be carried out for the case of wall velocity arising from demagnetizing torques only. In this case we ignore stray fields, as might be appropriate for a very thick film. One finds

$$V_{s0} = \gamma\Delta_0 M. \tag{17.20}$$

This velocity is a factor of $2\pi$ smaller than the corresponding peak velocity, just as in the case of Eq. (17.19). As discussed in Section 16,B, Eq. (17.20) describes surprisingly well the saturation velocity of many bubble films regardless of thickness but particularly in a high-drive regime ($H > H_{qs}$, see Section 16,B). No sound theoretical justification for this surprising agreement has yet been proposed. One can only speculate that in the high-drive regime, the effect of stray fields, which leads to the conventional Eq. (15.8) or (15.9), is somehow averaged out because of precession of the surface spins.

# IX

## Nonlinear Bubble Translation

Perhaps the most important topic in bubble dynamics is bubble translation because most devices rely on bubble translation for access to fixed reading and writing locations. We have already discussed bubble translation in the steady-state, linear region in Chapter VII. Here we discuss nonlinear bubble translation, based on the discussion of nonlinear dynamics in the previous chapter. This is one of the theoretically most complex topics, synthesizing most of the concepts we have considered in the rest of this review.

In Section 18 we consider the basic theory of nonlinear bubble translation, which is often called "dynamic conversion" because an initially simple bubble configuration is believed to convert dynamically into a complex one containing Bloch lines. Two basic results of the theory are that the bubble should move at a saturation velocity and should exhibit large ballistic effects, and these predictions are compared to experiment in Section 19. Other experiments such as the bias-jump effect and $S = 1$ state changes are discussed in the same light. In Section 20, a particular class of domain translations is considered, called "automotion," in which no gradient acts on the bubble. The concepts of the preceding sections are brought to bear on high-speed device operation in Section 21.

### 18. Theory of Nonlinear Bubble Translation

The theory of nonlinear bubble translation is analogous to the theory developed in Chapters VI and VIII for a quasi-planar wall. In Section 18,A we derive the steady-state Bloch-line structures that are stable for a given bubble velocity V. In Section 18,B we describe how the momentum concept can be generalized to the bubble geometry, and how each of the structures can be characterized by a vectorial momentum $\mathbf{\Psi}$. Thus we derive the $\mathbf{V}(\mathbf{\Psi})$ relation for bubble translation. Next, in Section 18,C we write the equations of motions for $\mathbf{\Psi}$ and $\mathbf{V} = \mathbf{V}(\mathbf{\Psi})$, based on the quasi-steady approximation. In Section 18,D we describe several general features of the solutions to the equations of motion, including the peak velocities, dynamic conversion, Bloch-line stacking and punch-through. Finally in Section 18,E we introduce a simple approximation for $\mathbf{V}(\mathbf{\Psi})$, called the "saturation velocity approximation" and use it to derive simple expressions for ballistic overshoot. In the interest of simplicity, Bloch points will be excluded from the mathematics in this section, although they enter qualitatively in the discussion of state changes in Section 18,D.

## A.  STEADY-STATE BLOCH-LINE STRUCTURE

Consider a cylindrical bubble with radius $r$ moving with velocity $V$ parallel to the $x$ axis, as shown in Fig. 18.1.[365] Let the origin of the moving coordinate system lie on the axis of the cylinder. The coordinate $z$ and the angular position $\beta = \sin^{-1}(y/r)$ specify a point of the cylinder surface. The normal velocity component at any point on the wall is

$$v_n(\beta) = V \cos \beta. \tag{18.1}$$

According to the basic torque equation of wall dynamics [Eq. (12.1)], internal restoring torques acting on the local wall-moment angle $\psi(\beta, z)$ must balance the reaction torque due to $v_n$ at every point on the wall. A complete numerical solution of this problem along the lines of Hubert's numerical

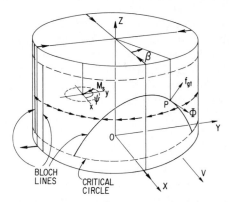

**Fig. 18.1.**  Schematic $S = 0$ bubble domain with dynamically generated Bloch line (after Malozemoff *et al.*[365]).

solution for the two-dimensional problem (Section 17,A) is not yet available, and so we base our discussion on a heuristic extrapolation of Hubert's results[245] plus some computations in the Bloch-line approximation.[365]

We consider first a static configuration ($V = 0$) with only two negative vertical Bloch lines at $\beta = -\pi/2$ ($S = 0$ bubble). These Bloch lines are in a stable position for $V > 0$ [as can be seen from the gyrotropic force expression (12.62)] and can be ignored in the subsequent discussion. $\psi$ is twisted by stray fields $H_r(z)$ as a function of $z$, as discussed in Section 8,E and shown in Fig. 8.9. One can define two critical circles corresponding to the critical points of the two-dimensional model, where $H_r(z) = \pm 8M$ (see Fig. 18.1). Between each of these circles and the nearest surface, the spins point normal to the domain wall (provided $\Lambda_0 \ll h$). If $V$ increases slightly, then $\psi$ increases or decreases slightly everywhere depending on the sign of $v_n$, until the restoring torque, primarily due to stray fields or magnetostatic energy, balances the dynamic reaction torque. A qualitative picture of $\psi(z)$ at any given $\beta$ (except at the vertical Bloch lines) can be obtained from Fig. 17.2 if the velocity parameter of those figures is taken to be $v_n(\beta)$. Clearly as long as $V < V_n$ in Fig. 17.2, the structure $\psi(\beta, z)$ is just a small perturbation of that for $V = 0$.

However if one attempts to drive the velocity above $V_n$, then Fig. 17.2 shows that instabilities occur, i.e., Bloch lines nucleate, at the front and back of the bubble where the normal wall velocity is maximum, but not at other points where the wall velocity is less. For the right-handed bubble chirality shown in Fig. 18.1, polarity considerations (Section 15,B) show that the Bloch lines nucleate at the lower critical circle in the front ($\beta = 0$) and at the upper critical circle on the back ($\beta = \pi$). The basic reason for this difference is that the front and back are moving in opposite directions relative to wall polarity. It is plausible that as the Bloch lines develop, they will form curves as illustrated in Fig. 18.1, terminating on the respective critical circles from which they originated and bulging farthest from the critical circles at $\beta = 0$ and $\beta = \pi$, respectively, where the absolute value of $v_n$ is maximum. In such a configuration it will still be true that over most of the wall surface the reaction to $v_n$ is balanced by a small displacement of $\psi$ from one or another magnetostatic equilibrium orientation. But in the Bloch-line regions $\psi$ "crosses over" from one equilibrium to another (e.g., Fig. 8.10b). In these regions the dynamic equilibrium is due to a combination of magnetostatic and exchange torques acting on the Bloch line. Mathematically it is conveniently described by balancing the dynamic reaction force of the Bloch line against the forces arising from the internal Bloch-line energy. We ignore any Bloch-line viscous-damping force.

The shapes of the Bloch-line curves can be derived using this basic principle and the Bloch-line approximation.[365] The dynamic reaction force $\mathbf{F}_{gL}$

is $2M\gamma^{-1}\Phi t \times V$ [Eq. (14.9)], where as usual $t$ is the unit vector tangent to the Bloch line [Eq. (8.10)] and $\Phi$ is the full angle of the Bloch-line rotation [Eq. (8.12)], which is a function of the stray field $H_r(z)$ and hence of the film-normal coordinate $z$. The direction of $F_{gL}$ in Fig. 18.1 is $+z$ at $\beta = 0$ and $-z$ at $\beta = \pi$. Except at $\beta = 0$ and $\beta = \pi$ there is a component of the Bloch-line reaction force normal to the wall, and it is always pushing outward in such a fashion as to distort the bubble elliptically perpendicular to the direction of motion (see Section 19,B for further details). For simplicity we assume here that the bubble remains cylindrical, as if it had infinite stiffness balancing the wall-normal component of $F_{gL}$.

The component $F_{gLt}$ of the reaction force which lies tangent to the bubble wall surface, by contrast, is balanced by the force arising from the energy of the Bloch-curve configuration. The latter force includes one term due to the Bloch-line curvature $\kappa$ (i.e., the inverse of the radius of curvature) within the tangent plane plus one due to the dependence of Bloch-line energy per unit length $E_L$ on $z$ owing to the dependence of stray field $H_r(z)$ on $z$. The force-balance equation is thus

$$F_{gLt} = E_L\kappa - (dE_L/dz)\cos\alpha, \tag{18.2}$$

where $\alpha$ is the angle between $F_{gLt}$ and the $z$ axis. The energy $E_L$ is given by Eq. (8.13) under the condition $|H_r| \leq 8M$. Equation (18.2) implicitly determines the shape of the Bloch curve and can be solved numerically for different values of $V$. Some results, computed using the plane-wall-model stray field of Eq. (8.33), are shown in Fig. 18.2a.[365] The figure shows one quadrant ($0 \leq \beta \leq \pi/2$) of the cylinder surface on the $+x$ or front side of the bubble. The Bloch lines are bounded on the bottom by the critical circle where $H_r(z) = 8M$, $\Phi = 0$, and $E_L = 0$. Symmetrically disposed Bloch lines on the $-x$ side would appear inverted from those in the figure and bounded by the upper critical circle where $H_r(z) = -8M$.

The family of curves in this figure range from the first member of infinitesimal arc length enclosing the nucleation point at $\beta = 0$ on the lower critical circle to the last and longest that barely touches the upper critical circle. The corresponding range of velocities is $0.6V_p \leq V \leq V_p$, where $V_p = 24.9\gamma A/hK^{1/2}$, which is only 4% greater than the maximum velocity $V_{po}$ for translation of a flat wall [Eq. (15.6)]. The minimum velocity equal to $0.6V_p$ is 40% higher than the flat wall minimum [Eq. (15.13)]. This greater difference is due to the greater curvature of the shorter Bloch curve. The total Bloch-line rotation angle $\Phi$ varies monotonically from 0 at the lower critical circle to $2\pi$ at the upper critical circle.

Thus we see that the velocity of domain translation is provided by two Bloch-line forces: one due to the gradient $dH_r/dz$ of the stray field, and one

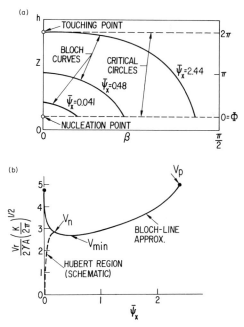

**Fig. 18.2.** (a) Flattened quarter of a bubble-domain wall showing critical circles and computed Bloch curves for a bubble with $r = h$ translating in the $\beta = 0$ direction and having the model stray field of Eq. (8.33). (b) $V(\bar{\psi})$ curve for the same case (after Malozemoff *et al.*[365]).

due to the line tension and curvature of the Bloch line itself. The fact that the allowed range of steady velocities $V$ does not differ exceedingly from that permitted for radial motion suggests that the more copious theoretical results for radial motion will provide reasonable guides (within a factor of 2) to the allowed velocities for translation. However, the additional dimension existing in the translational case does lead to different inertial phenomena as detailed in Section 8,E.

As noted in Section 14,A [Eqs. (14.8)–(14.10)], the presence of Bloch lines does not add much to the drag on a moving domain. Thus when Bloch lines of both twist senses are present in nearly equal numbers and maintain constant positions in the bubble, the domain mobility is not ordinarily greatly affected. The Bloch-line drag becomes important only when the density of Bloch lines is great enough to appreciably decrease the wall thickness. But when this condition is satisfied the wall energy and therefore the shape of the domain changes drastically, and the question of mobility becomes in any case exceedingly complex.

## B.  CANONICAL MOMENTUM

The concept of total canonical momentum was useful in the treatment of quasi-planar wall motion in Chapters VI and VIII. Here we discuss heuristically the total translational momentum of a domain of arbitrary shape and the special case of a cylindrical bubble of radius $r$. More exact discussions are given in the literature.[349,380] Throughout we assume the domain has downward magnetization. That is, the direction of magnetization outside the domain, specified by the unit vector $z_0$, is up. Only those formulas containing $z_0$ can be applied for either magnetization polarity.

We have previously shown that

$$\mathbf{p}_n = p_n \mathbf{n}, \qquad p_n = 2M\gamma^{-1}\psi \qquad (18.3)$$

is the momentum density vector normal to a domain wall at any point. Here $\mathbf{n}$ is the unit-vector normal, and it is directed from the $M_z = +M$ to the $M_z = -M$ side of the wall, in accord with the convention of Section 12. The wall-moment angle $\psi$ is measured in the counter-clockwise direction from a reference direction fixed in the medium. It is reasonable to assume that the total momentum vector $\mathbf{P}$ of the entire domain of general shape is given by the integral over $\mathbf{p}_n$

$$\mathbf{P} = \int \mathbf{p}_n \, dA, \qquad (18.4a)$$

carried over the wall surface area $A$. Let us consider the Cartesian components

$$P_x = \int dA \, p_{nx}, \qquad P_y = \int dA \, p_{ny}. \qquad (18.4b)$$

Within these integrals we can transfer direction cosine factors from $p_n$ to $dA$ thus

$$
\begin{aligned}
p_{nx} \, dA &= p_n \cos(p_n, x) dA = \pm p_n \, dy \, dz, \\
p_{ny} \, dA &= p_n \cos(p_n, y) dA = \pm p_n \, dx \, dz,
\end{aligned}
\qquad (18.4c)
$$

where $dy \, dz$ and $dx \, dz$ are the projections of the oblique area element $dA$ on the $yz$ and $xz$ planes, respectively. The sign is plus $(+)$ for the back and minus $(-)$ for the front $(x > 0$ or $y > 0)$ of the domain. In this way we find the expressions

$$\mathbf{P} = 2M\gamma^{-1}\mathbf{\Psi}, \qquad (18.5)$$

$$\Psi_x = -\int_0^h dz \oint \psi \, dy, \qquad (18.6a)$$

$$\Psi_y = \int_0^h dz \oint \psi \, dx. \qquad (18.6b)$$

$\oint$ represents a contour integral taken counterclockwise around the domain in a plane of constant $z$. These formulas can be written more succinctly in vector notation, using $\mathbf{x} = (x, y)$:

$$\mathbf{\Psi} = \mathbf{z}_0 \times \int_0^h dz \oint \psi \, d\mathbf{x}. \tag{18.7}$$

Problems arise in the evaluation of this expression for the case of nonzero winding number $S$. We will return to this case in the next section and proceed for the remainder of this section assuming $S = 0$. Integrating Eq. (18.7) by parts, one finds the equivalent expression

$$\mathbf{\Psi} = -\mathbf{z}_0 \times \int_0^h dz \oint \mathbf{x} \, d\psi. \tag{18.8}$$

Next we give several examples of the evaluation of $\mathbf{\Psi}$. Equation (18.8) reduces to useful special expressions whenever the Bloch-line model becomes valid. For then the spatial gradients of $\psi$ become concentrated within the Bloch-line regions. Using the properties of the Kronecker $\delta$-function we may write the approximate expression

$$d\psi/ds = \sum_i \Phi_i \delta(s - s_i) + d\psi_0/ds, \tag{18.9}$$

where $\psi_0(z, s)$ describes the stray-field twisted wall structure not having Bloch lines ($S = 1$), and where $\Phi_i(z, t)$ and $s_i(z, t)$ are the change in $\psi$ and the arc-length coordinates $s$ around the contour, for the $i$th point where a Bloch line intersects a plane of constant $z$ (see Fig. 12.1). The sign of $\Phi_i(z, t)$ is the same as that of $\mathbf{z}_0 \cdot \mathbf{t}_i$, where $\mathbf{t}_i$ is the tangent vector [Eq. (8.10)] of the $i$th Bloch line. Using the equivalence

$$\oint \mathbf{x} \, d\psi = \oint \mathbf{x}(d\psi/ds) \, ds, \tag{18.10}$$

and substituting Eqs. (18.9) and (18.10) in (18.8), one finds

$$\mathbf{\Psi} = -\mathbf{z}_0 \times \int_0^h dz \sum_i \Phi_i \mathbf{x}_i + \mathbf{\Psi}_0, \tag{18.11}$$

where $\mathbf{x}_i(z, t)$ is the position vector of the $i$th Bloch line and $\mathbf{\Psi}_0$ is given by Eq. (18.8) with $d\psi = d\psi_0$. The origin for $\mathbf{x}_i(z, t)$ must be taken to be the same as that used in calculating $\mathbf{\Psi}_0$, and it is usually taken to be the domain centroid. The term $\mathbf{\Psi}_0$, contributed by regions of slowly varying $\psi$, depends on the shape of the domain and is not always negligible. However it vanishes by symmetry in the case of a circular domain if $\psi$ can be replaced in these regions by its static equilibrium value and if the origin for Eq. (18.8) is the bubble center. Thus Eq. (18.11) shows that domain momentum can often be substantially concentrated within Bloch lines.

Let us consider the particularly simple case that the Bloch lines are all 180° lines ($\Phi_i = \pm\pi$), as if stray fields or other in-plane fields could be ignored. Then, according to Eq. (18.11), changes of $\Psi_x$ and $\Psi_y$ measure plane projections of areas swept out by the Bloch lines as the domain structure varies from one dynamic state to another, as shown in Fig. 18.3. More precisely we have

$$\Psi_x = \pi \sum_i A_{ix} \qquad \Psi_y = \pi \sum_i A_{iy}, \tag{18.12}$$

where $A_{ix}$ and $A_{iy}$ are signed areas associated with each Bloch line. In the case of a line beginning and ending on one film surface, $A_{ix}(A_{iy})$ is the area enclosed by the projection of the Bloch line and the film surface on the $yz$ ($zx$) plane (Fig. 18.3a). In the case of a line spanning the film thickness, $A_{ix}(A_{iy})$ is the area enclosed by the film surface and the projections of the Bloch line and the axis $y = 0$ ($x = 0$). The signs of $A_{ix}$ can be determined in any particular case by inspection from Eq. (18.11) and the twist sense $d\psi/ds$ of the Bloch lines. When only straight vertical Bloch lines are present, the area construction applies exactly because the effect of stray fields on $\Phi_i(z)$ averages to zero over $z$. Equation (18.11) then reduces to

$$\Psi_x = \pi h \sum_i (\mathbf{z}_0 \cdot \mathbf{t}_i) y_i \qquad \Psi_y = -\pi h \sum_i (\mathbf{z}_0 \cdot \mathbf{t}_i) x_i \tag{18.13}$$

where $(x_i, y_i)$ are coordinates of the $i$th Bloch line (relative to the center of the domain), and $\mathbf{t}_i$ is its unit vector tangent as defined in Eq. (8.10).

In the case of a circular cylinder, a dimensionless notation for momentum $\bar{\boldsymbol{\psi}} = (\bar{\psi}_x, \bar{\psi}_y)$ is convenient:

$$\bar{\psi}_x = (\pi r h)^{-1} \Psi_x \qquad \bar{\psi}_y = (\pi r h)^{-1} \Psi_y. \tag{18.14}$$

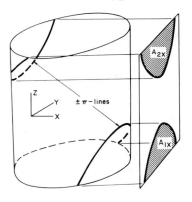
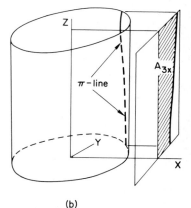

(a)                                                         (b)

**Fig. 18.3.** Schematic geometrical constructions for canonical momentum $2M\gamma^{-1}\boldsymbol{\Psi}$ of a bubble domain whose wall structure consists of $\pm\pi$ Bloch lines. $\Psi_x$ is given by $\pi \sum_i A_{ix}$, where $A_{ix}$ is the projection of the area enclosed by the $i$th Bloch line on the $yz$ plane (after Slonczewski[349]). (a) shows geometric construction of areas $A_{1x}$ and $A_{2x}$ when the ends of the line terminate on one film surface. (b) shows the construction when the ends terminate on opposite film surfaces.

In view of Eqs. (18.6), each of these quantities equals twice the surface mean of a projection of the vector $\psi\mathbf{n}$, and consequently represents a weighted-mean precession angle for the domain. For example, let us consider the Bloch-line configuration illustrated at the top of Fig. 13.5, in a bubble moving along the $+x$ direction with bias field up. According to the gyrotropic force expression $2M\gamma^{-1}\mathbf{\Phi t} \times \mathbf{V}$ [Eq. (14.9)], positive Bloch lines will collect on one flank of the bubble ($\beta = +\pi/2$) and negative Bloch lines on the other ($\beta = -\pi/2$). Assuming $n_+$ positive Bloch lines and $n_-$ negative ones, Eqs. (18.13) and (18.14) predict a dimensionless momentum

$$\bar{\psi}_x = n_+ + n_- \qquad \bar{\psi}_y = 0. \tag{18.15}$$

Thus, for this simple case the dimensionless momentum equals precisely the total number of Bloch lines. The velocity corresponding to such bubble states containing loosely spaced vertical Bloch lines is zero because there are no curvature or stray field forces acting on them. Thus these bubbles are metastable. It is worth emphasizing that such bubbles have what may be termed a "stored momentum." There is no mechanical analogy for this concept because we customarily think of an object with momentum as moving. However, as we shall see in Section 19, the "stored momentum" concept is very useful in the bubble analogy because such bubbles have a remarkable tendency for motion in the momentum direction once they are perturbed from their metastable equilibrium.

In contrast to the simple considerations of Eq. (18.12), correct determination of the momentum of the Bloch-curve states of Fig. 18.2a requires a numerical integration using Eq. (18.11) because the Bloch lines are not $\pi$-lines. Results are shown on this figure and a more complete plot of $V(\bar{\psi})$ is shown in Fig. 18.2b. Here $\bar{\psi}$ increases with the projected area of the Bloch loops, though not proportionally. At small values of $\bar{\psi}$, where the Bloch line is close to the originating critical circle (see Fig. 18.2a), the Bloch-line approximation is expected to fail, and one can expect an initial linear $V(\bar{\psi})$ region, shown schematically as a dashed line in Fig. 18.2b, in analogy with the $V(\bar{\psi})$ curve of Hubert's two-dimensional calculation described in Section 17,A.[245]

## C. EQUATIONS OF MOTION

Just as we integrated the momentum over the surface of the domain to obtain the total domain momentum, so we can integrate the forces acting on the domain wall to give the total vector force $\mathbf{F}$. We consider such a force to include not only external fields but also viscous and coercive drag. For example, a cylindrical bubble domain of radius $r$ with downward magnetization in a gradient $\mathbf{V}H_z$ has

$$\mathbf{F} = -2\pi rhM(r\mathbf{V}H_z + H_{cb}\mathbf{V}|V|^{-1} + \mu^{-1}\mathbf{V}), \qquad (18.16)$$

where $H_{cb}$ is the bubble coercivity and $\mu$ is the conventional linear wall mobility. Here we have ignored the small extra viscous drag of the Bloch lines.

Given the total force considered above and the total momentum for a domain of general shape, it is plausible to assume the basic equation of motion[349]

$$dP/dt \equiv 2M\gamma^{-1} \, d\Psi/dt = \mathbf{F}, \qquad (S = 0). \qquad (18.17)$$

This equation is another expression of the pressure equation of domain wall dynamics for the special case of vanishing winding number $S$. It can be derived directly from the Landau–Lifshitz equation, using the volume expression[349,380]

$$\Psi = -\tfrac{1}{2} \iiint \phi(\mathbf{V} \cos \theta)dx \, dy \, dz. \qquad (18.18)$$

For a cylindrical bubble of constant radius $r$ moving in a straight line, Eq. (18.17), combined with Eqs. (18.14) and (18.16), gives

$$d\bar{\psi}/dt = \gamma[H_g - H_{cb} - \mu^{-1}V], \qquad (S = 0, \; H_g > H_{cb}), \qquad (18.19)$$

where $H_g = |r\mathbf{V}H_z|$. This simple equation expresses the principle that the mean rate of spin precession within the bubble wall is given by the sum of effective fields due to the external gradient, coercivity, and viscous drag, in close analogy to the corresponding equation for a plane wall. If drive and coercivity are zero, we see that there is a simple relation between a change in $\bar{\psi}$ and the bubble displacement $X$, namely $dX = \Delta\alpha^{-1}d\bar{\psi}$. This relation is very useful in discussions of ballistic overshoot or automotion in subsequent sections.

Equation (18.17), or the more specialized Eq. (18.19), serves as a basis for transient bubble dynamics in the quasi-stationary limit as follows. Assume one has calculated the steady-state relation

$$V = V(\Psi) \qquad (18.20)$$

for constant velocity $\mathbf{V}$. We have seen several examples of this relation in Sections 18,A and 18,B. The key assumption, as discussed before in Section 12, is that this relation holds also in first approximation for nonsteady motions in which both $\mathbf{V}$ and $\Psi$ depend on time. Integration, typically by numerical means, of Eqs. (18.17) and (18.20) will in principle provide the bubble displacement $\mathbf{X}(t) = \int_0^t \mathbf{V}(t')dt'$ for a given force $\mathbf{F}(t)$. Particular highly simplified solutions of this problem are discussed in Section 18,E.

As expected from Hamilton's general equations of motion and borne out by analysis,[349] we may also write the steady-state relation

$$\mathbf{V} = dW/d\mathbf{P} = (\gamma/2M)(dW/d\mathbf{\Psi}),\qquad(18.21)$$

where $W(\mathbf{\Psi})$ is the domain energy minimized at constant $\mathbf{\Psi}$. This makes clear that $W(\mathbf{\Psi})$ may be considered the kinetic energy of a domain, with the same restrictions as in the two-dimensional case (Section 12).

For the remainder of this section we reconsider the definition of momentum and the equation of motion for the more general case of winding number $S \neq 0$.[349] Problems arise, for in this case $\psi$ cannot vary continuously with position across all surfaces in the medium, since $\psi$ increases by $2\pi S$ in one cycle around the domain (we do not consider Bloch points), and Eqs. (18.6) are consequently ambiguous. Since the observable wall-surface-moment vector is continuous everywhere on the wall, the discontinuity of $\psi$ cannot have any physical consequence. Its occurrence is simply a fault of the mathematical representation. In the expression (18.6a) for $\mathbf{\Psi}_x$ it is convenient to choose $\psi$ continuous at all points on the wall surface except along one curve (straight line for a cylinder), where the wall intersects a particular plane $y = Y(t)$ that may, if we choose, depend on time (see Fig. 12.1). Across this cut, $\psi$ has a discontinuous step-change $\delta\psi = 2\pi S$. The value of $\mathbf{\Psi}_x$ clearly depends on the location of $Y$.

If we choose $Y$ to be independent of time, then it is plausible that Eq. (18.17) is correct even for $S \neq 0$.[349] However, a constant $Y$ has the awkward nonphysical consequence that a simple rigid displacement of the distribution $\mathbf{M}(\mathbf{x})$ in the $y$ direction would change $P_x$. Moreover if the domain moves in the $y$ direction a distance greater than its maximum $y$-dimension, then clearly $Y$ must change with $t$ in order that the cut "keeps up" with the domain. On the other hand let us consider a fixed spin configuration but assume that the cut plane is moving at a velocity $\dot{Y}$. In this case, since $\psi$ changes by $2\pi S$ at every point the cut plane sweeps through, the time rate of change $\mathbf{\Psi}_x$ is $-2\pi h S \dot{Y}$ according to Eq. (18.6a). Yet it is obvious that the net force is zero since nothing has physically happened to the domain. Now if we consider the general case where the physical spin configuration changes and simultaneously the cut plane moves, $d\mathbf{\Psi}_x/dt$ must consist of both contributions, a "dynamic" term $(\gamma/2M)F_x$ causing the physical changes in the spin distribution, and a "kinematic" term arising from the time dependence of the cut at $Y(t)$. Evaluation of $\mathbf{\Psi}_y$ similarly requires consideration of a moving cut plane $x = X(t)$. The combined results are expressed vectorially by

$$d\mathbf{\Psi}/dt = (\gamma/2M)\mathbf{F} + 2\pi h S \mathbf{z}_0 \times \mathbf{u},\qquad(18.22)$$

where $\mathbf{u} = (\dot{X}, \dot{Y})$ is the velocity of the intersection of the cut planes, where $\mathbf{\Psi}$ is understood to be evaluated using these same cuts, and where $\mathbf{z}_0$ is the magnetization direction outside the domain. This equation is the fundamental equation of motion for a closed domain of general shape and general $S$. Note that the last term in this equation is precisely the gyrotropic deflection force discussed in Sections 13 and 14 if we consider the cuts to move with the uniform velocity of the domain.

Since cuts are cumbersome and confusing, it is fortunate that the following stratagem eliminates them from the equations: Equations (18.6) are unchanged by the translation of the coordinate axes; so we replace $\mathbf{x}$ in this equation by the position vector $\mathbf{x}'$ measured from the generally moving intersection $0'$ of the cuts (see Fig. 12.1). Then carrying out the contour integrals in Eqs. (18.6) from one side of the cut to the other by parts produces Eq. (18.8), with $\mathbf{x}$ replaced by $\mathbf{x}'$, where the vanishing of the "integrated parts" is assured by the special choice of origin. Note that in this equation the question of cuts does not arise because only the differential $d\psi$

and not $\psi$ itself appears, as made plain by Eq. (18.10) where $d\psi/ds$ is continuous for all integers $S$. It follows that $\boldsymbol{\Psi}$ as given by Eq. (18.8) satisfies the dynamical equation (18.22) provided $\mathbf{u}$ in this equation is taken to be the velocity of the now moving origin for a redefined $\mathbf{x}$. [Indeed, further consideration shows that Eqs. (18.8) and (18.22) are consistent no matter what the origin of $\mathbf{x}$, even if it lies outside the domain.] Thus the formulas of Section 18,B are correct even for $S \neq 0$, with the understanding that the origin of the $\mathbf{x}_i$ in Eq. (18.11) or the center of the cylinder in the case of Eq. (18.15) has velocity $\mathbf{u}$. The convenient location for the origin in most applications, and the one we use in the remainder of this review, will therefore be the domain centroid. For then not only are these formulas correct but $\boldsymbol{\Psi}$ will be unchanged ($d\boldsymbol{\Psi}/dt = 0$) by a rigid translation (steady-state motion).

## D. DYNAMIC STATE CONVERSION

As will be discussed in greater detail in Section 19, many experiments testify to the fact that the linear portion of the velocity-gradient relation for bubble transport is bounded by observed critical velocities $V_t$ that are generally an order of magnitude lower than the Walker velocity and of order $V_{po} = 24\gamma A/hK^{1/2}$. Various explanations of this effect have been given and the differences between these explanations have not yet been resolved in all cases.

Thiele[351] considered the possibility that the low-velocity state with no Bloch lines might convert dynamically to a more twisted state described by curved Bloch lines such as those in Figs. 18.1–18.4. His micromagnetic estimate of the threshold velocity for Bloch-line nucleation is of the same order as $V_w$. This result led him to conclude that the low-drive structure ideally persisted even for $H > \mu^{-1}V_t$, where $\mu$ is the linear mobility and $V_t$ is the observed critical velocity ($< V_w$). To account for the observed departures of the velocity from linearity, he proposed that normal modes of bubble wall vibration (see Section 22) are excited when the wall impinges on structural defects or phonons. The expenditure of energy in these processes would account for the increased drag and lowered apparent mobility

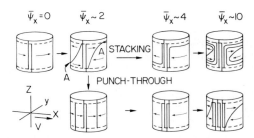

**Fig. 18.4.** Schematic Bloch-line wind-up for a propagating $S = 0$ bubble, illustrating the different possibilities of stacking and punch-through. This is the bubble analog of the plane-wall version shown in Fig. 15.1 (after Malozemoff and Maekawa[139]).

of the domain. An example of an experimental velocity versus drive curve exhibiting two apparent mobility regions is shown in Fig. 6.4. Because specific knowledge of the density and magnetic nature of the postulated defects is unavailable, the strength of this effect has not been estimated theoretically. His estimate of the characteristic velocity for the inception of these effects (see Section 22,B) is of the same order of magnitude as the Bloch-line instability velocity $V_{po}$ and thus can roughly account for the observed $V_t$. As will be discussed below, the reason for the velocity break and lower apparent mobility can also be explained more quantitatively by a Bloch-line model, and therefore, in the absence of a need to invoke Thiele's mechanism, we do not consider it further in this chapter.

Hagedorn[226] suggested a different explanation of the breakdown of linear mobility based on the great amount of scatter present in the experimental apparent velocities for $H > \mu^{-1}V_t$ (e.g., see Fig. 6.4). Like Thiele he supposed that the Bloch-line nucleation threshold was intrinsically of order $V_w$ and he also postulated randomly distributed defects in the form of surface inhomogeneities, for example, but having lower density and greater strength than those of Thiele. If a domain wall in a particular case should by chance not strike an imperfection during the course of observation, then its velocity lies on the extrapolated low-drive line up to a critical velocity in the range of Walker's velocity. If however the wall does, in the course of the propagation, strike one or more defects with sufficient force, then a Bloch line becomes nucleated and expands as indicated in the sequence shown in Figs. 18.2a and 18.4. Once nucleated, the Bloch line causes the domain to move at a velocity of order $V_{po}$, as discussed in Section 18,A. Repeated defect-induced nucleations of Bloch lines result in dynamic conversion of the domain into a many-Bloch-line state of lowered mobility.

The two-dimensional dynamic wall structure computations of Hubert[245] (see Section 17,A) impacted this problem by showing that for plane walls a Bloch line nucleates naturally at a peak velocity $V_n$ not far from $V_{po}$, as a natural consequence of the Landau–Lifshitz equations and contrary to Thiele's estimates.[351] It is natural to assume that Bloch lines can therefore also nucleate without defects during translation of a cylinder domain at comparable velocities. According to Eq. (18.19), drive fields $H_g = |r\nabla H_z|$ less than $H_{cb} + \mu^{-1}V_n$ will cause steady-state motion without Bloch-line nucleation, while greater drives will cause nucleation and growth of curved Bloch lines. In view of the steady-state relationship between the Bloch-line configuration and $\bar{\psi}$ described in the previous sections, the progress of these Bloch lines across the bubble-domain surface is described in the quasi-steady limit by Eqs. (18.17) or (18.19) and (18.20). A portion of the latter relation is shown in Fig. 18.2b. If $H_g$ is less than $H_{cb} + \mu^{-1}V_p$, then the Bloch line comes to dynamic equilibrium ($\dot{\bar{\psi}} = 0$) and steady bubble

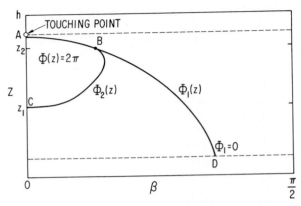

**Fig. 18.5.**  Flattened quarter of bubble-domain wall showing a computed complex Bloch-curve state representing the second cycle of Bloch-line wind-up. The momentum $\bar{\psi}_x$ is greater than any of the values shown in Fig. 18.2a. The velocity is $0.63V_p$. Here the Bloch curve segment having angle $\Phi_1$ may be thought of as the "original" Bloch curve shown in Fig. 18.2a. The Bloch curve with angle $\Phi_2$ could theoretically be nucleated at the point where the original Bloch curve touches the upper critical circle ($\bar{\psi} = 2.44$ in Fig. 18.2a) (after Slonczewski[381]).

motion can proceed. However if $H_g$ is great enough for $\dot{\bar{\psi}}$ to be always positive, then the pair of Bloch curves expands steadily with increasing $\bar{\psi}$ according to the sequence illustrated in Fig. 18.2a.

Thus far we have considered only the single pair of Bloch curves initially nucleated. What happens at and beyond the point where $\bar{\psi}$ reaches the critical value $\bar{\psi} = \bar{\psi}_p = 2.44$ where $V(\bar{\psi}) = V_p$? There are three possibilities with some experimental support for each one:

*1. Bloch-line "pile-up" or "stacking."* As discussed in Section 15,D, the barrier to Bloch line annihilation may be stronger than the barrier to nucleation of a second Bloch line in the two-dimensional model. This is particularly likely if the surface is capped by an ion-implanted or low-anisotropy layer. Embracing provisionally the postulate of second-Bloch-line nucleation and extending it to the case of bubble translation, we expect nucleation of a new Bloch line when the first Bloch line touches the second critical circle (see curve marked $V = V_p$ in Fig. 18.2a). An example of the resulting compound Bloch-line configuration appears in Fig. 18.5.[381] Its shape was computed according to the principles of Section 18,A. Note that in the figure the total Bloch-line twist angle $\Phi$ of segment $AB$ satisfies $\Phi(z) = 2\pi$, $BD$ satisfies $0 \le \Phi_1(z) \le 2\pi$, and $CB$ satisfies $\Phi_2(z) = 2\pi - \Phi_1(z)$. Thus, at $\bar{\psi} = \bar{\psi}_p$, $V$ changes discontinuously to a new value on a different curve $V = V_2(\bar{\psi})$, as shown schematically in Fig. 18.6a. Then $\bar{\psi}$ continues to increase as the Bloch-line shape continues to change with the point $C$ moving

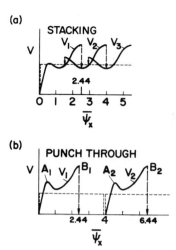

**Fig. 18.6.** Schematic $V(\bar{\psi})$ curves for bubble translation assuming (a) Bloch-line stacking or (b) punch-through. In (b) $A_i$ indicates a Bloch-line nucleation, while $B_i$ indicates punch-through. Dotted line indicates saturation velocity approximation.

downward in Fig. 18.5. When $C$ reaches the lower critical point a new Bloch line now nucleates. Repetition of this process leads to the sequence of wall configurations illustrated schematically in Fig. 18.4 and constitutes Bloch-line "pile-up" or "stacking."[226] Note that according to the computations the line tension is sufficient to prevent straightforward separation into vertical and horizontal segments and the lines remain curved. The corresponding schematic velocity-momentum relation appears in Fig. 18.6a, $\bar{\psi}$ increasing in rough proportion to the number of loops. The ever-presence of curved Bloch lines in pure pile-up would imply $V_i(\bar{\psi}) \neq 0$ for all $\bar{\psi}$ except $\bar{\psi} = 0$.

As long as the number of Bloch lines times the Bloch-line width $\sqrt{2\pi\Lambda_0}$ is much less than the domain dimensions, the velocity should remain comparable to what it was in the first cycle of Bloch-line nucleation and propagation. Thus an estimate of the average velocity during a wind-up can be obtained by averaging the velocity over the first cycle. Using the results of Fig. 18.2b, one finds that the average translational velocity should be $0.69V_{po}$. Evidence for pile-up in capped or triple-layer films is described in Section 19,D.

*2. Punch-through.* Hubert's plane-wall computations (Section 17,A) do not support the pile-up model.[245] Rather, his steady-state structures suggest that the wall moment at and near the film surface flips by about 360° when the wall is driven beyond its critical condition at $V_p$, which is close to $V_{po}$ for thick films. Naive extension of this result to the translation case implies that the curved Bloch line "punches through" to the film surface. There it severs into two Bloch curves extending from the critical circle where $\Phi = 0$, to the opposite film surface where $\Phi = \pm 2\pi$. The critical velocity for this process, estimated with the Bloch-line model, is typically also of the order $V_{po}$.[226] The Bloch lines are then expected to straighten up rapidly into vertical Bloch lines under the influence of line tension, as illustrated in Fig. 18.4. We assume $\bar{\psi}$ is conserved in this process (see Section 17,C). Repetitions of punch-through events with increasing $\bar{\psi}$ would lead to accumulation of oppositely-wound vertical-Bloch-line clusters on opposite flanks of the domain as shown in Fig. 18.4. If every Bloch curve punched through, no more than one pair of curved Bloch lines would exist at any one time. Therefore $V(\bar{\psi})$ falls to zero just after each critical punch-through point $B_i$, illustrated schematically in Fig. 18.6b. The average velocity, taken over one full cycle of $\bar{\psi}$, is now $0.42V_{po}$, using the curve of Fig. 18.2b. It is not implausible for some mixture of punch-through and pile-up to occur, in which case the abrupt velocity changes in the pure punch-through model would be smoothed out. Elliptical distortion of the bubble during propagation will also smooth out the velocity changes. The bias-jump and rocking experiments discussed in Section 19,C show that punch-through occurs in most as-grown films.

*3. Winding-state change.* The winding number $S$ can only be changed by the passage of Bloch points through the magnetic film. While $S$ cannot change if only pile-up occurs, the punch-through process may be altered in concept to effect a change of $S$ as follows: Punch-through of the original Bloch loop to the upper film surface may be accompanied by the nucleation of a Bloch point at the upper terminus of one or both resulting Bloch-line segments. The energy to nucleate the Bloch point may come from a portion of the energy liberated when the Bloch loop is replaced by two vertical Bloch lines. Following nucleation, the Bloch point moves rapidly down to the midplane of the film, because of its stray-field energy, and it changes the net twist value of the Bloch line from $\pm 1$ to 0 as described in Section 9,B. The affected vertical Bloch line now stretches between the two critical circles. Disturbance of the bubble sufficient to cause collision of the Bloch-point-containing Bloch line with any other vertical Bloch line will cause both to mutually annihilate (e.g., see Fig. 9.7). Thus one can see that symmetric generation of Bloch points in pairs leads ultimately to states with no Bloch points whose momentum is different than otherwise. Preferential generation of Bloch points on one side of the bubble leads alternatively to an ultimate change in $S$. Such a change may conceivably occur dynamically if, while the bubble is propagating, Bloch points nucleate on one side of a dynamically stable curved Bloch line, when the domain impinges on a material inhomogeneity, so that a state change persists only while the bubble is moving but disappears when it comes to rest and the curved Bloch line retracts.[226] A mechanism for the dynamic stabilization of Bloch points in this manner remains to be proposed.

Experimental evidence supporting each of the possibilities described above will be described in Section 19.

### E. SATURATION-VELOCITY APPROXIMATION AND THEORY OF BALLISTIC OVERSHOOT

As we have seen in the previous section, steady-state motion is no longer possible once the drive field $H_g$ exceeds $H_{cb} + \mu^{-1}V_p$. In this case one must integrate Eqs. (18.19) and (18.20) directly. This is difficult because of the complicated form of $V(\bar{\psi})$, illustrated in Fig. 18.2b or 18.6. It has proven useful for comparison with experiment to introduce a simple approximation for $V(\bar{\psi})$, permitting an analytic solution to the equations of motion.[115,379] One takes for motion at constant $Y$ the step function

$$V_x = V_s \, \text{sgn} \, \bar{\psi}_x, \qquad (18.23)$$

where $V_s$ is called the "saturation velocity" and the sgn function is defined in Eq. (5.2) (see also Section 17,B). This approximation ignores the linear region for small $\bar{\psi}$, which is plausible since this region gives rise to a wall mass comparable to the Döring mass, and this mass is negligible compared to the ballistic effects we shall consider below. The approximation also ignores the velocity peaks, which is plausible since according to the previous sections the theoretically calculated peaks are at most only a factor of two above the velocity minima in the nonlinear region of $V(\bar{\psi})$, and they may be further smeared by interaction with imperfections. If $\bar{\psi}_x$ exceeds $\bar{\psi}_{xp} = 2.44$, at which point the Bloch curve touches the critical circle (Fig. 18.2b), there are two possibilities, as discussed in the previous section. Either the Bloch curve will stack at the surface, in which case one can plausibly extend Eq. (18.23) beyond $\bar{\psi}_{xp}$, or it can punch through, in which case the velocity is assumed to drop to zero until the next Bloch line nucleates (see dotted line in Fig. 18.6).

Let us now introduce Eq. (18.23) into the equation of motion (18.19), assuming a cylindrical bubble of fixed radius $r$. We consider the case of a rectangular drive pulse of length $T$ and strength $H_g = |r\nabla H_z| > H_{cb} + \mu^{-1}V_s$, and we assume the bubble has $S = 0$ as in Fig. 18.1. Initially $\bar{\psi}_x = 0$ and $V = 0$ (we ignore the momentum $\bar{\psi}_x = 2$ of the original two Bloch lines).

When the gradient drive is applied, the equation of motion becomes

$$d\bar{\psi}_x/dt = \gamma(H_g - H_{cb} - \mu^{-1}V_s), \qquad (18.24)$$

for $0 < t < T$. Thus $\bar{\psi}_x$ increases linearly with time, representing the winding up of Bloch lines under the pressure of the drive. By the end of the pulse, $\bar{\psi}_x$ reaches

$$\bar{\psi}_{xT} = \gamma(H_g - H_{cb} - \mu^{-1}V_s)T. \qquad (18.25)$$

During this time $V = V_s$ according to the saturation-velocity assumption, and so the bubble moves linearly with time, reaching $X_T = V_s T$ by the end of the pulse.

After the gradient drive turns off at $t = T$, the equation of motion becomes

$$d\bar{\psi}_x/dt = -\gamma(H_{cb} + \mu^{-1}V_s). \qquad (18.26)$$

Now $\bar{\psi}_x$ decreases linearly with time, corresponding to the Bloch lines unwinding under the pressure of coercivity and viscous damping. This process occurs for a time

$$t_0 = (\bar{\psi}_{x\infty} - \bar{\psi}_{xT})/(d\bar{\psi}_x/dt)$$
$$= [(\mu H_g - \mu H_{cb} - V_s)T - \mu\gamma^{-1}\bar{\psi}_{x\infty}]/(\mu H_{cb} + V_s), \qquad (18.27)$$

which represents the overshoot time of the bubble. Here $\bar{\psi}_{x\infty}$ represents the final momentum of the bubble at the end of the overshoot. $\bar{\psi}_{x\infty}$ may be nonzero if, for example, punch-through has occurred and vertical Bloch lines remain at the flanks of the bubble as in Fig. 18.4. During the overshoot time the bubble continues to move forward at a velocity $V_s$ because of the torques from unwinding Bloch curves. Thus the overshoot displacement is $X_0 = V_s t_0$. If $H_g \gg \mu^{-1}V_s$, the overshoot displacement $X_0$ can be very large compared to the displacement $X_T$ during the pulse, since, if $H_g > \mu^{-1}V_s$ and $H_{cb} = 0$, one has $X_0/X_T = (\mu H_g/V_s) - 1$. It is also useful to write the overshoot displacement in terms of the change $\delta\bar{\psi}_x$ in $\bar{\psi}_x$ which occurred during the overshoot:

$$X_0 = -\mu\gamma^{-1}\delta\bar{\psi}_x V_s/(V_s + \mu H_{cb}). \qquad (18.28)$$

Thus, one can extract $\delta\bar{\psi}_x$ directly from the size of the overshoot. This equation also makes clear that provided $H_{cb} = 0$, $\mu\gamma^{-1} \equiv \Delta\alpha^{-1}$ is a distance characteristic of the overshoot due to one unit of momentum, corresponding to one vertical Bloch-line segment. For typical films, this distance can be as much as a micron.

Combining Eq. (18.28) with $X_T = V_s T$, one can find the position of the bubble during the overshoot as a function of $\bar{\psi}_x$ for a given $H_g$ and $T$:

$$X = (\mu H_g T - \mu\gamma^{-1}\bar{\psi}_x)V_s/(V_s + \mu H_{cb}). \qquad (18.29)$$

The final position of the bubble is simply this same equation with $\bar{\psi}_x = \bar{\psi}_{x\infty}$. If $H_{cb} = 0$ and $\bar{\psi}_{x\infty} = 0$, we find a remarkable result, namely the final position is $\mu H_g T$, where it would have been according to the simple linear theory, ignoring Bloch-line effects altogether. This is in fact a general result based merely on energy conservation, considering that the bubble began and ended at the same internal energy state and that the only loss was due to viscous damping through the linear mobility $\mu$. This result explains the tendency of bubble propagation experiments to show higher apparent velocities than other experiments such as dynamic bubble collapse. Compared to those experiments, it is simply the geometry of the translation experiment that makes the Bloch lines curved and thus increases the likelihood that the momentum they represent will be recovered in ballistic overshoot. By contrast, in an experiment such as dynamic bubble collapse, kinetic energy is irretrievably lost every time a Bloch-line ring punches through at a surface.

If we take $\bar{\psi}_{x\infty} = 0$ but $H_{cb} \neq 0$ in Eq. (18.29), we find

$$X_\infty/T = \mu H_g V_s/(V_s + \mu H_{cb}), \tag{18.30}$$

which is valid for $H_g > H_{cb} + \mu^{-1}V_s$. Below this drive, the linear mobility formula $V = \mu(H_g - H_{cb})$ applies. Combining these results, one can expect an apparent velocity-versus-drive curve with two linear regions. This effect, along with additional lowering of the $X_\infty$ curve due to punch-through ($\bar{\psi}_{x\infty} \neq 0$), may explain the characteristic two-region curves illustrated in Fig. 6.4.

This analysis also offers a way of determining a lower bound $\mu_{min}$ for the linear mobility in situations where the lower region is smeared out by static coercivity.[379] A straight line pivoting around the origin and leaning down against the scattered data points has a slope given by Eq. (18.30), which can be solved for $\mu_{min}$. An example of this analysis is given in Fig. 6.5, where, given $H_{cb} = 0.25$ Oe and $V_s = 900$ cm/sec, one finds $\mu_{min} = 2600$ cm/sec Oe.

Further discussion of the theory will be considered in the next section, in conjunction with experimental results.

## 19.  Comparison to Experiment

The comparison of the theory of the preceding section to experiment will reveal some difficulties in the prediction of the magnitude of the saturation velocity. But if one treats this velocity as an adjustable parameter in the saturation-velocity approximation, one can go on to make a host of predictions based on the domain-momentum concept. These predictions are beautifully borne out in experiments involving ballistic overshoot, bias jumps, in-plane field effects, and (as discussed in Section 20) automotion.

## A.   TRANSLATIONAL PEAK AND SATURATION VELOCITIES

In the previous section we have seen that steady-state bubble motion can be maintained up to some peak velocity, and if the drive is increased beyond that point, multiple Bloch-line nucleation can occur with cyclic velocity variations of different types. The average velocity should scale with $V_{po} = 24\gamma A/hK^{1/2}$. Here we review experimental data bearing on these points and find only partial agreement with theory, much as in the case of two-dimensional wall motion reviewed in Section 16. While many bubble propagation experiments have shown apparent nonlinearities, including peaks and saturation velocities[166,209,296,382,383] we consider here only those propagation experiments done by high-speed photography to determine the true average velocity. With one reported exception[266] the average velocity versus drive curves rise smoothly to saturation, as illustrated by the $X_T$ points for a EuGaYIG film in Fig. 6.5, and no velocity peak is observed *in zero in-plane field*.[189,310,365,384] By contrast, Fig. 19.1 shows velocity peaks appearing

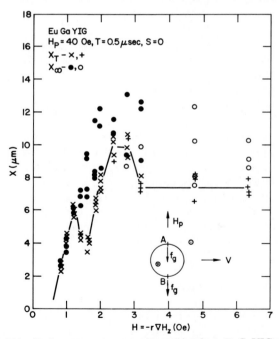

**Fig. 19.1.** Bubble displacement versus gradient drive in a EuGaYIG film in a bubble propagation experiment with high-speed photography (as in Fig. 6.5) and an in-plane field $H_p = 40$ Oe. Initial bubble state was $S = 0$. ×'s and ●'s indicate propagations without state changes, +'s and o's with state changes. Insert illustrates schematically the initial Bloch line distribution and the gyrotropic forces on the Bloch lines (after Malozemoff[299]).

with the application of an in-plane field.[299,310] Possible reasons for the lack of a peak in zero in-plane field include smearing due to coercivity, material inhomogeneities or excitation of wall modes. In particular, for high-mobility materials, the critical drive field $\mu^{-1}V_p$ is often less than the excess of the static coercivity over the dynamic coercivity, so that the linear region may be altogether suppressed in a pulsed-gradient experiment. An in-plane field increases the critical drive field and thus makes the peak observable.[94]

The degree to which the velocity is actually saturated as a function of time differs from study to study in a way that has not yet been resolved. Some data, like those shown in Fig. 19.2a for a GdTmYIG film,[385] show a dependence of bubble position on time that is linear within experimental scatter. Other data, like those shown in Fig. 19.2b for a LuGdAlIG film,[130] show noticeably greater scatter and, particularly if one draws a line through the upper end of the data scatter, an apparent initial higher speed motion some time during the first 100 nsec. In none of these data is there confirmation of the cyclic velocity variations predicted by theory.

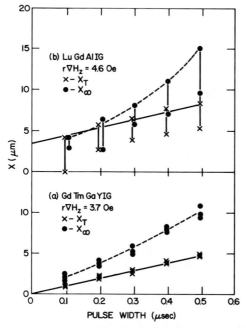

**Fig. 19.2.** Pulse-width dependence of bubble displacement in a bubble propagation experiment in (a) a GdTmGaYIG film (after De Luca *et al.*[385]) and (b) a LuGdAlIG film (after Vella-Coleiro[130]). $X_T$ and $X_\infty$ denote displacement at end of pulse and long after the pulse, respectively. Case (b) suggests a tendency to rapid initial wall motion.

From data like those of Figs. 6.5, 19.1, and 19.2 the translational saturation velocity $V_t$ can be determined. Compared to the predicted $0.69V_{po}$ for the stacking model of Section 18,D, the experimental results are generally within 30% for device quality samples.[365] For example, for the EuGaYIG film of Fig. 6.5, the observed saturation velocity is $\sim 800$ cm/sec compared to a calculated 650 cm/sec. For a SmCaGeYIG film, the observed value is 1760 cm/sec compared to a calculated 1900 cm/sec. The comparison of these two samples illustrates the $\gamma$ and $A$ dependence in the formula $V_{po} = 24\gamma A/hK^{1/2}$. $\gamma$ is increased by $\sim 30\%$ and $A$ by $\sim 100\%$ in going from EuGa to SmCaGe (see Sections 1 and 3), thus explaining the large increase in the saturation velocity. This increase is an important factor in the choice of material for device applications (see Section 21). The $K$ dependence in the $V_{po}$ formula was studied in a series of EuGaYIG films annealed to give different $K$ values while other parameters remained roughly the same.[362] The saturation velocity varied from 920 cm/sec in a film with $Q = 7.6$ to 1700 cm/sec in a film with $Q = 2.1$, in agreement with theory. The $K^{-1/2}$ dependence arises essentially from the fact that wall velocity is proportional to the wall width $\Delta_0 = (A/K)^{1/2}$.

Although the above results appear encouraging, they are probably fortuitous in view of the fact that the inverse thickness dependence in the saturation-velocity formula is not borne out in experiment. Although one study on ion-milled EuGdGaYIG films showed apparent velocity increases by a factor of 2 for a thickness decrease of a factor of 2,[386] this study was conducted without high-speed photography and thus the true velocities were not determined. By contrast, high-speed photography experiments on two series of EuGaYIG films, one grown to different thicknesses and the other ion-milled from a single sample to different thicknesses, showed saturation velocities varying by less than 30% over a thickness range from 2 to 10 $\mu$m.[365] As discussed in Section 16,B, this insensitivity to thickness raises serious and as yet unanswered questions for the entire Bloch-line theory of nonlinear wall motion. If the theory does not hold in the low-drive region of bubble translation, it is difficult to see why it should hold any better for describing straight-wall or bubble-collapse experiments. In fact the saturation velocities found by these different techniques appear to be not very different and to be given roughly by the empirical relation $1.2\gamma\Delta_0 M$. Further discussion of this relation has been given in Sections 16,B and 17,C. It should be noted that from a practical point of view, large saturation velocities are desirable. To achieve large saturation velocities, both the empirical relation and the formula for $V_{po}$ imply that $\gamma$ and $A$ should be increased while $K$ should be decreased. The two formulas differ chiefly in the effects of $h$ and $M$, which cannot be varied much in a device environment anyway, once the bubble size is prescribed.

Experiments involving stripe run-in (see Section 6,A) are closely related to bubble propagation and provide an alternative means of measuring the saturation velocity. Two studies have been reported, one on a EuTmGaYIG film[189] and another on a EuYbGaYIG film.[190] In both cases saturation velocities of 500–700 cm/sec were reported, both for ion-implanted and as-grown films, although the ion-implanted films gave slightly higher velocities. These velocities were close to but generally slightly higher than saturation velocities determined by dynamic bubble collapse or bubble propagation. A further point is that the shape of the stripe head became distorted during the motion, more so for the as-grown than for the ion-implanted film. Presumably this difference is due to the greater accumulation of Bloch lines in the as-grown case.

The mechanism of Bloch-line generation for a contracting stripe is like that for a bubble (see Section 18). A difference between the stripe and the bubble is that as the stripe contracts, it pushes the nucleated Bloch lines before it. The Bloch lines cannot keep accumulating ad infinitum. A likely alternative is that Bloch-line annihilation occurs at the front edge of the clump as fast as the lines are generated at the head.[189] The rate of energy loss by Bloch-line annihilation may be estimated from the Bloch-line energy per length $8AQ^{-1/2}$ [Eq. (8.8)] and the rate of Bloch-line generation $\dot{\psi} \sim \gamma H$, where $H$ is the net drive field acting on the stripe head. Balancing this rate of energy loss against $2MHVw$, the rate of energy gain for a stripe of width $w$ moving at velocity $V$, one finds $V \sim 4(2\pi)^{1/2}\gamma A/wK^{1/2}$. This result has the familiar form of the Bloch-line instability velocity $V_{po} = 24\gamma A/hK^{1/2}$, except that an inverse stripe-width dependence replaces the inverse film thickness dependence. This result can explain in a semiquantitative way the similarity of the stripe-head velocity to saturation velocities determined by other techniques. Velocities exceeding Walker's velocity have been reported in stripe runout experiments on YIG platelets,[387] although in this case $Q < 1$ and so the dynamic wall structure may differ considerably from the $Q \gg 1$ case treated in this review.


## B.   BALLISTIC OVERSHOOT

For some years the possibility of significant inertial effects during gradient propagation was overlooked because of the smallness of the Döring wall mass $m_D$ [Eq. (11.6)]. The corresponding time constant for wall relaxation $m_D/b$ is of the order of 10 nsec for high-mobility 5 $\mu$m garnets. For a typical velocity of $10^3$ cm/sec, the corresponding distance 0.1 $\mu$m is too small to be resolved in a microscope. Thus it was assumed in the conventional pulsed-gradient measurements of bubble velocity that the total displacement $X_\infty$ from the initial static position $X = 0$ before the pulse to the final static position $X = X_\infty$ following termination of the pulse took place during the width $T$ of the pulse. Thus $X_\infty/T$ was interpreted as the mean velocity for the duration of the pulse.

The results of such experiments were initially puzzling in that they showed a great amount of scatter both in displacement and propagation angle, particularly at high drives in high-mobility samples, as shown for example in Fig. 6.4 for a LuGdAlIG sample.[209] Sometimes the scatter showed several peaks in a plot of probability of occurrence versus displacement at a given drive.[209] The upper envelope of the scatter generally showed two regions of different apparent mobility.[209,328] Most puzzling of all, the maximum apparent velocities exceeded by far the saturation velocities observed by other techniques such as dynamic bubble collapse[168,328] or photometric stripe wall displacement,[369] but bubble propagation around a permalloy disk showed similar high velocities.[328]

One proposal to explain the scatter was that in the presence of a bias-field gradient, the average bias would drop during propagation. Therefore the bubble might run out and eventually contract back to a random position after the pulse turned off.[388,389] However, it was found that use of proper bias compensation did not eliminate the scatter.[128] Thus the scatter appeared to be due to some intrinsic domain-wall state change during the motion, given the generic term "dynamic conversion."[226,351]

The key oversight in the interpretation of the experiments was revealed when high-speed photography was used to study the translational bubble motion. It was discovered that contrary to earlier assumptions of a negligible mass effect, the propagated domains often continued to coast after the end of the drive pulse for distances up to several times the displacement during the drive pulse.[130,134a,210,211,266,357,365,384,390] For example, a plot of bubble positions at the end of the pulse ($X_T$) and long after the pulse ($X_\infty$) are shown in Fig. 6.5 for a EuGa film. The difference between $X_\infty$ and $X_T$ is termed the "ballistic overshoot" and it can be quite large. The bubble continues to move during this overshoot at velocities close to the velocity during the pulse.[130,365] Large scatter generally develops during the overshoot phase of the motion, while the scatter at $X_T$ is small,[210] although in one study the reverse was observed.[130] Furthermore, the true average velocity $X_T/T$ was observed to be saturated, as discussed in the previous section.

These results resolved the main discrepancies that were earlier apparent in the literature. The saturation velocities observed by different experimental techniques are now, broadly speaking, in qualitative if not quantitative agreement. The high velocities observed in bubble propagation around a permalloy disk are attributable to the stabilization of the wall structure by the in-plane field, which is always present in this experiment.

Next we consider various aspects of these experiments in the context of the saturation-velocity theory described in Section 18,E. Equation (18.29) shows that if coercivity is negligible and the bubble returns to its original momentum state ($\bar{\psi}_{x\infty} = 0$), the bubble displacement should be $\mu H_g T$, independent of the details of the motion. This result may help explain some results on a LuGdAlIG film,[130] in which the scatter in $X_T$ was significantly larger ($\sim 4\ \mu m$) than the scatter in $X_\infty$ ($\sim 2\ \mu m$). It should be noted that the scatter in $X_T$ cannot easily be explained in the context of the saturation velocity picture, and Hagedorn's picture[226] of material imperfections causing random Bloch-line nucleation over a high-velocity barrier seems more appropriate (Section 18,D). On the other hand, other experiments show more scatter in $X_\infty$ than in $X_T$.[210,365] The low scatter in $X_T$ is just what the saturation velocity model would predict. The scatter in $X_\infty$ can be partly explained by variability in the number of punch-throughs from one experiment to the next, causing variations in $\bar\psi_{x\infty}$ in Eq. (18.29). There may also be some randomness in the precise moment and location of the punch-through, giving rise to the "fuzziness" observed in rocking experiments to be discussed further on.[166] Furthermore since the effective drive $\mu^{-1}V_s$ coming from the Bloch-line torques is often small during the overshoot motion, particularly for high-mobility films ($\mu^{-1}V_s < 1$ Oe), nonuniformities in the coercivity can have a large effect on the equation of motion (18.26) and hence on the final bubble position. A remarkable example of the effect of static coercivity in causing scatter is a single-shot photometric trace of bubble displacement on a SmCaGeYIG film illustrated in Fig. 6.6, in which the bubble remained hung-up for 600 nsec out of a 1 $\mu$sec pulse.[211] It should also be noted that if there existed an effective "coercivity" holding back Bloch lines as they propagate within a wall, such a phenomenon would contribute to the scatter and the reduced velocities sometimes observed during overshoot.[130,365] However, the most likely mechanism for scatter is simply lack of reproducibility in the starting wall state of the bubble, particularly in the initial positions of Bloch lines. Evidence for the importance of initial wall state in translation experiments is provided by "bias jump" and "turnaround" experiments described in the next section.

A detailed application of the saturation velocity theory to the results of pulsed gradient field experiments is illustrated in Fig. 19.3a for a EuGaYIG film.[139] Final positions $X_\infty$ of bubbles are indicated by triangles while the dashed line at 2.4 $\mu$m represents the bubble position $X_T$ at the end of the pulse. The latter value is extrapolated from results like those of Fig. 6.5, which give a saturation velocity $V_s = 800$ cm/sec for this sample. Superimposed on the data in Fig. 19.3a are lines representing Eq. (18.29), plotted as $X$ versus $H_g$ for different values of $\bar\psi_x$ and fixed $T$. The number of Bloch lines in the bubble at any point in the overshoot phase of the motion in this figure is, therefore, represented by the value of $\bar\psi_x$ for the line running through that point. For example, at 4 Oe drive, $\bar\psi_x = 11$ at the end of the pulse, but it ranged from 6 to 10 long after the pulse. The hypothesis of complete punch-through described in Section 18,D would allow a maximum $\delta\bar\psi_x$ change of 2.44 after the end of the pulse (assuming the $V(\bar\psi)$ curve of Fig. 18.2b and ignoring bubble distortion). The hypothesis of complete stacking, on the other hand, implies a change of 11 ($\bar\psi_{x\infty} = 0$ at the end of the motion). The experimental results show that the change $\delta\bar\psi_x$ during the overshoot phase in the motion was 1 to 5, suggesting that some mixture of punch-through and

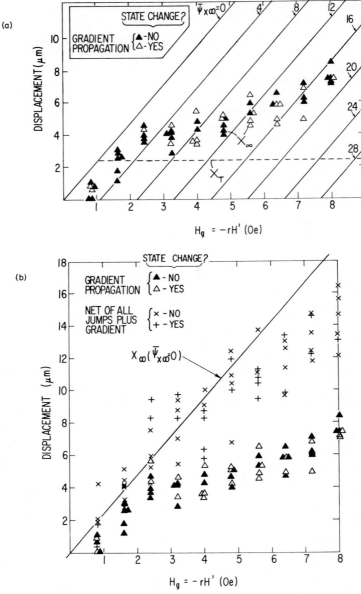

**Fig. 19.3.** (a) Bubble displacement as a function of drive $H_g = -rH'$ in a EuGaYIG film in a conventional pulsed-gradient experiment with a pulse length $T = 0.3$ $\mu$sec. Dashed line indicates extrapolated displacement at $t = T$. Diagonal lines represent Eq. (18.29) of text. (b) Same as (a) plus final positions of bubbles after repeated bias pulsing (after Malozemoff and Maekawa[139]).

stacking probably occurred, although an alternative explanation is offered by the bubble distortion effect described next. The results of Fig. 19.3a typify the data obtained so far in a variety of samples.[130,365,384]

Another kind of evidence for Bloch-line nucleation and dynamic conversion comes from the fact that bubbles propagating in a uniform field gradient tend to become elliptical perpendicular to their direction of motion.[365,385] This ellipticity is evidence of the gyrotropic reaction forces of the Bloch lines [see Eq. (14.9)]. Balancing this reaction force against the elliptical restoring force of Eq. (2.13), one finds a bubble distortion $r_2$, defined by $r(\beta) = r_0 - r_2 \cos 2\beta$, of

$$r_2 = \bar{\psi}_{xT} V_s r_0 / 6\pi M \gamma h [(l/h) - S_2(2r_0/h)], \qquad (19.1)$$

where $\bar{\psi}_{xT}$ is the total number of vertical Bloch lines and $S_2$ is defined in Eq. (2.13). Comparison with experiment on a GdTmGaYIG film[385] is shown in Fig. 19.4. Agreement is obtained for moderately short pulse times. The ellipticity implies an additional mechanism for ballistic overshoot, since Eq. (18.13) implies that a momentum change of $\pi h n r_2$ can arise simply from the relaxation back to circular shape of a bubble with $n$ vertical Bloch lines on its flanks. Preliminary estimates indicate that such a mechanism can account for most of the ballistic overshoot observed in typical experiments.[391] For longer pulses or larger drives than in Fig. 19.4, the distortion becomes very large and often triangular in shape.[392] In this case an additional mechanism for overshoot comes from the fact that the number of Bloch lines is so large that they are squeezed together, and then exchange torques provide the motive force for the overshoot. This mechanism has been described in the discussion of hard-wall mass in Section 13,C.

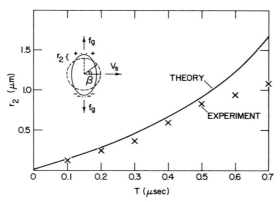

**Fig. 19.4.** Elliptical distortion $r_2$ of a bubble propagating under a gradient drive of strength $|r\nabla H_z| = 3.7$ Oe and pulse width $T$, in a GdTmGaYIG film. Theory represents Eq. (19.1) (after De Luca et al.[385]).

Further evidence for Bloch-line formation comes from the rocking experiment illustrated in Fig. 5.3 for a EuTmCaGeYIG film.[166] The images of the end points of the repetitive bubble trajectory determine the displacement $X_\infty$. For $S = 0$ bubbles, these images are generally somewhat fuzzy, presumably because of the two vertical Bloch lines rattling around in the wall of the bubble. However for $S = 1$ bubbles with no Bloch lines, the images are sharp up to a critical drive, and beyond that drive they become fuzzy as illustrated in Figs. 5.3a and 5.3b. Simultaneously with the onset of fuzziness, the curve of apparent velocity versus drive shows a break as illustrated in Fig. 5.3c, and state changes begin to occur, observed as random changes of the skew angle. Furthermore, above the break point one can observe bias-jump and turnaround effects that will be discussed in the next section and that indicate formation of vertical Bloch lines. Thus the onset of fuzziness can be interpreted as coinciding with Bloch-line punch-through. For any given drive, then, one can determine the gradient pulse width $T_p$ for punch-through. A plot of $T_p^{-1}$ versus $H_g$ is found to give a straight line with a slope $\gamma/2.5$ and an intercept $H_g = |r\nabla H_z| = 1.1$ Oe. This is in remarkable agreement with Eq. (18.25), where $\bar{\psi}_{xp} = 2.44$ from Fig. 18.2b and where $H_{cb}$, $\mu$, and $V_s$, determined in separate experiments, are 0.2 Oe, 1360 cm/sec Oe, and 1280 cm/sec, respectively for the EuTmCaGeYIG film.

Figure 5.3c also dramatizes the problems in interpreting bubble-translation data taken without high-speed photography. One curve is the reciprocal pulse time taken by adjusting the pulse width $T$ until the bubble displacement is a fixed value $X_\infty$, in this case 5.4 $\mu$m. The higher curve is reciprocal pulse time for half the displacement of the first curve, i.e., for $X_\infty = 2.7$ $\mu$m. Surprisingly the breakpoint appears almost twice as high in drive field. This demonstrates graphically that the observed breakpoint does not indicate the intrinsic critical drive and critical velocity for nonlinearity since the apparent values shift drastically with different experimental conditions. The shift may be understood in terms of Eq. (18.29) of the saturation-velocity theory, which may be rewritten as

$$X_\infty/T_p = V_s\{1 - \mu\gamma^{-1}\bar{\psi}_{xp}X_\infty^{-1}[V_s/(\mu H_{cb} + V_s)]\}^{-1}. \qquad (19.2)$$

If $X_\infty$ is large, then the apparent velocity at the breakpoint is just $V_s$. However, if $X_\infty$ is small enough, i.e., less than $\mu\gamma^{-1}\bar{\psi}_{xp}$ (assuming $H_{cb} = 0$), then the apparent velocity at the breakpoint goes to infinity and no nonlinearity will be seen at all. Since $X_\infty$ is often taken to be the bubble size[166,207] and $\mu\gamma^{-1}\bar{\psi}_{xp}$ is often of order a few microns, this apparent suppression of non-linearity can easily occur, particularly in experiments on bubbles smaller than a few microns. This concept is also significant for bubble-lattice device operation, for if drive pulses could be kept short enough, large drives could be used without risking punch-through and hence state changes.

Next let us consider the effects of ballistic overshoot on the deflection effect.[393] If we ignore coercivity ($H_{cb} = 0$), we can extend the treatment of Section 18,E to the case of a bubble with $S \neq 0$. In this limit, the final vector position $X_\infty$ does not depend on the nature of the function $V(\bar{\psi})$. For, integration of Eq. (18.22) between times $t = 0$ and $\infty$ is facilitated by the relation $\int_0^\infty V dt = X_\infty$. Assuming a cylindrical bubble of radius $r$, a drive $H_g = -r\nabla H_z$, a pulse time $T$ and ignoring coercivity, we obtain

$$\bar{\psi}_\infty = -\gamma r \nabla H_z T - \gamma \mu^{-1} X_\infty + 2r^{-1} S z_0 \times X_\infty, \qquad (19.3)$$

where $\bar{\psi}_\infty$ is the change in momentum from the beginning to the end of the experiment. If there is no Bloch-line reorientation or punch-through, $\bar{\psi}_\infty = 0$, in which case Eq. (19.3) is represented graphically by a right triangle, with legs $\gamma\mu^{-1} X_\infty$ and $2r^{-1} S X_\infty$, and hypotenuse $\gamma H_g T$. Thus the net deflection angle is

$$\rho_\infty = \tan^{-1}(2\mu S/\gamma r), \qquad (19.4)$$

and the apparent mobility in a conventional gradient experiment using still photography is

$$|X_\infty|/H_g T = \mu/[1 + (2\mu S/\gamma r)^2]^{1/2}. \qquad (19.5)$$

These are precisely the results that held in the absence of all inertial effects (see Section 13). Thus inertial effects are often obscured in conventional gradient propagation because corrections to Eqs. (19.4) and (19.5) are needed only to the extent that coercivity ($H_{cb} \neq 0$) and punch-through ($\bar{\psi}_\infty \neq 0$) play significant roles.

However while the pulse is on and assuming the velocity is saturated, the deflection angle, according to Eq. (19.3), is

$$\rho_{t<T} = \sin^{-1}(2SV_s/\gamma r H_g). \qquad (19.6)$$

In contrast to Eq. (19.4), $\sin \rho_t$ depends inversely on drive if $V_s$ is a constant. An example of this effect along with comparison to theory is shown in Fig. 14.4 for an $S = 1$ bubble in a EuGaYIG film.[362] Even more remarkable is the fact that after the drive pulse turns off, the angle increases noticeably, as indicated by the $\rho_\infty$ points in the figure.[357,362] A simple way of understanding this effect[384] is to note that during Bloch-line wind-up, the Bloch lines are oriented symmetrically around an axis parallel to $V$ (i.e., $\bar{\psi}\|V$), which points along the angle of Eq. (19.6). But after the pulse turns off, the Bloch lines themselves provide the driving force for the velocity, and the bubble will tend to deflect off from this direction, thus causing a gradual increase in skew angle and hence a spiraling motion. Coercivity and Bloch-line punch-through may prevent the bubble from reaching its final position, thus giving a reduced deflection angle from the simple prediction of Eq. (19.4).

## C.  Bias-Jump and Turnaround Effects

In this section we discuss several additional experiments that have given further evidence for Bloch lines and stored momentum in gradient-propagated bubbles. First we consider what happens when a uniform bias pulse is applied to a bubble that has previously been gradient-propagated. High-speed photography of the transient bubble shape during the pulse has revealed that for expanding or contracting pulses, the bubble becomes elliptical with its major or minor axis, respectively, along the original direction of motion.[138,139] These observations imply that lower velocity regions exist on the flanks of the bubble relative to the original direction of motion. Since wall regions containing Bloch lines are known to move slower, one concludes that Bloch lines have been left at the flanks of the bubble, as the punch-through model would predict.

Another remarkable effect in this experiment is that during the pulse the bubble tends to move forward a net distance along the original direction of translational motion.[139,384] The direction of motion is the same irrespective of the polarity of the bias-field pulse. This fact speaks eloquently for the plausibility of the "stored momentum" concept. The "bias jump" or "bubble creep" can be up to 3 $\mu$m in size, in the EuGa film of Fig. 19.3b, and if multiple pulses are applied the bubble jumps forward by decreasing amounts each pulse until it finally comes to rest. The rest positions after a series of pulses are shown as $\times$'s and $+$'s in Fig. 19.3b. Net displacements of 9 $\mu$m beyond the end position of the original gradient translation are seen to occur. The bias jump may be understood from the fact that the bias field causes unwinding Bloch-line clusters on opposite sides of the bubble to move towards each other and annihilate. The inward motion of the Bloch lines exerts a force to move the bubble forward, as follows from the gyrotropic force formula $2\pi M\gamma^{-1}\mathbf{t} \times \mathbf{v_L}$. If one ignores coercivity, radius changes, and skew deflection, Eq. (19.3) predicts a net displacement of $\mu\gamma^{-1}\delta\bar{\psi}$, where $\delta\bar{\psi}$ is the momentum change due to Bloch-line annihilation. This displacement, plus the initial gradient propagation distance, should give the total distance $\mu H_g T$ the bubble would have gone if no punch-through had occurred to start with. This prediction is represented by the line $\bar{\psi}_{x\infty} = 0$ in Fig. 19.3b and is in only rough agreement with experiment, presumably because of the nonnegligible effects of coercivity and radius changes.[139]

A related effect has been observed in the "deflectometer" geometry described in Section 6,C.[214] A bubble in a EuGaYIG film was propagated up to the potential well at the edge of a stripline, at which point it ran out into a stripe if the current in the stripline was kept on long enough. When the current was turned off, the stripe experienced a uniform bias field causing it to contract. But in addition to contracting, the domain was found to move forward along its original direction of propagation to a point as much as 14 $\mu$m beyond the original position of the potential well. It is likely that this motion is a result of momentum release from Bloch lines wound up during the original propagation phase of motion.

Remarkable momentum effects are also observed by comparing gradient propagation of different starting bubble states. In one study on a GdTmGa-YIG film, for example, three different kinds of $S = 1$ bubbles, namely $\sigma^+$, $\sigma^-$, and $\chi$, were identified by automotion (see Section 20) in an in-plane field along the $y$ direction.[393] Then the in-plane field was removed and the bubbles were propagated by 0.3 or 0.9 $\mu$sec pulses of bias field gradient in the $x$ direction, giving the results shown in Fig. 19.5. The propagation distances and deflection angles are noticeably different. These differences can be understood in terms of the different starting momenta $\bar{\psi}_x = -2$, 0, and 2 [Eqs. (18.13) and (18.14)] for the $\sigma^-$, $\chi$, and $\sigma^+$ bubbles that have the presumed starting states shown in the insert. The $\chi$ bubble moves straight along its skew direction because initially it has no momentum. The $\sigma^+$ bubble moves farther along $x$ and has a smaller deflection angle because its initial momentum was oriented along the $x$ direction. This effect can also be viewed as arising from the reorientation of the two Bloch lines of the $\sigma^+$ bubble from their original positions along $y$ to positions on the flanks perpendicular to the skew direction of motion. The $\sigma^-$ bubble moves the least far of the three bubbles. This effect can be understood from the gyrotropic forces on the two Bloch lines, which point inward in their starting configuration. Thus the

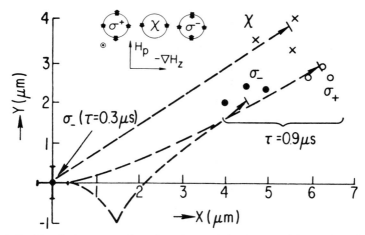

**Fig. 19.5.** Net displacements of $\chi$, $\sigma^+$, and $\sigma^-$ bubbles due to a field gradient of strength $|r\nabla H_z| = 0.8$ Oe and pulse width $\tau$ in a GdTmGaYIG film. Dashed curves show theoretical trajectories (after Maekawa and Dekker[393]).

Bloch lines are in unstable positions and they will rotate around the wall of the bubble and interchange positions (assuming they do not collide with each other and annihilate). The momentum change for this process is $\delta \bar{\psi}_x = 4$, which implies a reduction in displacement along the $x$ direction of approximately $4\mu\gamma^{-1}$ [Eq. (19.3), ignoring the deflection effect and coercivity]. Since $\sigma^+$ bubbles do not incur this momentum change, one may compare $4\mu\gamma^{-1}$, which is 4 $\mu$m for the film in question, to the observed 2.5 $\mu$m difference in dispacement between $\sigma^+$ and $\sigma^-$ bubbles. A more complete theoretical treatment, using equations to be described in Section 20, gives the dashed lines shown in the figure.

These momentum changes are presumably also responsible for the "turn-around" effect, which occurs when gradient polarity is changed from one pulse to the next. Instead of changing direction smoothly, bubbles are often observed to deviate to the side by several microns or even to keep going in their initial direction.[175,389,393,394] All these observations emphasize the need for careful and consistent preparation of bubbles in gradient propagation experiments, if scatter in $X_T$ and $X_\infty$ is to be minimized. One common procedure is to select a bubble on the basis of skew angle, to bring it up to the center-line of the gradient with pulses as little as possible above the coercive threshold, and to propagate it in the same direction as the preparatory motion.[310]

## D.   EFFECT OF CAPPING LAYERS AND IN-PLANE FIELDS

Next we describe the effect of capping layers on velocity saturation and ballistic overshoot. High-speed photography studies of ion-implanted garnet films have so far shown very little difference from the saturation velocity of as-grown films.[365,384] Even a triple-layer film with two low-anisotropy layers of GdYIG sandwiching a EuGaYIG bubble film[376] showed a saturation velocity of roughly 1300 cm/sec, not much larger than velocities of similar single layer films.[395] These results suggest that capping layers do not suppress Bloch-line nucleation (contrary to earlier belief based on the remarkable improvements capping layers gave in device operation). The effect of capping layers on overshoot is illustrated in Fig. 19.6 by the measurements of apparent velocity in "rocking" experiments (Section 6,C) on EuGaYIG films.[382] The overshoot increases, presumably because Bloch loops pile up at the capping layers rather than punching through. Similar results using the conventional gradient-propagation experiment have been reported on permalloy-coated GdTmGaYIG, multilayer GdBiGaYIG and ion-implanted EuTmGaYIG films, where apparent velocities $X_\infty/T$ increased and scatter in $X_\infty$ diminished.[308–309b] However, contradictory results have been reported in other measurements of EuGa, EuTmCaGe, and other YIG films,[365,384,395a] which show that a capping layer may *decrease* $X_\infty$. A possi-

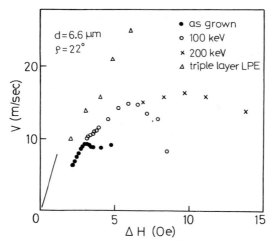

**Fig. 19.6.** Apparent bubble translation velocity $V = |\mathbf{X}_\infty|/T$ in a "rocking" experiment on as-grown, ion-implanted and triple-layer EuGaYIG films as a function of drive $\Delta H = 2r|\nabla H_z|$ (after Konishi et al.[382]).

ble explanation is that punch-through occurs with Bloch-point nucleation leading to Bloch-line annihilation and consequent reduction in stored momentum. One may speculate that the experimental differences described above are in some way connected with the "strength" of the capping layer, determined by its thickness and magnetization. In-plane domain structure can also give rise to apparent asymmetries in forward and backward motion.[309] In one dramatic case a closure domain in an ion-implanted EuYbCa-Ge YIG film was pinned at the edge of a conductor. When an in-plane field was applied, instead of the closure domain contracting towards the bubble, the bubble was dragged 6 μm towards the pinning point before the closure domain broke loose from that point.[396] The effect of capping layers on device performance will be considered further in Section 21,B.

Now we turn to the effect of in-plane fields on bubble propagation. Results from a high-speed photography study of an $S = 0$ bubble in a EuGaYIG film are shown in Fig. 19.1.[299,310] While no velocity peak in $X_T$ exists for $H_p = 0$ (see Section 19,A), two peaks are observed with $H_p = 40$ Oe. These peaks can be understood with reference to the insert of Fig. 19.1. As long as the bubble velocity is sufficiently small, the static in-plane field force $2MH_p\pi\Delta_0$ [Eq. (8.11)] prevents the Bloch lines from rotating away from their static equilibrium positions on opposite sides of the bubble at $A$ and $B$. If in addition, no new Bloch lines are nucleated, the velocity should follow the simple linear relation $V = \mu H_g$ and no significant overshoot should occur. This can explain the low drive ($H_g \leq 1.2$ Oe) region of Fig. 19.1. However the gyrotropic force $2\pi M\gamma^{-1}\mathbf{t} \times \mathbf{V}$ [Eq. (14.9)] points down on both

Bloch lines in the figure. Therefore the Bloch line at $B$ is dynamically stable, while the Bloch line at $A$ becomes unstable if the gyrotropic force becomes large enough to overcome the in-plane field force. Balancing the two forces, one finds the critical velocity and field[250]

$$V = \gamma \Delta_0 H_p \qquad H_g = |r \nabla H_z| = \alpha H_p. \qquad (19.7)$$

Given $H_p = 40$ Oe and an FMR damping constant $\alpha = 0.003$, one predicts a velocity peak at 1.2 Oe above the coercive field, in reasonable agreement with the experimental results in Fig. 19.1. Experiments varying $H_p$ and $\alpha$ also bear out these simple predictions.[299]

Above the peak, the unstable Bloch line at $A$ rotates around the back side of the bubble and joins the other Bloch line at the dynamically stable position $B$. Then the velocity can rise again until new Bloch lines are nucleated, giving rise to the second peak. In this intermediate region, overshoot occurs because after the drive pulse ends, the Bloch lines relax to their original static position. At drives above the threshold for generation of new Bloch lines, the velocity becomes saturated. The predicted threshold [Eq. (15.16)] is 2.5 Oe, in good agreement with experiment. [299,378] Careful bias compensation is required in these experiments because the threshold can be shifted by bias pulses.[213]

A similar high-speed photography study has not been reported for $S = 1$ bubbles. However, in conventional bubble-propagation experiments on EuGa and SmCaGeYIG films, the apparent velocity $(X_\infty/T)$ of $S = 1$ bubbles did not increase as rapidly as that of $S = 0$ bubbles, when an in-plane field was applied.[261,397] The apparent velocity difference was as much as a factor of two for a EuGaYIG film with 80 Oe in-plane field. The reason for this difference has not been established, although it may be related to momentum changes arising when 1H bubbles convert to the (1, 2) or $\sigma$ state.

A demonstration of such state changes for $S = 1$ bubbles has been made in the following way. A static in-plane field of sufficient magnitude transforms chiral bubbles into 1H bubbles (Section 8,E) with Bloch lines on different sides, depending on the chirality, as shown in Fig. 19.7. A pulse of bias field causes precession and hence motion of the Bloch loop in a direction dependent on the $z$ field polarity. The direction can be determined from $\mathbf{f}_g \sim \mathbf{t} \times \mathbf{V}$ in the usual way [Eq. (14.9)]. If punch-through occurs, one ends up with $\sigma^+$ or $\sigma^-$ states as shown. Next, if the in-plane field is removed and a bias pulse applied, the vertical Bloch lines can be made to rotate towards each other and annihilate. This is the familiar bias-jump effect described in the previous section. According to Eqs. (18.13) and (18.14), there is a change in momentum of $\delta \bar{\psi}_x = 2$ between the $\chi$ and $\sigma$ states. Thus one expects displacements of the bubbles of order $\Delta \alpha^{-1} \delta \bar{\psi}_x$, and in studies of GdTmGa and SmCaGeYIG films,[229,250] displacements of order 1 $\mu m$ were indeed observed.

We have earlier seen (Section 16, B) that $\chi^+$ and $\chi^-$ states can be transformed from one to the other by "chiral switching" with bias pulses.[229] We

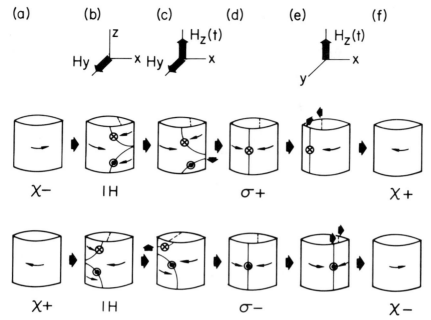

**Fig. 19.7.** Sequence of schematic bubble states in dynamic interconversion of $\chi^+$, $\chi^-$, $\sigma^+$, and $\sigma^-$ states, using steady in-plane and pulsed bias fields.

will see in Section 20 that the phenomenon of automotion provides an exceedingly simple means of distinguishing $\sigma^+$ and $\sigma^-$ states. These effects, plus the processes for $\chi \rightarrow \sigma$ and $\sigma \rightarrow \chi$ conversions described in the previous paragraph, offer a complete system for controlling and identifying $S = 1$ substates. In particular one can now identify whether a $\chi$ bubble is $\chi^+$ or $\chi^-$ by converting it to $\sigma$ by a bias pulse of known polarity and then using automotion to distinguish whether the $\sigma$ state is $\sigma^+$ or $\sigma^-$. Such a procedure underlies the chiral-switching results on a GdTmGaYIG film shown in Fig. 16.3.[229]

An interesting situation arises when a film has a capping layer on one surface and an in-plane field is simultaneously applied. Returning to Fig. 19.7 and assuming the capping layer is on the top surface, one concludes that the $\chi^- \rightarrow \sigma^+$ transition occurs, but the $\chi^+ \rightarrow \sigma^-$ transition is prevented if punch-through cannot occur at the top surface. A similar situation arises if gradients are applied to move the bubble in a direction perpendicular to the in-plane field. If the motion is in the $+x$ direction in Fig. 19.7, the 1H state is stabilized by the cap, irrespective of the original chirality. However, if the motion is in the $-x$ direction, the $\sigma^+$ state is stable. Thus the bubble state depends on the direction of bubble motion.[250] These predictions have been confirmed in measurements on SmLuCaGeYIG films. To distinguish the 1H and $\sigma^+$ states, a quasi-static cap switch was induced by raising the in-plane field (Section 9,D). As indicated in Fig. 9.6, this process generates the $\frac{1}{2}*$ and (0, 2) states, respectively, which can then be distinguished in propagation skew-angle measurements.

We have seen that in-plane fields have a strong effect on the dynamics of different bubble states. Because of the interest in using these states for coding a bubble-lattice file, their dynamic stability has been studied, both in the idealized environment of a pulsed gradient propagation experiment and in a more realistic device environment. The effect of in-plane field on the stability of some of these states was shown in Fig. 9.6 for a constant gradient drive. The drive dependence of the stability of some of these states is shown in Fig. 19.8, as determined in a "rocking" experiment.[165,398,399] The results in these two figures suggest that there are several different processes that limit the state stability.

Stability of many states is limited by static "cap-switch" processes discussed in Section 9,D. Such transitions are essentially independent of gradient drive or direction of motion relative to the in-plane field. For ex-

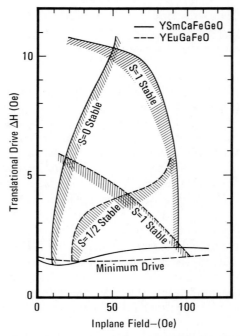

**Fig. 19.8.** Stability diagram for $S = 1$, 0, and $\frac{1}{2}$ states, as determined by a rocking experiment, as a function of gradient drive $\Delta H = |2r\nabla H_z|$ and in-plane field $H_p$, for ion-implanted SmCaGe-YIG and EuGaYIG films (after Brown and Hsu[398]).

ample, such a process can account for the vertical stability threshold for $S = 1$ bubbles in SmCaGeYIG at $H_p \sim 90$ Oe in Fig. 19.8. Stability of other states is limited by Bloch-line annihilation. For example, an $S = \frac{1}{2}$ state has two Bloch lines, one with, the other without a Bloch point. If these two lines collide, they annihilate and the state converts to $S = 1$. The threshold for this process is determined by Eq. (19.7), which determines the field at which a Bloch line can rotate around the bubble wall.[250] Such a process is sensitive, therefore, to the orientation and strength of the gradient drive and may account for the $S = \frac{1}{2}$ contour in the EuGaYIG sample of Fig. 19.8. Yet another limitation on state stability is set by nucleation of new Bloch lines, for which the threshold is given by $H_g = \mu^{-1}[V_n + (\pi/2)\gamma\Delta_0 H_p]$ from Eq. (15.16). Taking $V_n \sim V_{po}$ and $H_p \sim 50$ Oe and using known material parameters, one predicts a maximum drive of $\Delta H = |2r\nabla H_z| = 21$ and 11 Oe for the SmCaGe and EuGa samples in Fig. 19.8, respectively. These values are more than double the observed maximum thresholds at $\sim 10$ and $\sim 4$ Oe. Considering other experiments discussed earlier (Sections 16,A and 19,A), this discrepancy is no surprise since Bloch-line nucleation thresholds generally seem to be lower than predicted by theory. The difference in the maximum drive for the two samples may be traced primarily to the factor of two difference in damping constant, SmCaGe having $\alpha = 0.1$ and EuGa having $\alpha = 0.05$. Thus high damping appears to be favorable for state stability. Of course high damping also reduces the various dynamic effects like skew-deflection angle [Eq. (8.26)] and bias-jump size [Eq. (8.27)] on which discrimination of the states could be based. Thus some practical compromise must be made if these effects are to be used in a bubble-lattice device.

Like in-plane fields, in-plane anisotropy should have strong effects on bubble propagation, although much less work has been done on this topic. One example is afforded by a study of EuGa, GdTmGa, and SmGaYIG films with small misorientations ($<1°$) of the surface normal from the [111]-direction. Such films have orthorhombic in-plane anisotropy (see Section 1,C). Propagation of bubbles in different directions in the plane revealed no variation of the saturation velocity, but there were maxima in the size of the overshoot when the direction of motion was along the easy anisotropy axis.[365] This effect can be understood from the fact that the in-plane anisotropy provides a force on the Bloch lines (Section 8,A), which pushes them towards the nearest point of intersection of the bubble circumference with the diameter lying in the easy direction.

The effects of in-plane field and in-plane anisotropy are particularly evident in the runout of stripes in chevron expanders of field-access bubble devices (Section 6).[191-193,400,401] In one study the stripe-head velocity in a SmGaYIG film with a 1.4° substrate misorientation was found to be 25% larger for expansion perpendicular to the easy axis than parallel to it.[400] The velocity ratio in the two directions was in approximate agreement with the predicted ratio of critical velocities given by Eq. (15.15) but the absolute velocities were considerably higher. This discrepancy presumably arises from the in-plane fields $H_p$, either externally applied or emanating from the permalloy structure, which are expected to increase the critical velocities by $(\pi/2)\gamma\Delta_0 H_p$ [Eq. (15.16)]. In a study of stripe expansion in a chevron expander in SmLuCaGeYIG, a linear-mobility region culminated in a peak velocity that increased with applied in-plane field at a rate of 130 cm/sec Oe.[193] This compares reasonably well to the predicted $(\pi/2)\gamma\Delta_0 = 180$ cm/sec Oe.

Similar experiments on GdCoMo amorphous films also gave a linear mobility except when the SiO$_2$ spacer between the amorphous film and the permalloy drive elements was made small (1500Å).[191] Under this latter circumstance a peak expansion velocity of 50,000 cm/sec and a velocity saturation of 10,000 cm/sec were observed, which may be compared to a predicted Walker velocity $V_w = 60,000$ cm/sec and a Bloch-line instability velocity $V_{po} = 16,000$ cm/sec for this material. The agreement between the peak velocity and Walker's velocity is doubtless fortuitous because of the in-plane fields present in the device.

## 20. Gradientless Propulsion or Automotion

### A. EXPERIMENTS

In the previous sections we have described phenomena such as bubble overshoot and the bias-jump effect in which bubbles move for a time without external gradient driving forces. This "inertial" motion arises from internal forces within the domain wall usually related to Bloch lines. In these cases the accumulated bubble displacement is finite (as long as $\alpha \neq 0$). By contrast, in this section we describe a class of phenomena in which essentially unlimited bubble displacement can occur in the absence of external gradient driving force. Such motion is called "gradientless propulsion" or "automotion." It relies both on the presence of internal forces within the domain wall, excited by a time-dependent *homogeneous* external field, and in addition on a non-linear loss mechanism such as coercivity.

We have already encountered examples of such an effect in Section 13,B. Pulses of bias field, which by symmetry can exert no external force on the orientation of a dumbbell-shaped hard bubble, cause rotation of the dumbbell ad infinitum, as long as bias pulses continue to be applied (see Fig. 13.4). We saw there that the effect relied on the simultaneous presence of Bloch lines and coercivity. Similarly, clumps of Bloch lines could move ad infinitum along stripe walls in the presence of repetitive bias pulsing (Section 13,A).

A related effect, but which *does* involve gradients, is motion of bubbles along a one-dimensional bias-field trough created by a straight conductor carrying a constant current. When gradient pulses are applied perpendicular to the axis of the trough, bubbles move along the trough, in a direction depending on the wall state.[402] A similar effect can be achieved by merely applying rf currents to a straight conductor, without any dc currents, whereupon bubbles move steadily along the edge of the conductor.[403] The necessity of a gradient drive in this experiment was not established. With respect to both of these trough experiments, the fact remains that the average motion was orthogonal to whatever gradient was present.

Also remarkable is the observation that a hexagonal bubble lattice can be made to rotate in a rotationally symmetric environment.[404] For example, a loose lattice is stabilized by a continuous train of current pulses in a pancake coil shaped and located so as to create a broad bias-field well containing the bubbles. Alternatively the lattice may be more firmly confined by a circular etched groove and pulsed with currents in the pancake coil. In both of these cases, bias-field pulses cause lattice rotation around the center of the confinement pattern. Rotations of up to 0.7° per bias pulse have been achieved for a lattice of 5 $\mu$m bubbles, in an as-grown EuGaYIG film with a circular groove 250 $\mu$m in diameter and subjected to bias pulses of 9 Oe.[404] Rotation can also be induced by sinusoidal rf bias fields, and its sense depends on the frequency. In the case of lattice rotation, the large scale of the drive coil (e.g., 1 mm I.D.) implies smaller (radial) gradients than in the typical straight-conductor experiments. But again, the possibility of an essential role for gradients could not be ruled out, and indeed figured in attempts at an explanation.[404]

Gradients were entirely ruled out in a study of isolated bubbles subjected to different combinations of spatially uniform bias or in-plane fields. For example, uniform in-plane field pulses with fast rise and slow fall (peak of 30 Oe) were applied to $S = 1$ bubbles in a EuGaYIG film.[405] Three groups of bubbles were observed, one group that did not move, and two others that moved roughly 0.5 $\mu$m per pulse in opposite directions at a roughly 45° angle to the in-plane field.

In another configuration, a uniform dc in-plane field was applied in conjunction with bias-field pulses to a GdTmGaYIG film. Three groups of $S = 1$ bubbles were also found, one that did not move, but two that moved in opposite directions nearly perpendicular to the in-plane field.[406] The size of the displacement per pulse is shown in Fig. 20.1 as a function of the strength of the in-plane field and bias pulse field.[249,406] It is noteworthy that except for very short pulses, the displacement is independent of pulse width. In the same environment of a uniform dc in-plane field and a pulsed bias field, $S = 0$ bubbles were found not to move, while $S = +\frac{1}{2}$ bubbles moved at 45° degrees to the in-plane field and $S = -\frac{1}{2}$ bubbles moved along the in-

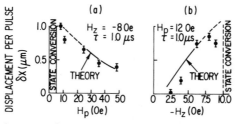

**Fig. 20.1.** Net displacement of a $\sigma$ bubble in a GdTmGaYIG film subjected to a constant in-plane field $H_p$ and a uniform bias pulse field of strength $H_z$ and length $\tau$. Dashed lines indicate theoretical calculations based on Eqs. (20.5)–(20.8) (after Argyle et al. [249]).

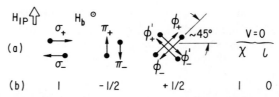

**Fig. 20.2.** Overall pattern of automotion observations for free isolated bubbles in the presence of a constant in-plane field and pulsed bias field. (a) Scheme of automotion velocity vectors. (b) Measured effective winding number $S$ (after Argyle et al.[249]).

plane field. Figure 20.2 shows the qualitative pattern of the observations and the symbols assigned to the respective wall states.[249] This effect has been termed "automotion."

Yet another configuration (known as "type II") utilizes a dc in-plane field and an asymmetric (fast-rise slow-fall) pulsed in-plane field in some other direction. Displacements of 0.8 $\mu$m/per pulse were observed with a dc field of 8 Oe and a pulsed field of 20 Oe at right angles to the dc field.[407]

These phenomena make clear that isolated bubbles with appropriate internal wall states can be moved in any direction at will, provided uniform in-plane fields are available with any orientation in the plane of the film. Since such motion does not require small driving structures such as permalloy elements, it has the potential of alleviating lithographic constraints in the construction of bubble devices, particularly of the lattice type. Gradientless propulsion is also extremely useful in identifying bubble states, particularly when coupled with skew propagation to identify the $S$ state.[229,249,408] As we shall see here, the two $\sigma$ or (1, 2) states are easily identified, and this has made possible the detailed studies of state interconversions described earlier in Section 19,D.

## B. MECHANISMS OF GRADIENTLESS PROPULSION

Next we turn to mechanisms of gradientless propulsion. Since these effects can continue ad infinitum, it is clear that the energy required must somehow be supplied by the time-dependent external fields. The internal bubble wall state acts like a transducer, converting part of the external field energy into the viscous and coercive dissipation attendant on bubble translation.

Essentially, gradientless propulsion can be viewed as a partial rectification of primarily oscillatory motion. The possibly complex wall state of the bubble somehow reacts to the periodic drive in such a way as to cause bubble displacement (or rotation in the case of the dumbbell domain) having alternating sign. In the absence of coercivity and other nonlinear loss mechanisms, the time-average displacement vanishes, as proved below for the free-bubble case. However, in the presence of nonlinear drag, a positive mean displacement arises if one arranges for the drag force retarding the motion to be less effective during the positive- than during the negative-velocity phase.

Let us first consider those cases in which pulsed or ac gradients are impressed along one axis ($y$) and the automotive progression follows a zigzag course along the orthogonal axis ($x$) as shown in Fig. 20.3. For example, in the case of bubbles in a field well created by a current-carrying conductor, gradients act perpendicular to the conductor. Similarly in the rotating lattice, gradients pointing radially inward arise from the containment well. Let us assume the bubble has a nonzero winding number. Then as each applied pulse of $dH_z/dy$ impels the bubble in the $y$ direction, it deflects in the $x$ direction. The restoring gradient causing the $y$ coordinate of the bubble to relax back to its initial value is weaker. For example, in the case of lattice rotation, the restoring gradient arises from the weak internal restoring forces of the lattice. Equations developed earlier for the skew angle of gradient-propagated bubbles show that the greater the drive relative to coercivity, the greater the deflection angle [e.g., see Eqs. (13.9) or (14.11)], assuming that the velocity is subcritical. Thus the bubble zigzags along the $x$ direction in steps proportional to the difference between the forward and backward skew angles. In this case, coercivity is the particular nonlinear drag mechanism indispensible to automotion. This model cannot explain the automotion occurring when the applied gradient is sinusoidal in time. The latter mode may be explained by invoking the dependence of viscous drag on bubble radius. Lack of knowledge of the actual gradients and bubble states has prevented quantitative comparison of experiment with the theory.

The experimental configurations considered in the previous paragraph exert in-plane fields on the bubbles. As the studies on isolated bubbles have made clear, these fields can also give rise to gradientless propulsion. Let us now turn for the remainder of this section to the free-bubble case in which no in-plane gradients of either static or dynamic character exist, and for which the most detailed understanding has been achieved.

To prove that a nonlinear drag mechanism is indispensible for linear automotion of free bubbles, consider the time average of the force-balance

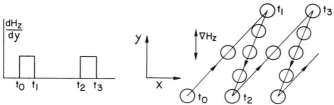

**Fig. 20.3.** Schematic gradient drive versus time and corresponding bubble motion, illustrating a possible mechanism for bubble-lattice rotation (after Argyle *et al.*,[249] copyright 1976 by International Business Machines Corporation, reprinted with permission).

equation (18.22), which holds in the absence of Bloch points, over one period $T$ of the applied force. One thus finds

$$\Psi(T) - \Psi(0) = 2\pi Sh\mathbf{z}_0 \times \bar{\mathbf{V}}T + (\gamma/2M) \int_0^T \mathbf{F} dt. \qquad (20.1)$$

Here $\bar{\mathbf{V}}$ is the mean domain-center velocity (the automotion velocity) that we are trying to find, and $\mathbf{F}(t)$ is the total instantaneous force on the domain. On the assumption that the motion is truly periodic, the left-hand side of Eq. (20.1) must vanish. By the hypothesis of no gradients, the only term in $\mathbf{F}$ is the dissipative drag force $\mathbf{F} = \mathbf{F}_d$ of whatever origin. This equation implies that automotion requires a nonlinear $\mathbf{F}_d$. For, suppose $\mathbf{F}_d$ had the simple linear form

$$\mathbf{F}_d = -C\mathbf{V}(t), \qquad (20.2)$$

where $C \, (>0)$ is a constant. Then Eq. (20.1) reduces to

$$4\pi M \gamma^{-1} Sh\mathbf{z}_0 \times \bar{\mathbf{V}} - C\bar{\mathbf{V}} = 0. \qquad (20.3)$$

Since the vectors $\bar{\mathbf{V}}$ and $\mathbf{z}_0 \times \bar{\mathbf{V}}$ are orthogonal, the only solution of Eq. (20.3) is $\bar{\mathbf{V}} = 0$, implying that automotion does not occur.

But suppose that $\mathbf{F}_d$ is replaced by some more complicated form of drag, e.g., due to coercivity or perhaps still of viscous character but taking into account the proportionality of $\mathbf{F}_d$ to a time-dependent bubble radius. Then the form of Eq. (20.3) is generally changed and $\mathbf{V} \neq 0$ is possible.

To illustrate how coercivity gives rise to automotion of bubbles having appropriate wall states, consider the example of a dc in-plane field $H_p$ and a pulsed bias field $H_z(t)$.[249] Assume the bubble has a $\sigma$ or $(1, 2)$ state as illustrated in Fig. 20.4. The in-plane field stabilizes the Bloch lines on the diameter along the in-plane field direction. A bias pulse acting to expand the bubble, as shown schematically in phase 1 of the figure, also causes spin precession within the Bloch lines so that they circle around the domain toward each other in the first phase of the motion. These Bloch-line displacements toward the $y = 0$ plane decrease the momentum $\bar{\psi}_x$ of the bubble according to Eq. (18.13) and cause a corresponding reaction of the bubble position towards the $x$ direction. This tendency may also be understood in terms of the $x$ component of force $\pm 2\pi M \gamma^{-1} \dot{y}_i$ [Eq. (14.9)] arising from the $y$ component of the Bloch-line motion. As in the case of ballistic overshoot, the bubble displacement is of order $\Delta\alpha^{-1}\delta\bar{\psi}_x$, where the maximum value of $\delta\bar{\psi}_x$ is 2 if Bloch-line annihilation is to be avoided. The rapid radial motion of the wall simultaneously reduces the coercivity during phase 1. During phase 2 of the motion, the bubble has nearly reached its equilibrium diameter while the bias field pulse is still on, but the Bloch lines are now relaxing back toward their original field-stabilized positions. During this phase the momentum increases and the bubble tends to move backwards. However, the coercivity-reducing effect of the radial bubble motion is weaker; so the coercivity is

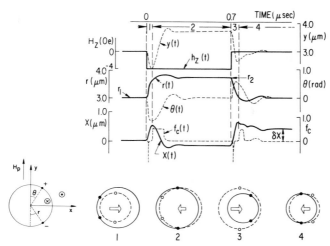

**Fig. 20.4.** Theoretical calculations, using Eqs. (20.5)–(20.8), of bubble displacement $X$, radius $r$, and Bloch line position $\theta$ of a $S = (1, 2)$ bubble as a function of time, in response to a pulse of bias field $H_z(t)$, with a dc in-plane field $H_p$. The sketches below illustrate the definitions of the dynamical variables for the theory and also the four phases of bubble and Bloch line motion during the pulse (after Argyle et al.[249]).

higher and prevents most of the backward motion (step 2 of Fig. 20.4). A similar two-phase process occurs following on the trailing edge of the pulse. Thus we see that the gradientless propulsion is the net of displacements associated with the leading and trailing edges of the pulse. This explains why the displacement per pulse is essentially independent of pulse width for widths greater than the relaxation time for positions of Bloch lines.

To construct a more detailed model of this process, we first quantify the coercivity-reduction effect we have mentioned above.[249] The net coercive reaction pressure that opposes wall motion (Fig. 20.5) is expressible as

$$P_c = -2MH_c \operatorname{sgn} V_n, \qquad (20.4)$$

where the coercive field $H_c$ ($>0$) is a constant and $V_n = \dot{r} + \dot{X} \cos \beta$ is the normal wall velocity at any point $\beta$ on the wall. By integrating over the cylinder surface we find the total $x$ component of coercive reaction force (for unit film thickness)

$$F_c^x = -8MH_c r(\operatorname{sgn} \dot{X})[1 - (\dot{r}/\dot{X})^2]^{1/2}, \qquad |\dot{r}| < |\dot{X}| \qquad (20.5)$$

$$F_c = 0, \qquad |\dot{r}| > |\dot{X}|.$$

An interesting property of $F_c^x$ is that its value depends strongly on the ratio of two kinds of motion, $\dot{X}$ and $\dot{r}$. In particular, if $|\dot{r}|$ exceeds $|\dot{X}|$, translational coercivity is completely suppressed. The physical reason for this is made

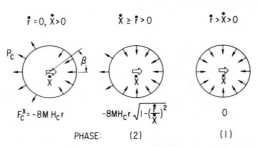

**Fig. 20.5.** Local wall-coercivity force (arrows) depending on relative radial ($\dot{r}$) and translational ($\dot{X}$) wall velocities. $F_c^x$ is the net translational coercive reaction force (after Argyle et al.[249]).

clear in Fig. 20.5, where the coercive pressure is indicated by arrows. The strength of the coercive pressure is in this case directed inward and independent of the local wall velocity; therefore its vectorial surface integral vanishes, giving $F_c^x = 0$.

Now we can write the complete equations for nonlinear bubble motion in terms of the bubble displacement $X$ perpendicular to the dc in-plane field $H_p$, the bubble radius $r$, and the angular deviation $\theta$ of the two Bloch lines from their static equilibrium positions as defined in Fig. 20.4. This model neglects the component of motion parallel to $H_p$ caused by gyrotropic deflection. Following the formalism of Section 14, one balances the components of conservative force derived from the total static energy $W$ against the dynamic reaction forces derived by superposing Bloch-line reaction forces (with Bloch-line damping neglected) onto normal-wall forces originating from viscous drag and coercivity [see Eqs. (14.8) and (14.10)]:

$$\partial W/\partial r = -\gamma^{-1}4\pi M(\alpha\Delta_0^{-1}r\dot{r}) \pm \gamma^{-1}4\pi M(\dot{X}\cos\theta + r\dot{\theta})$$
$$- 8\,MH_c r \arcsin'(\dot{r}/|\dot{X}|), \qquad (20.6)$$

$$\partial W/\partial \theta = \mp\,\gamma^{-1}4\pi Mr(\dot{X}\sin\theta + \dot{r}), \qquad (20.7)$$

$$\partial W/\partial X = \mp\,\gamma^{-1}4\pi M(\dot{r}\cos\theta - r\dot{\theta}\sin\theta) - 4\pi M\alpha r(2\Delta_0\gamma)^{-1}\dot{X}$$
$$+ F_c^x(\dot{r}, \dot{X}), \qquad (20.8)$$

where the function arcsin′ $u$ is defined as $\frac{1}{2}\pi$ sgn $u$, for $|u| \geq 1$ and arcsin $u$ for $|u| \leq 1$. Equation 20.6 includes the radial force due to the tangential component of Bloch-line motion. Equation (20.7) gives the tangential Bloch-line reaction to its normal velocity component. Equation (20.8) gives the bubble-displacement reaction ($\dot{X}$) to the Bloch-line motion ($\dot{\theta}$ and $\dot{r}$). The energy of the bubble is (per unit film thickness)

$$W = 2\pi r^2 MH_z(t) - 4\pi r M\Delta H_p \cos\theta + \tfrac{1}{2}C_0(r - r_0)^2, \qquad (20.9)$$

where $r_0$ is the equilibrium radius with $H_z(t) = 0$ and $H_p = 0$. The usual potential energy of the bubble with a constant bias $H_b$ is assumed to have a parabolic form with a Hooke's law constant $C_0$. This equation tells us that the equilibrium radius depends on $H_p$ and is expressed as

$$r = r_0 + 4\pi M H_p \Delta_0 C_0^{-1}. \tag{20.10}$$

This in-plane field dependence of the static radius has been confirmed experimentally. The static forces are obtained by differentiating the energy (20.9), and their substitution into the left-hand sides of Eqs. (20.6)–(20.8) gives three simultaneous differential equations governing $\theta(t)$, $r(t)$, and $X(t)$.

An example of numerical computations of $\theta(t)$, $r(t)$, and $X(t)$ is shown in Fig. 20.4, and the predicted dependences of the net displacement per pulse on in-plane field, drive field, and pulse time are shown in Fig. 20.1.[249] The detailed behavior is more complex than the simple intuitive model described at the beginning of this section. In particular, the change of $X$ following the rise of the pulse (end of phase 2) is negative, rather than positive as stated above. This discrepancy arises from the quite reversible effect of the radial shift of stable Bloch-line positions between field-on and field-off, neglected in the qualitative explanation of automotion given above. The agreement between theory and experiment in Fig. 20.1 gives considerable confidence in the proposed mechanism and in the identification of the $\sigma$-states as (1, 2) states. That identification is further confirmed by state-conversion experiments as described in Section 19,D.

The more recently observed in-plane-pulse automotion (type II) of $\sigma$ bubbles[407] has a simpler explanation. In this experiment the dc in-plane field $H_y$ provides restoring forces orienting the Bloch lines in the direction $\theta = 0$ (see lower left corner of Fig. 20.4, but visualize both Bloch lines staying opposite each other on a diameter). In the presence of the superposed in-plane pulse field having amplitude $H_x$, the vector sum $\mathbf{H}_y + \mathbf{H}_x$ provides the new equilibrium direction $\theta = \theta_1$, which is the angle between $\mathbf{H}_y$ and $\mathbf{H}_y + \mathbf{H}_x$. The net change of dimensionless momentum component along the bisector of this angle is $\delta\bar{\psi} = 4\sin(\theta_1/2)$, according to Eqs. (18.13)–(18.14). Let us assume that the rise time and the Bloch-line motion are too rapid for coercivity to have an appreciable effect. Then in analogy with the pulsed-bias jump (Section 19,C) the asymptotic displacement along the bisector after the rise of a long pulse is, by Eq. (18.19)

$$X = \int V \, dt = 4\Delta\alpha^{-1} \sin(\theta_1/2). \tag{20.11}$$

If the pulse fall time is infinite, coercivity prevents further bubble displacement while $\theta$ falls back to zero, and Eq. (20.11) represents the maximum possible net displacement. Type II automotion of this order of magnitude up to 1 $\mu$m per pulse) has been observed.[407]

More recent work has revealed a type of $S = 0$ bubble that does move under the conditions described above.[407a] A configuration of four vertical Bloch lines grouped in clusters of one and three has been proposed to explain this effect. Indeed, other work shows by means of pulsed laser photography that even chiral bubbles can be caused to automote under the above-described nominal conditions.[407b] Clearly the presence of static Bloch lines is not a prerequisite for this form of automotion.

It has been proposed that chiral bubbles $[S = (1, 0)]$ would "automote" if subjected to an in-plane field pulse instead of the conditions described above.[405] Assuming a simple plane-wall model [e.g., Eq. (10.10)], the in-plane field pulls the wall spins by an angle $d\psi$ whose magnitude is $\arcsin(H_p/8M)$. The senses of this in-plane-field-induced rotation are obviously opposite for the two possible wall chiralities. In the absence of coercivity or external drive, one has from the pressure equation (18.19) the usual relationship $dq = \Delta\alpha^{-1}d\psi$ between the wall displacement $dq$ and the rotation angle $d\psi$. Thus one expects a wall displacement whose direction depends on the wall chirality. If the in-plane field is applied rapidly and removed slowly, coercive suppression of the reverse motion may be expected. Experiments done to demonstrate this effect as described in Section 20,A have revealed groups of bubbles moving in opposite directions.[405] However, subsequent experiments indicate that the active bubbles are most likely the (1, 2) bubbles engaging in type II automotion with some kind of in-plane anisotropy providing the static restoring force. The response of the pure chiral bubbles to the in-plane field is unaccountably weak.

One may summarize by saying that only in the case of the free-bubble automotion shown in Figs. 20.1 and 20.4 is there a convincing comparison of theory and experiment. The fact that free-bubble automotion is so strong opens up the question whether gradients are actually necessary for the conductor-line and lattice-rotation forms of automotion discussed above. Beyond the free-bubble case, work on the phenomenon of gradientless propulsion has been very preliminary, and no report has yet been made of efforts to use this remarkable phenomenon in a device environment.

## 21. Device Dynamics

### A. Mobility and Critical-Velocity Limitations

In this section we review the implications of bubble-domain dynamics for field-access device operation. As discussed in Section 6,B, such devices operate by means of in-plane fields $H_p$ that rotate at some frequency $f$ and that act on some in-plane structure with the spatial period $P$ to create a moving potential well. As the well moves, the bubble lags in the well until it experiences a gradient sufficient to maintain the externally imposed velocity. The maximum gradient drive $H_g = |r\nabla H_z|$ that the well can provide is roughly proportional to $H_p$ as long as the drive layer is not saturated. Thus

$$H_{max} = \rho H_p, \tag{21.1}$$

where the proportionality constant $\rho$ has been estimated for typical devices to be in the range 0.1–0.3.[198,199] The behavior of the device is often characterized by a "margin plot" in which the conditions for some fixed probability of error, say 50%, are determined as a function of bias field and in-plane drive field, for different frequencies and a given number of steps of operation. An example is shown in Fig. 21.1 for a T and I bar device operated at different frequencies on a SmGaYIG film. The upper and lower margins are primarily determined by static considerations, namely bubble collapse and runout, respectively, although there is a weak frequency effect also.[410] On the other hand, the dependence of the minimum in-plane drive field $H_{pmin}$ on frequency is quite strong and can be related to the bubble dynamics theory we have described in the rest of this review.

To construct a simple model, let us consider a bubble material with mobility $\mu$, bubble coercivity $H_{cb} = 4\pi^{-1}H_c$, and critical velocity

$$V_p(H_p) = V_p + \mu_p H_p, \tag{21.2}$$

where $\mu_p$ is the rate of increase of the critical velocity with in-plane field. $V_p$ is typically of order $V_{po} = 24\gamma A/hK^{1/2}$ [Eq. (15.6)], while $\mu_p$ is of order $(\pi/2)\gamma\Delta_0$ [Eq. (15.16)]. The simplest case to analyze is the one in which the well velocity is constant, as in the experiment of bubble rotation around the edge of a permalloy disk (see Section 6,B). In this case at low frequencies $f$, the mini-

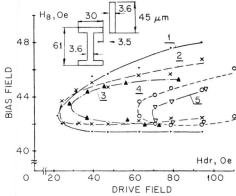

**Fig. 21.1.** Margin plot of TI-bar device on a SmGaYIG film at the frequencies: 1–50 kHz, 2–100 kHz, 3–125 kHz, 4–170 kHz, 5–300 kHz (after Kleparsky *et al.*[409]).

mum in-plane drive is proportional to coercivity and viscous drag and is

$$H_{\text{pmin}} = \rho^{-1}(H_{\text{cb}} + \mu^{-1}Pf) \qquad \text{(permalloy disk, small } f\text{)} \qquad (21.3)$$

for the model parameters defined above, where $P$ is the circumference of the disk. At higher frequencies, the bubble may reach its saturation velocity, so that by inverting Eq. (21.2) we have

$$H_{\text{pmin}} = \mu_{\text{p}}^{-1}(Pf - V_{\text{p}}) \qquad \text{(permalloy disk, large } f\text{)}. \qquad (21.4)$$

These relations for minimum in-plane drive field versus frequency are plotted schematically in Fig. 21.2, where the break in the curve denotes the change-over from the mobility-limited to the critical-velocity-limited region. The critical frequency at which this changeover occurs is

$$f_{\text{crit}} = \rho P^{-1}(V_{\text{po}} + \mu_{\text{p}}\rho^{-1}H_{\text{cb}})(\rho - \mu_{\text{p}}\mu^{-1})^{-1}. \qquad (21.5)$$

If $\mu\rho < \mu_{\text{p}}$, that is, for low mobility materials, it follows that the break to nonlinear behavior will be suppressed.

In a practical device with discrete permalloy elements, the dependence of $H_{\text{pmin}}$ on $f$ is modified in two main ways. First, the minimum drive field is generally dominated by the magnetostatic forces that tend to hold back a bubble from crossing the gap between permalloy elements.[198,199,411] The above model can be generalized to include this effect approximately by replacing Eq. (21.3) by

$$H_{\text{pmin}} = H_{\text{pmino}} + \rho^{-1}\mu^{-1}Pf \qquad \text{(device with gap, small } f\text{)}, \qquad (21.6)$$

where $P$ is the device period, and where the constant $H_{\text{pmino}}$ includes a coercivity contribution $\rho^{-1}H_{\text{cb}}$ and a gap contribution. The gap contribution is usually proportional to the magnetization of the bubble material. For example, for a 2 $\mu$m T- and I-bar device on amorphous GdCoMo films,[411] the gap contribution was 0.06 $(4\pi M)$ Oe, while the coercivity contribution was 10 Oe. Secondly, well velocity is generally not constant over a single period in a device. The ratio $r$ of maximum to average velocities in various device structures including T and I bars, T and X bars, and chevrons ranges from

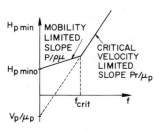

**Fig. 21.2.** Schematic plot of minimum in-plane drive field $H_{\text{pmin}}$ of a device versus frequency $f$.

3.5 to 4.8, as determined in a high-speed stroboscopic study.[133] The high velocity occurs at some place other than the gap, which is reasonable since at the gap the drive is smallest. Thus to account for the velocity variation in the theory, one can replace Eq. (21.4) by

$$H_{pmin} = \mu_p^{-1}(Pfr - V_p) \qquad \text{(device with gap, large } f\text{)}, \qquad (21.7)$$

which is determined by the critical velocity limit at the fastest point in the circuit. The $r$-factor is not introduced into Eq. (21.6) since this equation represents the mobility limit at the gap, where the potential well is assumed to translate at approximately the average velocity $Pf$. Equations (21.6) and (21.7) are plotted in Fig. 21.2 and again show the cross-over from mobility limited to critical-velocity limited device operation.

Comparison with permalloy disk experiments has shown mobility-limited behavior in those cases investigated so far.[202,203] Such behavior has also been seen in an 8 $\mu$m-period T- and I-bar device on a GdCoMo film with $\mu = 200$ cm/sec Oe, where $H_{pmin}$ increased by 13 Oe as $f$ increased to 1 MHz, in agreement with Eq. (21.6)($\rho \sim 0.3$).[191] More detailed calculations have been performed to determine the full time evolution of the potential well under various permalloy structures, and the shape and phase lag of the bubbles as a function of time or position in the circuit have been calculated assuming mobility-limited behavior.[133a,199,411a] The agreement with stroboscopic observations on devices is good provided the operating frequencies are not too high.

Is there any evidence for critical-velocity limits in devices? A break of the type illustrated in Fig. 21.2 can be observed in the study of a 37 $\mu$m period T- and I-bar device on a SmGaYIG film, shown in Fig. 21.1, although the frequency dependence above the break is nonlinear.[409] In various other device studies[390,412,413] operation failed at lower frequencies than expected from the mobility limitation alone. For example, in one important study the maximum frequency of failure-free multistep device operation was determined for a series of SmCaGeYIG films at a given drive.[413] The results were in reasonable agreement with Eq. (21.7), taking $V_p = V_{po} = 24\gamma A/hK^{1/2}$ (plus stray-field corrections) and $r = 2.5$ but $\mu_p = 0$. (However, the theoretically more preferable value $\mu_p = (\pi/2)\gamma\Delta_0$ would predict significantly higher operating frequencies.) One interesting feature of these results was that significantly higher frequencies were achieved in thinner than in thicker films, in agreement with the $h^{-1}$ dependence of the formula for $V_{po}$. This result contrasts with bubble-propagation measurements described in Section 19,A, which show essentially no dependence on $h$. A possible reason for this discrepancy is that in the device there are stray in-plane field components from the drive elements, concentrated near the surface. These components would have a larger effect, through $\mu_p \neq 0$, on a thin film than on a thick one. Considering typical values for $\mu_p = (\pi/2)\gamma\Delta_0$, one finds that $V_p(H_p)$ can be strongly increased for typical device operating values of $H_p$.[133] Such an effect can also explain the apparent discrepancy between a maximum velocity of 2400 cm/sec observed in bubble rotation around a permalloy disk in a EuGaYIG film as compared to a bubble-collapse saturation velocity of only 1000 cm/sec.[314] Further evidence for a saturation-velocity limitation in device operation has come from a careful study of the temperature dependence of the bias-field margin during high-speed operation of half-disk devices on SmLuCaGeYIG films.[413a] Although mobility increases by a factor of almost three from $-60$ to $+60°C$, the size of the margin remains constant, even in a region where the margin is strongly dependent on frequency. This effect is consistent with the approximate temperature independence of the saturation velocity. A saturation velocity was also observed in a stroboscopic study of bubble propagation by charged walls at ion-implanted pattern edges.[75d]

Awareness of the simultaneous need for high mobility and high critical velocity has been important in the development of adequate bubble materials. For example, the CaGe-based garnet systems have typically higher exchange stiffness $A$ than Ga systems,[32] and this correlates with an improved high-frequency performance attributed to the effect of $A$ in $V_{po}$. The dynamic properties have been an important factor in making CaGe compositions such as SmLuCaGe and EuTmCaGe the most widely used materials for bubble devices today. Angular-momentum-compensated materials have also shown superior device operation with little margin degradation out to 2 MHz.[84,85] This is explained by the large value of $\gamma$ in the formula for $V_{po}$ and correlates with the lack of ballistic overshoot in pulsed gradient propagation experiments on these materials.[130] So far, however, they have not been used in practical devices primarily because of temperature sensitivity problems.

## B.   MULTISTEP TESTING AND CAPPING-LAYER EFFECTS

Another aspect of bubble-device performance has been revealed in tests of the bias margin at a fixed drive and frequency but for different numbers of steps $N$.[414] Typical results are shown in Fig. 21.3 for a SmGaYIG film with three different ion-implantation conditions.[414,415] One finds that often the bias margin decreases with log $N$ according to [414,415]

$$\Delta H_b = \Delta H_{bo} - \varepsilon \log N, \tag{21.8}$$

where $\Delta H_{bo}$ and $\varepsilon$ are empirical constants. This behavior is unexpected in the light of the simple model of the last section and indicates that statistical processes reduce the bubble stability range. One possible cause of the randomness comes from coercivity or structural variations in the permalloy drive layer, but this topic is beyond the scope of our review.[412,416] Another

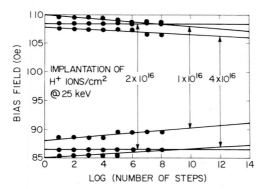

**Fig. 21.3.** Bias-field margin versus the log of the number of steps in a T- and I-bar circuit on a SmGaYIG film for three different ion-implantation conditions (after Shumate et al.[414,415]).

possible cause is randomness in the dynamic-conversion process, which is presumably related to the dramatic effect of capping layers on device performance. While uncapped films may operate for a few steps with as wide a margin as capped films, they fail for even very moderate values of $N$ and at low values of $f$. A frequent concomitant of failure in uncapped films is the appearance of hard bubbles.[235] By contrast, properly capped materials along with properly designed devices can yield essentially flat margin-versus-log $N$ plots as shown in Fig. 21.3.[414,415] Particularly high operating velocities have been achieved in triple-layer films (Section 16,D).[376,417]

Next we discuss the mechanism by which capping layers give this dramatic improvement in device performance. As discussed in Section 19,D, various physics experiments have been done to compare capped and uncapped films. The results indicate that Bloch lines can nucleate and propagate in capped as well as uncapped materials with little change in the critical or average velocity because the stray fields are only modified partially at one surface. In addition, a plausible hypothesis is that a "good" capping layer is one that prevents punch-through of Bloch lines at that surface because of the extra energy of in-plane domains that would have to form. Thus punch-through can occur in a capped film, but only at the uncapped surface. Examination of the multiple Bloch curve structures (e.g., see Fig. 18.4) reveals that only the first punch-through can yield vertical Bloch lines, whereas all subsequent loops pile up at the capped surface, leading to an increased overshoot and reduction in scatter.[382] The in-plane fields present in a device will substantially modify the shapes of the Bloch curves, but there will still be at most only two vertical Bloch lines.

Considering these results, a likely model for the effect of capping layers on device performance is as follows: Clearly the criterion for failure-free device operation is not simply that the instantaneous velocity always remain below the critical velocity for Bloch-line nucleation, since that velocity appears to be comparable in both capped and uncapped films and cannot explain the large performance difference between the two. Instead one must invoke a "jerkiness" in device operation, arising both from the discreteness of the drive-element pattern and from static and dynamic coercivity fluctuations both in the permalloy elements and the bubble film. This jerkiness is presumed to cause sudden drives on the bubble that are long enough and strong enough to cause Bloch-line nucleation and punch-through, even though the average bubble velocity may be far below $V_p$. In an uncapped film, such punch-throughs can cause a gradual pile-up of vertical Bloch lines along with state changes on a random statistical basis. Thus there is a finite probability of forming hard bubbles, which eventually escape from the device potential well because of their low mobility or large deflection effect, or simply because they expand and deform into "dumbbells" (Section 8,D). Another possibility is that the large momentum of the piled-up Bloch lines can cause bubbles to stray from their proper trajectories wherever the bubble must change direction, for instance at corners in the device propagation structure. Since the formation of a hard bubble requires the concatenation of a long sequence of random events, multistep testing is required to reveal dynamic conversion failures. By contrast in capped layers the maximum state change is $\Delta S = \pm 1$ if a maximum of only one pair of vertical Bloch lines can be formed as discussed above. The effectiveness of capping layers can then be explained if it is supposed that such a small state change does not materially affect device performance. It would be of interest to investigate this supposition by determining the effect of different bubble states on field-access device performance, and work in this direction has only recently begun.[399]

Theoretically, device designers of field-access devices need to make but small allowances for Bloch-line states when the winding number $S$ and the Bloch-line number $n$ are small. The transverse field $r|\nabla H_z|$ required to balance the gyrotropic force on a typical $r = 2$ $\mu$m bubble moving at the critical velocity is on the order $S$ Oe, according to Eqs. (18.16), (18.22), and (15.6). (The effect will increase rapidly with decreasing radius, however.) The inertial overshoot effects may be gauged by estimating the bias-field equivalent of the bubble kinetic energy represented by Bloch-line energy [Eq. (8.8)]. One equates $2\pi r^2 M H_z = 8 A Q^{-1/2} n$ and finds $H_z \sim 0.1 n$ Oe for $r = 2$ $\mu$m. Since both of these fields are small (unless $n$ is large and the bubble is hard) compared to the fields provided by the permalloy elements, it is clear that bubbles should stray little from the "track" they would otherwise follow, unless there are weak wells at critical points in the circuit.

# X
# Wall Waves and Microwave Effects

It is well known that an infinite ferromagnetic domain has a continuous spectrum of propagating plane-wave excitations, whose circular frequency $\omega$ depends on wave vector $\mathbf{k}$. For large $Q$, the minimum frequency is approximately the "anisotropy gap" $\omega(0) = \gamma H_K = 2\gamma K/M$. In the presence of film surfaces, the spectrum becomes complicated by the existence of "magnetostatic modes." The addition of domain walls further complicates matters by introducing the following features: (1) the bulk spin waves may reflect from the walls; (2) new "magnetostatic interface waves" may appear, which are domain spin-precession modes concentrated near the walls but have such a symmetry that the wall remains stationary;[350] and (3) the walls may themselves support loaded-membrane waves, leaving the adjoining domain moments largely stationary. Much of the spectrum for the latter wall waves lies far below $\gamma H_K$ and therefore is substantially uncoupled from the domain precessions. The wall-wave spectrum forms one subject of this chapter (Section 22), and their coupling to domains is considered only to the extent required to understand microwave excitation of domain-wall displacement (Section 23).

Also discussed in Section 22,B are the normal modes of oscillation for an isolated bubble. Still another subject of this chapter is the propagating wave motion that can occur in a hexagonal lattice of bubbles when they are regarded as interacting moving bodies (Section 22,C). One then has both "acoustic" modes, in which the bubble centers move, and "optic" modes, in which the bubble radii oscillate. Very little is known experimentally about either membrane or lattice waves. The theories are included here to stimulate work in these potentially fruitful areas. Finally we discuss, both theoretically and experimentally, the extraordinary phenomena connected with the coupling of microwave radiation to domain walls in bubble films (Section 23).

## 22. Wall-Wave Spectra

In Section 12,E we suggest heuristically the velocity relation $u_f = (\sigma_0/m)^{1/2}$ [Eq. (12.36)] for propagation of a wall deflection along the wall. This relation rests on the analogy of a uniform "loaded" membrane, which is valid if only the "surface tension" $\sigma_0$ of the wall contributes restoring forces to the wall of effective-mass density $m$.[350] Detailed studies show that additional restoring forces are needed to provide an accurate theory. Section 22,A considers the infinite medium, in which long range demagnetizing fields from poles in the *wall* surface make a significant contribution. Section 22,B considers how magnetic poles on the *film* surface affect both stability and wave dispersion of a plane wall. It also covers the normal-mode frequencies arising from the finite dimensions of a bubble. Section 22,C discusses the propagating modes of a hexagonal bubble lattice.

### A. INFINITE PLANE WALL

There exist many papers on the theory of spin-wave spectra in the presence of domain walls, particularly in the Soviet literature from which we give some recent citations.[334,419-421] Here we restrict our attention to the lowest branches of this spectrum that represent oscillations of the wall about its equilibrium position. Our simplified treatment utilizes a method previously applied to the problem of stability of uniform wall motion.[287]

Consider the infinite ferromagnetic medium with two domains statically separated by a domain wall with no Bloch lines. To discuss the linear excitation spectrum, we use the notation of Section 12,A and linearize the wall-motion equations (12.1), (12.2), (12.5), and (12.6) about the values $q = 0$ and $\psi = 0$ under the special assumption that the external field term generally included in $H_y$ vanishes. We find

$$\dot{\psi} = \gamma H_z + (\sigma_0\gamma/2M)\nabla^2 q - \alpha\Delta_0^{-1}\dot{q}, \tag{22.1}$$

$$\Delta_0^{-1}\dot{q} = [(\pi\gamma/2)H_x + 4\pi\gamma M]\psi - (2\gamma A/M)\nabla^2\psi + \alpha\dot{\psi}, \tag{22.2}$$

where

$$\nabla^2 \equiv (\partial/\partial x)^2 + (\partial/\partial z)^2. \tag{22.3}$$

Here $q(x, z, t)$ is the wall displacement in the $y$ direction, $\psi(x, z, t)$ the wall-moment angle measured from the $x$ axis, and $H_x$ is an applied homogeneous field component in the unperturbed wall plane. For the $z$ component of field intensity $H_z(x, z, t)$ acting at the point $y = q(x, z, t)$, we write

$$H_z = -H_z'q + H_{zd}(x, z), \tag{22.4}$$

where $H_z'$ is an external field gradient serving to define the static wall position.

Since the demagnetizing field $H_{zd}(x, y, t)$ is due only to deformation of the wall from a plane $y = q$, $H_z$ vanishes at the equilibrium condition described by $q = 0$ for all $x$ and $z$. If we neglected the quantities $\nabla^2\psi$ and $H_{zd}$, then the remaining equations above would represent simply the motion of a damped mass-loaded membrane with the propagation velocity of Eq. (12.36). When included in Eq. (22.2), the term $-(2\gamma A/M)\nabla^2\psi$ effectively represents a wave-vector dependent mass, which becomes important at large wave vectors and is easily taken into account. A more significant quantity is that part of $H_{zd}$ that arises from magnetic charges appearing dynamically when $dq/dz \neq 0$[287,422] because it affects the spectrum strongly at small wave vectors.

To calculate $H_{zd}$, consider the $yz$ section of the wall undulation depicted in Fig. 22.1a. As made clear by the enlarged view in Fig. 22.1b, the discontinuity in $\mathbf{M}$ at the wall, considered infinitely thin, produces a magnetic pole density on the wall surface. Applying Gauss's law to the infinitesimal rectangle $ABCD$, we find

$$(H_n{}^+ + H_n{}^-)ds + 2 \cdot 4\pi M \, dq = 0, \tag{22.5}$$

where $H_n{}^\pm$ are $yz$ plane projections of the wall normal component of $\mathbf{H}(x, y, z)$ on the two sides ($y \neq q$) of the wall, and $ds$ is an element of wall length in the $yz$ plane. In the limit of small displacements, we have $ds \to dz$, $H_n{}^+ \to H_n{}^- \to H_{yo}^+$, where $H_{yo}^+$ is the $y$ component of $\mathbf{H}$ on the $y > q$ side of the wall, i.e., evaluated at $y = q^+$. Equation (22.5) reduces to

$$H_{yo}^+ = -4\pi M \, \partial q/\partial z. \tag{22.6}$$

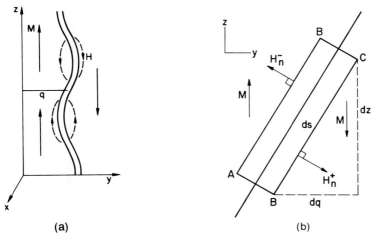

(a)　　　　　　　　　　　　　　　　　　　　(b)

**Fig. 22.1.** (a) Section of undulating wall indicating origin of the demagnetizing field component $H_{zd}$. (b) A geometric construction used to derive $H_{zd}$.

Note that $H_z(x, y, z)$ is continuous across the wall to first order in $q$. To find $H_{zo} = H_z(x, 0, z)$, consider the fundamental Maxwell equation

$$\partial H_z/\partial y = \partial H_y/\partial z. \tag{22.7}$$

Evaluating this at $y = q^+$ and substituting Eq. (22.6) we find

$$\partial H_{zo}^+/\partial y = -4\pi M(\partial^2 q/\partial z^2). \tag{22.8}$$

Further progress in evaluating $H_{zo}$ from $q$ cannot be made without specializing the function $q(x, z)$ because of the long-range character of the demagnetizing field. Consider a wall wave of form

$$q = \text{Re } q_0 \exp i(k_x x + k_z z - \omega t), \tag{22.9}$$

having wave vector $(k_x, k_z)$ and circular frequency $\omega$. This wave will generally entrain a wave of magnetostatic potential $V(x, y, z)$, which decays as a function of distance from the wall and from which $\mathbf{H}(x, y, z) = -\nabla V$ may be derived, except at $y = q$:

$$V = \text{Re } V_0 \exp[-\kappa|y - q| + i(k_x x + k_z z - \omega t)] \quad (\kappa > 0), \tag{22.10}$$

in the continuous regions $y \neq q$. In the limit of large $Q$ we can neglect the tilting of $\mathbf{M}$ within domains away from the easy axis by $\mathbf{H}$. Then $\nabla \cdot \mathbf{M}$ vanishes within the domains, and we have

$$\nabla \cdot \mathbf{H} = -\nabla^2 V = 0 \quad (y \neq q). \tag{22.11}$$

The expression (22.10) only satisfies this relation under the condition

$$\kappa = (k_x^2 + k_z^2)^{1/2} \equiv k. \tag{22.12}$$

From Eqs. (22.10) and (22.12) we have

$$\partial H_{zo}^+/\partial y = -kH_{zd}, \tag{22.13}$$

where $H_{zd} \equiv H_{zo}^+$ is the demagnetizing field sought, and Eq. (22.8) reduces to

$$H_{zd} = -4\pi M k_z^2 k^{-1} q, \tag{22.14}$$

which is the desired relation. Particularly noteworthy is the fact that for a given wave direction specified by $k_x/k_z$, $H_{zd}$ is proportional to the first power of $k$. It follows that for sufficiently small $k$ this restoring force will dominate the one due to surface tension that is proportional to $\nabla^2 q = -k^2 q$, except in the special direction $k_z = 0$.

To find the wave dispersion, we let $\psi(x, z, t)$ also have a form like (22.9) and substitute Eq. (22.14) for $H_{zd}$ in Eq. (22.4). The resulting quadratic secular equation derived from Eqs. (22.1) and (22.2) has the solution

$$\omega = \gamma(H_1 H_2)^{1/2} - \tfrac{1}{2}i\alpha\gamma(H_1 + H_2), \tag{22.15}$$

$$H_1 \equiv 4\pi M + (\pi/2)H_x + (2A/M)k^2, \tag{22.16}$$

$$H_2 \equiv \Delta_0 H_z' + 4\pi M \Delta_0 k_z^2 k^{-1} + (2A/M)k^2. \tag{22.17}$$

Equation (22.15) is the exact solution of Eqs. (22.1) and (22.2) for $\alpha = 0$, but the imaginary term representing attenuation is given only to first order in $\alpha$. In the harmonic-oscillator analogy, $H_1^{-1}$ corresponds to the mass and $H_2$ to the restoring-force coefficient, except for dimensional factors. In actuality $H_1$ measures the stiffness of the wall-moment with terms due, respectively, to the wall demagnetization, external field and exchange stiffness. $H_2$ measures the restoring-force coefficient on $q$ with terms due, respectively, to external-field gradient, magnetic charges on the wall surface, and wall-surface tension. All of the equations in this section are correct to leading order in $Q^{-1}$ and $\Delta_0 k$. The latter condition means that the wavelength is large compared to wall thickness. Equations (21.15)–(21.17) agree, in appropriate limits, with results of more rigorous calculations.[422-424]

Figure 22.2 illustrates schematically the dispersion relation given by Eqs. (22.15)–(22.17) for propagation parallel ($k_x = 0$) and perpendicular ($k_z = 0$) to the easy axis, and in the absence and presence of the external gradient $H_z'$. Note that $H_z'$ creates a "gap" in the spectrum and that the power dependence of $\omega$ on $k$ is different in all four cases, as indicated in the figure. Owing to the character of our approximations, only the portion of the wall-wave spectrum, which lies well below the anisotropy gap of the single-domain bulk spin-wave spectrum, at the frequency $2\gamma K/M$, is given correctly by Eqs. (22.15)–(22.17).

Noteworthy is the generally strong attenuation of wall waves on a device scale. We have considered above $k$ to be real and $\omega = \omega_1 - i\omega_2$ to be complex. Instead write now $\mathbf{k} = \mathbf{k}_1 + i\mathbf{k}_2$ and let $\omega$ be real. Then by expanding the formal relation $\omega = \omega_1(\mathbf{k}) + i\omega_2(\mathbf{k})$ about $\mathbf{k}_2 = 0$, we find the first-order spatial attenuation coefficient

$$\mathbf{k}_2 = \omega_2/(\partial\omega_1/\partial\mathbf{k}_1)_{k_2=0}. \tag{22.18}$$

For example, in the special case of "surface tension" waves with $k = k_x$, we

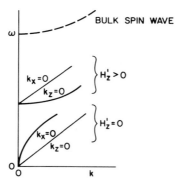

**Fig. 22.2.** Qualitative dispersion relations for domain-wall waves in an infinite ferromagnetic medium with ($H_z' > 0$) and without ($H_z' = 0$) external field gradient.

would have $H_1 = 4\pi M$, $H_2 = (2A/M)k^2$, so that for $k < \Lambda_0^{-1}$, we have the attenuation

$$k_2 = \alpha/2\Lambda_0 = \alpha M(\pi/2A)^{1/2},\tag{21.19}$$

where $\Lambda_0$ is the Bloch-line width parameter. Garnet bubble films generally satisfy $\alpha \geq 10^{-3}$ so that in materials having $M \geq 15$ Oe, we have $k_2^{-1} \lesssim 0.025$ cm. Moreover, even for $\alpha = 0$ a coercivity on the order of 0.1 Oe would produce at least as strong an attenuation as $\alpha = 10^{-3}$. This means that, without amplification or great improvement of materials, wall waves cannot be used for carrying signals farther than a few times 0.025 cm, which is smaller than the usual dimensions of bubble-circuit chips.

It is worth noting that wave propagation in a hard wall does not obey Eq. (12.36). Moreover, it cannot be derived from Eqs. (22.1) and (22.2) because of the condition $\sigma \gg \sigma_0$. In a hard wall the restoring force due to wall demagnetization is essentially absent. It becomes meaningful then to neglect magnetostatic energy altogether. In this limit the dispersion is given, in the absence of external fields, by[425]

$$\omega_h = (2\gamma A/M)[(1 + \Delta_0^2 k_s^2)^{-1} k_x^2 + k_z^2],\tag{22.20}$$

under the condition

$$|k_x| < |k_s|(1 + \Delta_0^{-2} k_s^{-2}),\tag{22.21}$$

where $k_s \equiv d\psi/dx$ is the static pitch ($\pi \times$ Bloch-line density) assumed to be wholly oriented in the $x$ direction (vertical Bloch lines). Note that since the dispersion is quadratic it initially lies below that of a soft wall. In this sense, the term "hard wall" is misleading, and its high degree of compliance has formed the basis for an interpretation of Bloch-line annihilation experiments.[425] On the other hand, nondecaying wall oscillations with $|k_x|$ exceeding the bound (22.21) do not exist, so that the number of hard-wall modes is comparatively small.

## B.  WALL WAVES IN A FILM

The presence of boundary conditions at the free surface of a film having finite thickness implies some discrete character to the spectrum of flexed-wall modes. The usual procedure for similar problems would be to construct the normal-mode wave functions from linear combinations of the infinite-medium solutions. This procedure is generally invalid in bubble materials, for several reasons: (1) The stray fields arising from the *dynamic* charges on the *film* surfaces make an essential and nonlocal negative contribution to the wall-restoring coefficient. (2) The long-range character of the stray field arising from *dynamic wall-surface* charges implies that Eq. (22.14) is incorrect in a bounded medium. (3) The stray fields originating from the *static* magnetic charges on the *film* surfaces decrease the wall-moment stiffness (inverse mass) inhomogeneously, making it smallest at the critical points close to the film surfaces, as explained in Section 17,B.

For these reasons, a quantitative theory of wall-mode spectra in bubble films will inevitably require extensive numerical computations that have not yet been carried out. We will describe here the results of rather approximate theories.

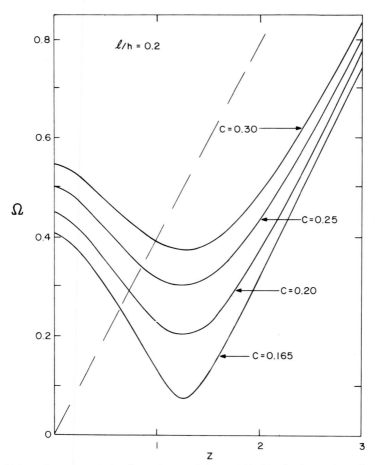

**Fig. 22.3.** Dispersion relation for a plane wall in a bubble film for four values of reduced external field gradient $C = H_z'h/8M$. The reduced circular frequency is $\Omega = (\omega/4\pi\gamma M)(\pi h/2)^{1/2} \cdot (K/A)^{1/4}(m/m_D)^{1/2}$, and the reduced wave vector is $z = kh$ (after Schlömann[61]). Here $m/m_D$ is the ratio of mean actual wall mass to the Döring mass of Eq. (11.6).

The effect of consideration (1) above on the dispersion of a plane soft wall was computed numerically as a function of wave vector $k_x$ parallel to the film plane.[61] Consideration (2) was avoided by the simplifying assumption that $q$ does not depend on $z$, which would be correct in the limit of thin films. Consideration (3) can then be incorporated in principle by means of a correction factor obtained from a separate computation of mean wall mass, as described in Section 17,B.

Figure 22.3 shows examples of computed dispersion relations[61] for four values of applied gradient $H_z'$. The effects (1) and (3) above lower the dispersion curve in comparison with that of an infinite medium having the same material parameters and $H_z'$. The dynamic surface poles reduce the restoring force on wall displacement at all $k_x$ except $k_x = 0$ and $k_x = \pm\infty$, producing a minimum in $\omega(k_x)$. The critical value of $H_z'$ for which the minimum value of $\omega(k_x)$ reaches zero represents a static stability limit below which ripples form spontaneously.[60] The negative sign of the dynamic surface-pole effect is understood from the fact that rippling the wall has the effect of bringing positive and negative charges on the film surfaces closer to one another, thereby decreasing their magnetostatic interaction energy. To date, there exist no experimental studies of wall-wave propagation although the quasi-static onset of rippling as $H_z'$ is reduced has been observed in good agreement with theory, as discussed in Section 2,D.

Extension of the above considerations to the wall vibrational spectrum of an isolated bubble domain yields only discrete frequencies because of the finite boundary conditions. Let us neglect again the quantitatively important long-range magnetostatic considerations discussed in the beginning of this section. [The importance of the wall-charge consideration (1) may be judged by setting the last two terms of Eq. (22.17) equal to each other, assuming $k_x = 0$. We thus see that for $k = |k_z|$ less than the critical value $2/l$, wall charges are more important than surface tension.] We may think of normal-mode frequencies $\omega(n_\theta, n_z)$, where $n_z$ is the number of half waves in the $z$ direction (free boundary condition) and $n_\theta$ is the number of full waves around the bubble of diameter $d$. Some of these frequencies are[424]

$$\omega(0, 0) = \{(h/l)[-(l/h) + S_0(d/h)]\}^{1/2}\omega_d,$$

$$\omega(1, 0) = 0,$$

$$\omega(0, 1) = (\pi d/2h)\omega_d,$$

$$\omega(2, 0) = \{(3h/l)[(l/h) - S_2(d/h)]\}^{1/2}\omega_d, \tag{22.22}$$

where

$$\omega_d \equiv 4(2\pi A)^{1/2}\gamma/d \tag{22.23}$$

and $S_n$ are the magnetostatic stability functions for the bubble distortions represented by the respective wave modes.[51] Note that $\omega(1, 0)$ must vanish in any approximation because the corresponding mode is equivalent to rigid bubble translation for which there is no restoring force.

According to one proposal (see Section 18,D), the coupling of these modes to structural defects in the magnetic film can introduce losses leading to decreased mobility.[424] For a given mode frequency $\omega(n_\theta, n_z)$, define a corresponding critical wall velocity (translational or radial) $V_c(n_\theta, n_z) = \Delta_0 \omega(n_\theta, n_z)$ at which the rate of spin flipping within the wall equals $\omega(n_\theta, n_z)$. Then if $V$ is markedly smaller than $V_c(n_\theta, n_z)$, the mode $\omega(n_\theta, n_z)$ is not excited and has no influence on the mobility. If, on the other hand $V$ is as great as, or greater than $V_c(n_\theta, n_z)$, then defects will begin to excite the mode $\omega(n_\theta, n_z)$, whose energy represents additional loss and therefore diminution of the mobility.

It follows that, with increasing $V$, there should be a decrease in differential mobility each time $V$ approaches one or another $V_c(n_\theta, n_z)$. It is interesting to observe that the velocity corresponding to the first axial mode is given by

$$V_c(0, 1) = 16\gamma A/hK^{1/2}, \tag{22.24}$$

which differs from characteristic velocities of the Bloch-line model (Section 15) only with respect to the numerical factor standing in front. This model thus provides an alternate explanation of diminished mobility at higher drive, and for the often higher apparent $\alpha$ values observed in mobility as compared to ferromagnetic resonance (Section 11,A). In addition, the absence of anomalously high $\alpha$ values derived from hard-bubble mobility has been attributed, according to this model, to the unavailability of modes outside of the range given by the inequality (22.21). Unfortunately, the present lack of any estimate for the rate of mode excitation by the defects prevents a more definitive comparison with experiment. Also, the mechanism does not seem to afford an explanation of ballistic effects (Section 19).

## C. BUBBLE-LATTICE DYNAMICS

A regular hexagonal lattice of bubbles (Fig. 2.1d,e,f) will remain stable over a wide range of bias fields.[54] Unless the static bias field $H_b$ is very close to the collapse point, at which the lattice distance is large, the bubbles interact appreciably via magnetic dipole interactions. The exchange interactions between different bubbles are comparatively miniscule because the tail of the Bloch-wall profile decays exponentially with distance. The repulsive bubble–bubble dipole interactions have the consequence that vibrations propagate along the lattice. We discourse briefly here on the theory of such bubble-lattice waves. There is no experimental literature to speak of on this subject, except for the radial oscillations with vanishing wave vector (see Section 16), because these are the only ones excited by a uniform drive field, and studies of overdamped lattice waves in bubble-lattice storage devices.[27,426]

The lattice vibrational spectrum lies in bands analogous to those of crystals. At large lattice distances, we have the discrete isolated-bubble spectrum $\omega(n_\theta, n_z)$, $(n_\theta = 0, 1, 2, \ldots, n_z = 0, 1, 2, \ldots)$ discussed in the previous section. When the bubbles are placed closer to one another, then each discrete frequency broadens into a band of plane-wave modes, with one mode $\omega(\mathbf{k}, n_\theta, n_z)$ corresponding to each two-dimensional wave vector $\mathbf{k}$, which is continuous for an infinite lattice. The domain of $\mathbf{k}$ lies within a so-called Brillouin zone,[427] which is a unit cell in the lattice that is reciprocal to the bubble lattice (see Fig. 22.4a). The models investigated thus far consider only three bands or branches. Two of these branches arise from bubble translation ($n_\theta = 1$, $n_z = 0$) and one from radial oscillation ($n_\theta = 0$, $n_z = 0$).

Let us regard the $i$th magnetic bubble having radius $r_i$ as a small dipole of moment $2\pi Mhr_i^2$. Then for large lattice parameters the potential energy of the bubble lattice is

$$V = 2\pi^2 M^2 h^2 \sum_{i \ne j} r_i^2 r_j^2 R_{ij}^{-3} + \sum_i [V(r_i) + 2\pi MhH_b r_i^2], \quad (22.25)$$

where $R_{ij}$ is the distance between bubble centers and $V(r_i)$ is the single-bubble self-energy (Section 2). Allowing $x_i$ and $y_i$ to represent the bubble-center

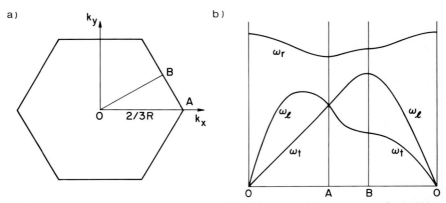

**Fig. 22.4.** (a) Brillouin zone for a hexagonal bubble lattice. All wave vectors for bubble-lattice waves may be considered to lie within this zone. The line segment $OA$ has length $2/3R$ where $R$ is the distance between nearest-neighbor bubble centers. (b) Schematic plot of frequency-versus-wavevector of the hexagonal bubble lattice along the symmetry axes indicated in (a).

position coordinates and $d$ the static bubble diameter, the kinetic energy in first-order approximation is

$$\varepsilon = (\pi mhd/4) \sum_i (2\dot{r}_i^2 + \dot{x}_i^2 + \dot{y}_i^2), \qquad (22.26)$$

where $m$ is the mass per unit area of the domain walls. Linearized equations of bubble motion based on these energy expressions have been solved to find the natural frequencies.[62,63,428-430] The scheme of the results, whose details depend on many parameters, along lines of symmetry $OA$, $AB$, and $BO$ in the Brillouin zone is illustrated in Fig. 22.4b. The two lower branches are predominantly translational, corresponding to the acoustic modes of an atomic lattice. In one of these $(\omega_l)$ the bubbles vibrate longitudinally with respect to $\mathbf{k}$, in the other $(\omega_t)$ transversally. The longitudinal and transverse characterizations are strictly valid only if $\mathbf{k}$ is small or if $\mathbf{k}$ lies on axes (like $OA$ and $OB$ in Fig. 22.4a) of reflection symmetry. The predominantly radial branch $\omega_r$ decreases monotonically with increasing $k$. It often has a minimum at the corner $A$ and a maximum at the origin $O$ of the Brillouin zone.

Note that the maximum frequency of the three-branch spectrum is attained by the radial mode at $k = 0$. The frequency band is in the order of tens of megahertz, typically, for 5 $\mu$m garnets.

In principle, an inhomogeneous field applied near the edges of a finite lattice can be used to constrain the lattice to a constant predetermined density. Increasing the bias field $H_b$ then has the effect of lowering the branch $\omega_r$ relative to the others. The particular mode $\omega_r(A)$, which is usually the minimum of $\omega_r$, is by symmetry purely radial. At some critical value of $H_b$, one has the condition $\omega_r(A) = 0$, which indicates an instability of the lattice.[431] According to this instability mode one out of every three bubbles, arranged on a periodic sublattice of the complete lattice will collapse simultaneously, as has been observed experimentally.[431a]

For bubble diameters comparable to the nearest-neighbor bubble–bubble distance $R$, Eq. (22.25) is inaccurate. A better approximation utilizing a Fourier expansion of the exact magnetostatic energy of the set of circular cylinder bubbles forms the basis of computations limited to infinitesimal $k$. From these results one obtains "acoustic" velocities of longitudinal ($C_l = \partial\omega_l/\partial k$) and transverse ($C_t = \partial\omega_t/\partial k$) character, evaluated at $k = 0$. These velocities are of the order $\gamma A^{1/2}$ or greater for film thickness $h = 4l$ and bias fields in the range $H_b/4\pi M \le 0.2$ when the bubble diameter is at least half the bubble–bubble distance.[428] Since these velocities ($\ge 100$ m/sec) are predicted to be an order of magnitude higher than saturation velocities of bubble translation (Section 19), they appear to possess potential as information signal carriers.

A refinement of the theory based on Eq. (22.25) and (22.26) takes into account the deflection effect acting on a lattice of hard bubbles all of which have the same winding number.[432] The effect of the Lorentz-like gyrotropic force [Eq. (12.64)] changes the polarization of the waves from linear to elliptic. It also raises the frequency of one of the two acoustic modes at $k = 0$ from zero to a finite value. In such applications one must take care to use the large hard-bubble mass discussed in Section 13,C. Experimental preparation of a lattice with identically hard bubbles appears to be difficult. Probably important in practice is the effect of viscous damping, which was taken into account by one of the published dispersion computations.[429] The propagation of disturbances in an *overdamped* lattice is important in bubble-lattice storage.[426] A published prediction of a parametric mechanism for the excitation of lattice waves having nonvanishing wave vector by means of a uniform drive field raises the hope for future experimental investigations of the spectrum.[433] The one-dimensional bands describing the vibrational structure of a set of plane-parallel walls in a film have been calculated.[433a]

The energy coupling a bubble domain to a strain field is proportional to the wall area. Theoretically this interaction leads to coupled magnetoacoustic waves.[433b] Although the resulting dimensionless coefficient coupling radial oscillations to the acoustic field is small ($\sim 10^{-3}$), there are the possibilities of generating acoustic waves by magnetic excitation or bubble-breathing oscillations by acoustic excitation.[433b]

## 23. Microwave Effects

The primary response of a ferrimagnet to an applied field oscillating with a frequency in the microwave range is by coherent spin precessions that take place throughout the entire medium, whether single- or multidomain, and which are referred to as ferromagnetic-resonance and spin-wave precessions. We refer the reader to literature for such effects, which are studied in both single[418,434] and multidomain[435-437b] specimens. The natural resonances of domain-wall displacement, which actually represent three-dimensional spin waves localized to inhomogeneous regions near wall-center planes, are too low to be excited at microwave frequencies (see Fig. 22.2).[437] Nonetheless, there still exist significant phenomena relating microwave fields and the motion of domain walls. A summary of the experiments in this area has been published.[438] They fall into the following categories, which we discuss in this section:

*a. Bubble detection.* The presence of a bubble domain changes the absorption of microwave energy from that in a single-domain ferromagnetic medium. This phenomenon provides in principle a means of detecting the presence of a bubble.

*b. Bubble nucleation.* A high-intensity microwave field can cause the moment of a single-domain medium to precess locally at such a large angle that it flips to the other easy direction, thus nucleating a bubble domain.

*c. Wall pressure.* The presence of the microwave field so alters the energy of an existing domain configuration that effective pressures, which are static on the microwave time scale, act on the walls. There is a variety of such effects depending on rf polarization and wall geometry.


## A. BUBBLE DETECTION

An experimentally proven scheme for detecting bubbles by means of microwaves is based on the homogeneous-precession mode of the single-domain magnetic film that exists in the absence of bubbles.[439] The moment in this case precesses about a net field provided by the combination of anisotropy ($H_K$), demagnetization ($-4\pi M$), and external dc bias ($H_b$), with all field contributions acting normal to the film plane. Hence the fundamental small-amplitude natural circular resonant frequency is

$$\omega_r = \gamma(H_K - 4\pi M + H_b). \tag{23.1}$$

Higher resonant spin-wave modes exist but only this mode couples to a homogeneous microwave field and absorbs power from the supply under the simplest assumptions.

Suppose that the supply frequency is equal to the frequency $\omega_r$, so that the precession amplitude is maximal. Then the signal in an output circuit coupled inductively to this magnetic precession is also maximized. Introduction meanwhile of a single bubble domain by external means will disturb the uniformity of the fundamental precession and reduce the output signal.

Figure 23.1 shows schematic top (a) and front (b) sectional views of a coplanar microwave structure used successfully to detect bubbles in this way.[439] The shaded region in Fig. 23.1a represents a gold layer evaporated on a glass substrate. In one mode of operation, microwave input current flows along the central member 1 of the coplanar transmission line, which is sometimes tapered (not shown), generating a circulating microwave magnetic field that threads through the etched slots on both sides. This line terminates at point 2 in a short circuit to the ground provided by the mass of gold material surrounding it.

Slot-line 3 in Fig. 23.1a, which is also short-circuited at point 2, carries output microwaves. The output microwave current flows in opposite directions on opposite sides of the slot and is concentrated near the slot, which is threaded by the microwave magnetic field as indicated in Fig. 23.1b. By symmetry, the current and field patterns of the input and output lines are

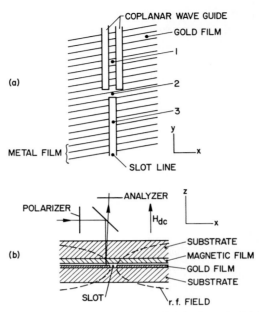

**Fig. 23.1.**  Schematic top (a) and front (b) sectional views of an experimental arrangement for observing microwave effects in bubble films. Depending on the particular experiment, microwave input power flows in along the coplanar or slot-line guides. Optics for double transmission Faraday observation of the magnetic film are indicated in (b) (after Dötsch[440]).

orthogonal to one another so that ideally no power is transmitted from input to output in the absence of magnetic material.

However, when this gold structure is pressed against a single-domain magnetic film, whose moment is normal to the plane, power begins to flow in the output line. The linearly polarized film-tangential input field $H_x$, which has its maximum amplitude in the magnetic film near a point above point 2, induces a circular precession of the moment **M** about the film normal. The consequent rf stray magnetic field generated by the rf component $M_y$ clearly links the output slot-line, thus transmitting power. Subsequent introduction of a bubble domain above point 2 by external means disrupts the uniformity of the fundamental precession mode. Its amplitude diminishes, thus reducing the measured output power; and this reduction indicates the presence of the bubble.

An actual experiment used a 0.3 $\mu$m thick gold layer whose tapered coplanar-guide input line was 12 $\mu$m wide at the point 2.[439] An input power of about 1 mW at a frequency of between 1 and 2 GHz applied to a GdGaYIG film produced an output of several hundred millivolts. Bubbles 5 $\mu$m in diameter were introduced by means of an attractive chevron bar of permalloy, whose magnetic state was varied by an external field, and were then detected with zero-to-one ratios as high as $6:1$.

Other garnet compositions that are more nearly like those used in practical bubble devices that have higher losses, were also used. Their resonance lines were too broad to transmit microwaves by the above scheme. However, placing a low-loss garnet film on one side of the microwave structure and the high-loss bubble material on the other side overcame this problem. For, in this arrangement the stray field of the bubble acts on the low-loss film to detune its resonance and thus diminish its transmission. The best detection, with a $3:1$ signal ratio, was found using, not the linear uniform precession, but another mode which resonated at 300 MHz. Interesting results involving a magnetoacoustic effect were also obtained.[439]

## B. BUBBLE NUCLEATION

With increasing applied rf power, the angle $\theta$ of the uniform-precession cone increases. Consequently, the film-normal component of magnetization $M_z = M \cos \theta$ decreases, as verified by means of the optical Faraday apparatus indicated schematically in Fig. 23.1b. In this experiment, the slot line carried the rf input while the coaxial line served to monitor the resonance transmission.[440] Remarkably large angles, as high as $\theta = 150°$ were reported at high bias fields. Such large cone angles are not observed in bulk specimens, where $\theta$ is limited to small values by a saturation phenomenon involving parametric excitation of spin waves.[91,441] When the bias field is decreased below the collapse field for bubble domains, the value that $\theta$ can be excited to is limited by domain nucleation thresholds. In particular, as soon as $\theta$ exceeds a critical value $\theta_{c1}$, the spins tip over and create a bubble. In a particular LaGaYIG film, $\theta_{c1}$ falls in the range $44°$ to $52°$, depending on bias field $H_b$. If the bubble diameter is large compared to the excitation region, the bubble remains centered on the excitation maximum. The moment of the new domain now also precesses, albeit with a larger central precession angle. Further increase of power increases the diameter of the bubble and raises $\theta$ at its center to another critical value $\theta_{c2}$, which ranges up to $83°$. At this point the central moment again reverses, and a bubble domain is created within the first one, thus producing a ring domain that is essentially a circular endless stripe domain. By appropriate progressive lowering of the bias field, this process may be repeated many times to ultimately generate a large system of concentric rings as illustrated by the photograph in Fig. 23.2.[438,442,442a]

As mentioned above, in bulk materials of low linewidth, $\theta$ is strongly limited by parametric excitation of short-wavelength spin waves; typically $\theta$

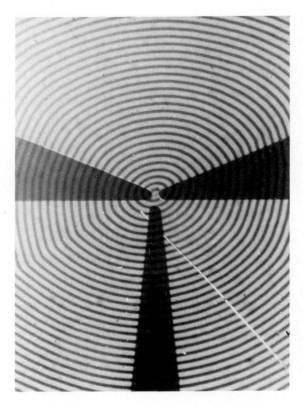

**Fig. 23.2.** Concentric circular domains generated by high-intensity microwaves. The somewhat different microwave structure from that of Fig. 23.1 also appears in the photograph (after Dötsch[440]).

does not exceed a few degrees. In thin films, however, the instability threshold is much higher.[443] Nonetheless, attainment of such large precession angles as those reported above is puzzling owing to the fact that for large $\theta$, effective anisotropy and demagnetizing fields diminish greatly. A simple analysis shows that the equation

$$\omega_r = \gamma[(H_K - 4\pi M)\cos \theta + H_b] \tag{23.2}$$

must replace Eq. (23.1) when $\theta$ is large. Thus, for example, at $\theta = 90°$ neither anisotropy nor demagnetization exert any torque on **M**, and they can contribute nothing to $\omega_r$, which reduces to $\omega_r = \gamma H_b$. In the LaGaYIG specimens investigated, $H_K - 4\pi M$ was about 100 times $H_b$.[440] Thus it is clear that $\omega_r$ vanishes near $\theta = 90°$, according to the uniform-precession model. The origin of the effective field maintaining precession at the experimental drive frequency given by Eq. (23.1) remains to be explained. It is interesting that large cone angles can also be excited over a broad band of input frequencies $\omega$ below $\omega_r$, but less efficiently than at $\omega = \omega_r$.

## C. WALL PRESSURE

We consider two types of wall pressure arising in the presence of microwave fields: For convenience we dub them (1) "dipolar" wall pressure and (2) "intrinsic" wall pressure.

*1. Dipolar pressure.* We saw how $|M_z|$ diminishes with increasing cone angle of ferromagnetic resonance. Due to the inhomogeneity of microwave excitation which exists in the presence of a domain wall, the resulting stray fields originating from the magnetic poles on the film surfaces produce a "dipolar" pressure on the wall.[444] If the gradient of $|M_z|$ is sufficiently smooth, this pressure pushes the wall in the direction of increasing $|M_z|$ as is evident by inspection of Fig. 23.3. Clearly the demagnetizing field $H_d$ at the wall opposes the domain with the larger moment.

Since $|M_z|$ decreases with increasing rf power, it is clear from the above that walls will ordinarily be repelled from regions of maximum rf intensity. This mechanism readily explains the expansion of ring domains away from the maximum of a sharply peaked inhomogeneous rf field distribution (point 2 in Fig. 23.1) to provide space for the nucleation of new domains as described above in connection with Fig. 23.2. Indeed the equilibrium radius of either a bubble or a ring domain centered on the rf maximum (point 2) is readily adjusted by varying the applied power.[440,444] The stability range of such microwave-stabilized rings may be as large as 20% of $4\pi M$, compared to circa 0.1% for static rings. The possibilities of measuring wall mobility of ring domains by modulating the rf power, e.g., by dynamic pulse collapse or ac response, have been demonstrated.[440]

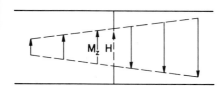

**Fig. 23.3.** Schematic view of a film with a smooth spatial gradient $|M_z|$, making clear that the demagnetizing field presses the wall toward the region of greater $|M_z|$.

Widening the input slot-line enlarges the region of high rf field intensity so that it is greater than a bubble diameter. Instead of expanding symmetrically about the point of rf maximum, each nucleated bubble domain in this case becomes repelled as a whole from the point of rf maximum, thus "stepping aside" so that the next bubble can nucleate. When excited by pulse-modulated microwave energy this process rapidly creates an array of bubbles whose equilibrium density qualitatively reflects the spatial distribution of the in-plane component of rf intensity. Comparison of the spatial distribution of $M_z$ indicated by the shading of the Faraday-effect photograph in Fig. 23.4a, with the bubble distribution of Fig. 23.4b illustrates this effect.[445] The precession amplitude vanishes directly above the center line of the input slot and has equal maxima on both sides. The bubbles accordingly accrete along the center line and at large distances but are absent from the maxima.

*2. Intrinsic pressure.* According to theory,[446-448] microwave excitation produces what one might call an "intrinsic" wall pressure to distinguish

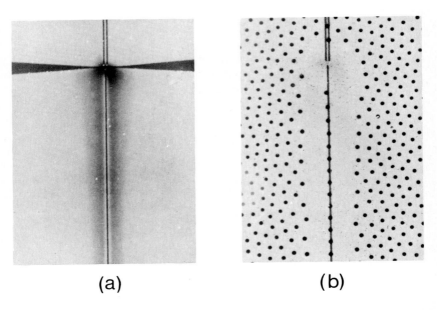

(a)                                        (b)

**Fig. 23.4.** Photographs showing effects produced with rf input following the slot line of Fig. 23.1. Shading of the Faraday effect (a) maps the amplitude of magnetic precession. The distribution of bubbles generated by higher intensity rf (b) accords with precession amplitude. In (a) an additional passive slot separates the input and coplanar guide shown in Fig. 23.1. The input slot is 2.5 $\mu$m wide (after Dötsch[445]).

from that of stray-field or dipolar origin discussed above. The most fundamental effect occurs in response to a circularly polarized drive field. Its
mechanism is as follows: Consider domain moments $\mathbf{M}_+$ oriented parallel
to $\pm z$ as indicated in Fig. 23.5a. They may be regarded as stabilized by static
effective fields $\mathbf{H}_{\mathrm{eff}\,\pm}$ due to anisotropy and demagnetization oriented parallel
to $\pm z$, respectively. According to the Landau–Lifshitz equation (Section
3,A)

$$\dot{\mathbf{M}} = \gamma \mathbf{H} \times \mathbf{M}, \tag{23.3}$$

each domain is capable of a small-amplitude precession at the natural Larmor
frequency $\omega_r = \gamma |\mathbf{H}_{\mathrm{eff}\,\pm}|$ in a right-hand-screw sense with respect to its
stabilizing field $\mathbf{H}_{\mathrm{eff}\,\pm}$.

Now suppose a circularly polarized externally supplied field $\mathbf{h}_{\mathrm{rf}}$ rotates at
frequency $\omega \gg \omega_r$ in the right-hand-screw sense about the $z$ axis. In this
case domains with moment $\mathbf{M}_+$ statically oriented parallel to $z$ are called
*Larmor domains* because their natural sense of precession is the same as
$\mathbf{h}_{\mathrm{rf}}$. Those with moment $\mathbf{M}_-$ are called *anti-Larmor* domains.

Consider now the motion of $\mathbf{M}_\pm$ purely in the field $\mathbf{H} = \mathbf{h}_{\mathrm{rf}}$. One substitutes

$$\mathbf{M} = \pm M\mathbf{z}_0 + \mathbf{m} \tag{23.4}$$

where $\mathbf{z}_0$ is a unit vector parallel to the $z$ axis and $\mathbf{m}$ is a deviation of $\mathbf{M}$ in
the $xy$ plane, into Eq. (23.3). Treating $\mathbf{m}$ as small and neglecting $\mathbf{H}_{\mathrm{eff}\,\pm}$, one

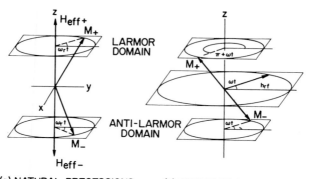

(a) NATURAL PRECESSIONS    (b) DRIVEN PRECESSIONS

**Fig. 23.5.** Schematic comparison of natural (a) and driven (b) precessions in domains, relating to the origin of intrinsic domain-wall pressure caused by circularly polarized microwave
field in high-frequency limit.

finds the solutions

$$\mathbf{m} = \mp \gamma M h_{rf}/\omega \tag{23.5}$$

for the Larmor and anti-Larmor domains, respectively. The dynamic domain energy densities, which are $-\mathbf{M} \cdot \mathbf{H} = \pm \gamma M h_{rf}^2/\omega$, respectively, differ in sign because of the opposing phases of $\mathbf{m}$. The energy difference

$$P = 2\gamma M h_{rf}^2/\omega \tag{23.6}$$

represents a domain-wall pressure tending to expand the anti-Larmor domains at the expense of the higher-energy Larmor domains.[446] Since $\mathbf{H}_{eff\pm}$ is neglected, Eq. (23.6) is correct only in the limit $\omega \gg \gamma H_{eff\pm}$. Since $|M_z|$ is the same in leading order for the two types of domain, this effect is not inherently connected with the gradients of $|\mathbf{M}_z|$ that are essential to the dipolar pressure and the experiments discussed above.

A detailed theory of intrinsic pressure for arbitrary drive frequencies takes into account the effects of anisotropy, dipolar fields from magnetic poles on domain walls, and damping, for an infinite array of parallel domains in a thick demagnetized platelet. Effects of magnetic poles on platelet faces are not included. The results are as follows:[447,448]

(i) *Elliptic polarization.*   When the rf field with frequency $\omega$ is elliptically polarized with principal axes parallel and perpendicular to the walls, the sign of the pressure examined as a function of $\omega$ changes at each of two resonant frequencies

$$\omega_1 = \gamma H_K \qquad \omega_2 = \gamma [H_K(H_K + 4\pi M)]^{1/2}, \tag{23.7}$$

where $H_K = 2K/M$ is the uniaxial-anisotropy field. The full dependence of pressure on $\omega$ is shown in Fig. 23.6 for the special case $Q = H_K/4\pi M = 1$ and resonance linewidth $\delta = \Delta\omega/\gamma H_K = 0.1$.[447] Experimental observations of remanent-moment changes induced by microwaves in high-power phase-shifters utilizing garnet toroids lend qualitative support to the existence of this pressure.[446] Evidence for this kind of wall pressure in bubble films does not exist.

(ii) *Linear polarization.*   Application of a linearly polarized rf field $h_{rf}$ either parallel or perpendicular to the wall creates no pressure by symmetry. More generally the predicted pressure is proportional to $\sin 2\phi$, where $\phi$ is the angle between $\mathbf{h}_{rf}$ and the wall plane. This effect, therefore, averages to zero for bubble or ring domains. The pressure is maximal for $|\sin 2\phi| = 1$ at the resonance $\omega = \omega_2$, where the pressure is equivalent to an effective

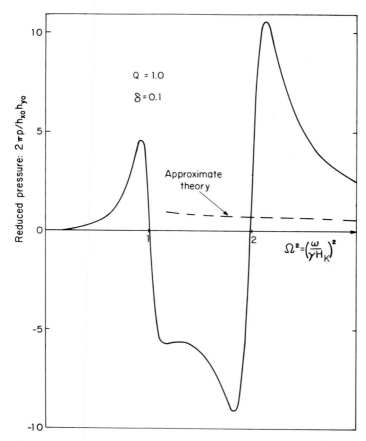

**Fig. 23.6.** Pressure on domain walls (in reduced units) induced by elliptically polarized magnetic field plotted as a function of square of reduced frequency, according to exact theory. The dashed curve is the high-frequency limit given by Eq. (23.6). Here $\delta = \Delta\omega/\gamma H_K = 0.1$ and $Q = K/2\pi M^2 = 1$ (after Schlömann[447]).

field $H_z = h_{rf}^2(1 + Q^{-1})/8\pi\delta M$.[447] No experimental test of this effect has yet been made.

No attempt has yet been made to extend the "intrinsic" energy considera-tion to the case of a gradient of rf excitation. However, a naive argument can be made if the ferromagnetic medium is assumed to be in a resonant condi-tion at all points. For then the applied rf field is always orthogonal (leading by 90°) to the instantaneous **m** and the stored applied-field energy vanishes.

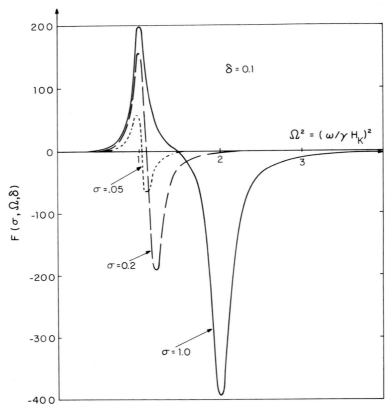

**Fig. 23.7.** Frequency dependence of wall-alignment energy for $\delta \equiv \Delta\omega/\gamma H_K = 0.1$ and three values of $\sigma \equiv 2\pi M^2/K$ (after Schlömann[447]).

The  principal  internal  energy  (assuming  $Q \gg 1$)  is  then  the  anisotropy $K \sin^2 \theta$  connected  with  the  constant  precession-cone  angle  $\theta$.  If  there  is a  gradient  of  $\theta$  because  of  inhomogeneous  excitation  or  whatever,  then  the wall pressure is roughly

$$P \approx K(\partial \sin^2 \theta/\partial n) = 2\pi QM^2(\partial \sin^2 \theta/\partial n) \tag{23.8}$$

where $n$ is normal to the wall. It thus appears that the pressure due to internal energy should dominate that due to stray-field (dipolar) energy, that is of order $M^2\nabla|\sin \theta|$ and that was invoked above in the discussion of experiments, whenever the inequality $Q|\sin \theta| > C$ is satisfied. Here $C (>0)$ is some constant of order unity.

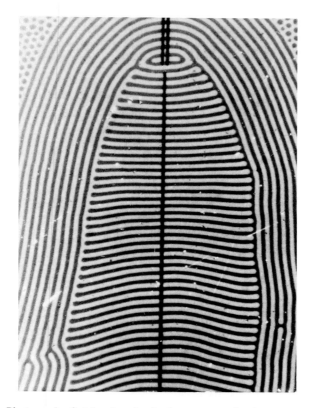

**Fig. 23.8.** Photograph of stripe domains in the presence of pulse-modulated microwave power applied to the slot line (see Fig. 23.1). The central domains align parallel to the rf field or its intensity gradient. The outer domains appear to suffer repulsion along the negative gradient of rf intensity (after Dötsch[445]).

*(iii) Torque on walls.* A torque tending to rotate the walls occurs generally when the rf field is linearly polarized. It is derivable from an orientational energy $(h_{rf}^2/16\pi Q)F \sin^2 \phi$, where the factor $F$, plotted in Fig. 23.7, has extrema at the resonant frequencies of Eq. (23.7) and a change of sign between them. The sign of $F$ is such as to align walls parallel to $\mathbf{h}_{rf}$ at $\omega = \omega_1$ and perpendicular to $h_{rf}$ at $\omega = \omega_2$.[447,448] This prediction accords qualitatively with the experimental observation of parallel alignment to a pulse-modulated microwave field whose frequency is below the resonance.[437,445] This effect, shown by the domains in the central portion of the photograph in Fig. 23.8, might alternatively be connected with the *gradient* of the micro-

wave intensity.[445] The stripe domain walls closer to the edge of this photo-graph show evidence of repulsion from the high-intensity regions (i.e., along the intensity gradient) in the magnetic film. More recently, a reorientation of domain walls from parallel to perpendicular to $h_{rf}$ has been observed with increasing microwave frequency.[442a]

# References

1. F. Bloch, *Z. Physik* **74**, 295 (1932).
2. L. Landau and E. Lifshitz, *Physik A, Soviet Union* **8**, 153 (1935).
3. L. Néel, *Cahiers Phys.* **25**, 1–20 (1944).
4. C. Kittel and J. K. Galt, *in* "Solid State Physics," Vol. 3 (F. Seitz and D. Turnbull, eds.), p. 437. Academic Press, New York, 1956.
5. J. F. Dillon, Jr., *in* "Magnetism," Vol. III (G. Rado and H. Suhl, eds.), p. 415. Academic Press, New York, 1963.
6. S. Middelhoek, "Ferromagnetic Domains in Thin Ni-Fe Films." Drukkerij Wed. G. Van Soest N. V., Amsterdam, 1961.
7. M. Prutton, "Thin Ferromagnetic Films." Butterworth, Washington, D.C., 1964.
8. R. F. Soohoo, "Magnetic Thin Films." Harper and Row, New York, 1965.
9. M. S. Cohen, *in* "Magnetism and Magnetic Materials," 1968 Digest (H. Chang and T. R. McGuire, eds.), p. 126. Academic Press, New York, 1968.
10. D. J. Craik and R. S. Tebble, "Ferromagnetism and Ferromagnetic Domains." North-Holland, Amsterdam, 1965; R. S. Tebble and D. J. Craik, "Magnetic Materials." Wiley-Interscience, London, 1969.
11. A. Hubert, "Theorie der Domänenwände in Geordneten Medien." Springer-Verlag, Berlin, 1974 (in German).
12. C. Kooy and U. Enz, *Philips Res. Repts.* **15**, 7 (1960).
13. A. H. Bobeck, *Bell Syst. Tech. J.* **46**, 1901 (1967).
14. M. Nalecza, ed., "Cylindrical Magnetic Domains in Digital Technology." State Educational Publishing Office, Poland, 1973 (in Polish).
15. T. H. O'Dell, "Magnetic Bubbles." Macmillan, London, 1974.
16. A. B. Smith, "Bubble Domain Memory Devices." Artech House, Dedham, Massachusetts, 1974.
17. A. H. Bobeck and E. Della Torre, "Magnetic Bubbles." North-Holland, Amsterdam, 1975.
18. H. Chang, "Magnetic Bubble Technology: Integrated-Circuit Magnetics for Digital Storage and Processing." IEEE Press and Wiley, New York, 1975.
19. H. Lachowicz, M. Maszkiewicz, and H. Szymczak, eds., Magnetic Bubbles, *Procceedings of the Winter School on New Magnetic Materials.* Polish Scientific Publishers, Warsaw 1976.
20. F. B. Hagedorn, *AIP Conf. Proc.* **5**, 72 (1972).
21. S. Konishi, *Butsuri* **30**, 178 (1975) (in Japanese).
22. K. Oteru, ed., "Handbook of Bubble Technology." Ohm Book Company, Tokyo, 1976 (in Japanese).
23. F. H. de Leeuw, *Physica* **86–88B**, 1320 (1977).

24. J. C. Slonczewski and A. P. Malozemoff, *in* "Physics of Magnetic Garnets," *Proc. Int. School of Physics "Enrico Fermi,"* Course LXX (A. Paoletti, ed.), p. 134. North Holland, New York, 1978.

25. O. Voegeli, B. A. Calhoun, L. L. Rosier, and J. C. Slonczewski, *AIP Conf. Proc.* **24**, 617 (1975).

26. L. L. Rosier, D. M. Hannon, H. L. Hu, L. F. Shew, and O. Voegeli, *AIP Conf. Proc.* **24**, 620 (1975).

27. B. A. Calhoun, J. S. Eggenberger, L. L. Rosier, and L. F. Shew, *IBM J. Res. Develop.* **20**, 368 (1976).

27a. H. L. Hu, T. J. Beaulieu, D. W. Chapman, D. M. Franich, G. R. Henry, L. L. Rosier and L. F. Shew, *J. Appl. Phys.* **49**, 1913 (1978).

28. W. F. Brown, Jr., "Magnetostatic Principles in Ferromagnetism." North-Holland, New York, 1962; "Micromagnetics." Interscience, New York, 1963.

29. S. Chikazumi, "Physics of Magnetism." Wiley, New York, 1964.

30. A. H. Morrish, "The Physical Principles of Magnetism." Wiley, New York, 1965.

31. "Physics of Magnetic Garnets," *Proc. Int. School of Physics "Enrico Fermi,"* Course LXX (A. Paoletti, ed.). North Holland, New York, 1978.

32. J. W. Nielsen, S. L. Blank, D. H. Smith, G. P. Vella-Coleiro, F. B. Hagedorn, R. L. Barns and W. A. Biolsi, *J. Elec. Mat.* **3**, 693 (1974).

33. P. Chaudhari, J. J. Cuomo, and R. J. Gambino, *IBM J. Res. Develop.* **17**, 66 (1973).

34. C. H. Bajorek and R. J. Kobliska, *IBM J. Res. Develop.* **20**, 271 (1976).

35. D. Treves, *Phys. Rev.* **125**, 1843 (1962); *J. Appl. Phys.* **36**, 1033 (1965).

36. R. L. White, *J. Appl. Phys.* **40**, 1061 (1969).

37. A. B. Harris and S. Kirkpatrick, *Phys. Rev.* **B16**, 542 (1977).

37a. J. T. Carlo, D. C. Bullock, and F. G. West, *IEEE Trans. Magn.* **MAG-10**, 626 (1974).

38. J. C. Slonczewski, A. P. Malozemoff, and E. A. Giess, *Appl. Phys. Lett.* **24**, 396 (1974).

38a. R. D. Henry and D. M. Heinz, *AIP Conf. Proc.* **18**, 194 (1973).

39. A. C. Gerhardstein and P. E. Wigen, *Phys. Rev.* **B18**, 2218 (1978).

39a. D. R. Krahn, P. E. Wigen, and S. L. Blank, *J. Appl. Phys.* **50**, 2189 (1979).

40. P. W. Shumate, Jr., *J. Appl. Phys.* **44**, 5075 (1973).

41. A. H. Bobeck, D. H. Smith, E. G. Spencer, L. G. Van Uitert, and E. M. Walters, *IEEE Trans. Magn.* **MAG-7**, 461 (1971).

42. A. Rosencwaig and W. J. Tabor, *AIP Conf. Proc.* **5**, 57 (1972).

43. W. J. Tabor, G. P. Vella-Coleiro, F. B. Hagedorn, and L. G. Van Uitert, *J. Appl. Phys.* **45**, 3617 (1974).

44. W. T. Stacy, A. B. Voermans, and H. Logmans, *Appl. Phys. Lett.* **29**, 817 (1976).

45. R. Wolfe, R. C. LeCraw, S. L. Blank, and R. D. Pierce, *Appl. Phys. Lett.* **29**, 815 (1976).

46. D. J. Breed, A. M. J. van der Heijden, H. Logmans, and A. B. Voermans, *J. Appl. Phys.* **49**, 939 (1978).

47. D. J. Breed, W. T. Stacy, A. B. Voermans, H. Logmans, and A. M. J. van der Heijden, *IEEE Trans. Magn.* **MAG-13**, 1087 (1977).

48. A. P. Malozemoff and J. C. DeLuca, *J. Appl. Phys.* **45**, 3562 (1974).

49. R. A. Abram, R. J. Fairholme, M. D. R. Tench, and K. A. Gehring, *J. Phys.* **D8**, 94 (1975).

50. J. Kaczér and R. Gemperle, *Czech. J. Phys.* **B11**, 510 (1961).

51. A. A. Thiele, A. H. Bobeck, E. Della Torre, and U. F. Gianola, *Bell Syst. Tech. J.* **50**, 711 (1971); A. A. Thiele, *Bell Syst. Tech. J.* **48**, 3287 (1969); **50**, 725 (1971); *J. Appl. Phys.* **41**, 1139 (1970).

52. J. A. Cape and G. W. Lehman, *J. Appl. Phys.* **42**, 5732 (1971).

53. Y. O. Tu, *J. Appl. Phys.* **42**, 5704 (1971).

54. F. A. de Jonge and W. F. Druyvesteyn, *Festkörperprobleme* **XII**, 531 (1972).

55. P. V. Cooper and D. J. Craik, *J. Phys. D: Appl. Phys.* **6**, 1393 (1973).

56. E. Della Torre, *IEEE Trans. Magn.* **MAG-4**, 822 (1970).
57. L. Gál, *Phys. Stat. Sol. (a)* **28**, 181 (1975).
58. H. Callen and R. M. Josephs, *J. Appl. Phys.* **42**, 1977 (1971).
59. J. Kaczér and I. Tomáš, *Phys. Stat. Sol. (a)* **10**, 619 (1972).
59a. T. G. W. Blake and E. Della Torre, *J. Appl. Phys.* **50**, 2192 (1979).
60. F. B. Hagedorn, *J. Appl. Phys.* **41**, 1161 (1970).
61. E. Schlömann, *IEEE Trans. Magn.* **MAG-10**, 11 (1974).
62. M. H. H. Höfelt, *IEEE Trans. Magn.* **MAG-9**, 621 (1973).
63. I. Tomáš, *Phys. Stat. Sol. (a)* **21**, 329 (1974).
64. A. H. Bobeck, S. L. Blank, and H. J. Levinstein, *Bell Syst. Tech. J.* **51**, 1431 (1972).
65. A. Rosencwaig, *Bell Syst. Tech. J.* **51**, 1440 (1972).
66. A. H. Bobeck, S. L. Blank, and H. J. Levinstein, *AIP Conf. Proc.* **10**, 498 (1973).
67. T. L. Hsu, *AIP Conf. Proc.* **24**, 624 (1975).
68. P. Hansen, *Appl. Phys. Lett.* **25**, 241 (1974).
69. W. J. DeBonte, *AIP Conf. Proc.* **10**, 349 (1972).
70. J. Haisma, G. Bartels, W. F. Druyvesteyn, U. Enz, J. P. Krumme and A. G. H. Verhulst, *IEEE Trans. Magn.* **MAG-10**, 630 (1974).
71. Y. S. Lin and P. J. Grundy, *J. Appl. Phys.* **45**, 4084 (1974).
72. H. Uchishiba, H. Tominaga, and K. Asama, *IEEE Trans. Magn.* **MAG-9**, 381 (1973).
73. J. J. Zebrowski, *Phys. Stat. Sol. (a)* **37**, 407 (1976).
74. R. Wolfe and J. C. North, *Appl. Phys. Lett.* **25**, 122 (1974).
75. G. S. Almasi, E. A. Giess, R. J. Hendel, G. E. Keefe, Y. S. Lin and M. Slusarczuk, *AIP Conf. Proc.* **24**, 630 (1975).
75a. Y. Okabe, *IEEE Trans. Magn.* **MAG-14**, 502 (1978).
75b. B. E. Argyle, M. H. Kryder, R. E. Mundie, and J. C. Slonczewski, *IEEE Trans. Magn.* **MAG-14**, 593 (1978).
75c. Y. S. Lin, D. B. Dove, S. Schwarzl and C. C. Shir, *IEEE Trans. Magn.* **MAG-14**, 494 (1978).
75d. I. L. Sanders and M. H. Kryder, *J. Appl. Phys.* **50**, 2252 (1979).
75e. C. C. Shir, *J. Appl. Phys.* **50**, 2270 (1979).
76. R. Suzuki, M. Takahashi, T. Kobayashi, and Y. Sugita, *Appl. Phys. Lett.* **26**, 342 (1975).
77. I. B. Puchalska, H. Jouve, and R. H. Wade, *J. Appl. Phys.* **48**, 2069 (1977).
78. I. B. Puchalska and H. Jouve, *IEEE Trans. Magn.* **MAG-13**, 1178 (1977).
79. T. Obokata, H. Uchishiba, and K. Asama, *IEEE Trans. Magn.* **MAG-13**, 1181 (1977).
80. H. Uchishiba, T. Obokata, and K. Asama, *J. Appl. Phys.* **48**, 2604 (1977).
81. T. L. Gilbert, *Phys. Rev.* **100**, 1243 (1955).
82. R. K. Wangsness, *Phys. Rev.* **91**, 1085 (1953).
83. R. C. LeCraw, J. P. Remeika, and H. Matthews, *J. Appl. Phys.* **36**, 901 (1965).
84. R. C. LeCraw, S. L. Blank, and G. P. Vella-Coleiro, *Appl. Phys. Lett.* **26**, 402 (1975).
85. R. C. LeCraw, S. L. Blank, G. P. Vella-Coleiro, and R. D. Pierce, *AIP Conf. Proc.* **29**, 91 (1976).
86. G. P. Vella-Coleiro, S. L. Blank, and R. C. LeCraw, *Appl. Phys. Lett.* **26**, 402 (1975).
87. N. Ohta, T. Ikeda, F. Ishida, and Y. Sugita, *J. Phys. Soc. Japan* **43**, 705 (1977).
88. D. C. Cronemeyer, *AIP Conf. Proc.* **18**, 85 (1974).
89. R. Hasegawa, *J. Appl. Phys.* **46**, 5263 (1975).
90. R. C. LeCraw and R. D. Pierce, *AIP Conf. Proc.* **5**, 200 (1972).
90a. T. Ikeda, N. Ohta, F. Ishida, and Y. Sugita, *J. Appl. Phys.* **49**, 1598 (1978).
91. M. Sparks, "Ferromagnetic Relaxation Theory." McGraw-Hill, New York, 1964.
92. G. P. Vella-Coleiro, D. H. Smith, and L. G. Van Uitert, *Appl. Phys. Lett.* **21**, 36 (1972).
93. G. P. Vella-Coleiro, *AIP Conf. Proc.* **10**, 424 (1973).
94. F. H. de Leeuw, *IEEE Trans. Magn.* **MAG-9**, 614 (1973).

95. F. H. de Leeuw and J. M. Robertson, *J. Appl. Phys.* **46**, 3182 (1975).
96. H. Nakanishi, K. Minegishi, K. Hoshikawa, N. Inoue, K. Takimoto and Y. Suemune, *Japan J. Appl. Phys.* **15**, 2267 (1976).
97. C. H. Tsang, R. L. White, and R. M. White, *J. Appl. Phys.* **49**, 1838 (1978).
98. G. P. Vella-Coleiro, *IEEE Trans. Magn.* **MAG-13**, 1163 (1977).
99. R. W. Shaw, R. M. Sandfort, and J. W. Moody, "Characterization Techniques Study Report: Magnetic Bubble Materials," ARPA-Contract Report No. 1533; Contract No. DAAH01-72-C-0490. Advanced Research Projects Agency, Washington, D.C., July 1972.
100. P. W. Shumate, Jr., *IEEE Trans. Magn.* **MAG-7**, 586 (1971).
101. R. M. Josephs, *AIP Conf. Proc.* **10**, 286 (1973).
102. B. E. Argyle, R. J. Gambino, and K. Y. Ahn, *AIP Conf. Proc.* **24**, 564 (1975).
103. A. P. Malozemoff and K. Papworth, *J. Phys.* **D8**, 1149 (1975).
104. M. Takahashi, T. Suzuki, O. Kobayashi, and T. Miyahara, *Japan J. Appl. Phys.* **14**, 415 (1975).
105. R. V. Telesnin, A. G. Shishkov, E. N. Ilicheva, N. G. Kanavina, and N. A. Ekonomov, *Phys. Stat. Sol.* (*a*) **12**, 303 (1972).
106. G. R. Woolhouse and P. Chaudhari, *Phys. Stat. Sol.* (*a*) **19**, K3 (1973).
107. K. R. Papworth, *Phys. Stat. Sol.* (*a*) **22**, 373 (1974).
108. G. R. Woolhouse and P. Chaudhari, *AIP Conf. Proc.* **18**, 247 (1974).
109. B. R. Brown, G. R. Henry, R. W. Koepcke, and C. E. Wieman, *IEEE Trans. Magn.* **MAG-11**, 1391 (1975).
110. J. M. Nemchik and S. H. Charap, *Metal. Trans.* **2**, 635 (1971).
111. J. A. Seitchik, W. D. Doyle, and G. K. Goldberg, *J. Appl. Phys.* **42**, 1272 (1971).
112. B. E. Argyle and A. P. Malozemoff, *AIP Conf. Proc.* **10**, 344 (1973).
113. B. E. Argyle and A. Halperin, *IEEE Trans. Magn.* **MAG-9**, 238 (1973).
114. D. C. Fowlis and J. A. Copeland, *AIP Conf. Proc.* **10**, 393 (1973).
115. A. P. Malozemoff, *J. Magn. and Magnetic Mat.* **3**, 234 (1976).
116. A. Marsh, R. J. Fairholme, and G. P. Gill, *IEEE Trans. Magn.* **MAG-7**, 470 (1971).
117. J. A. Cape, *J. Appl. Phys.* **43**, 3551 (1972).
118. E. I. Il'yashenko, I. G. Avaeva, V. B. Kravchenko, and S. N. Matveyev, *Phys. Stat. Sol.* (*a*) **36**, K1 (1976).
119. F. C. Rossol, *J. Appl. Phys.* **40**, 1082 (1969).
120. G. E. Moore, *IEEE Trans. Magn.* **MAG-7**, 751 (1971).
121. G. P. Vella-Coleiro and T. J. Nelson, *Appl. Phys. Lett.* **24**, 397 (1974).
122. T. Ikuta and R. Shimizu, *Rev. Sci. Inst.* **44**, 1412 (1973).
123. F. C. Rossol, *AIP Conf. Proc.* **10**, 359 (1973).
124. A. P. Malozemoff, *IBM Tech. Disclosure Bull.* **15**, 2756 (1973).
125. J. C. Slonczewski, A. P. Malozemoff, and O. Voegeli, *AIP Conf. Proc.* **10**, 458 (1973).
126. F. B. Humphrey, *IEEE Trans. Magn.* **MAG-11**, 1679 (1975).
127. M. H. Kryder and A. Deutsch, *Proc. of the SPIE Technical Symposium on High Speed Optical Techniques, San Diego, California, August 26–27, 1976.*
128. G. P. Vella-Coleiro, F. B. Hagedorn, and S. L. Blank, *Appl. Phys. Lett.* **26**, 69 (1975).
129. G. P. Vella-Coleiro, *Appl. Phys. Lett.* **29**, 445 (1976).
130. G. P. Vella-Coleiro, *J. Appl. Phys.* **47**, 3287 (1976).
131. F. C. Rossol, *IEEE Trans. Magn.* **MAG-7**, 142 (1971).
132. K. Yoshimi, S. Fujiwara, F. Yamauchi, and T. Furuoya, *IEEE Trans. Magn.* **MAG-8**, 669 (1972).
133. T. Kobayashi, P. K. George, and F. B. Humphrey, *IEEE Trans. Magn.* **MAG-12**, 202 (1976).

133a. V. G. Kleparsky, M. A. Rozenblat, and A. M. Romanov, *IEEE Trans. Magn.* **MAG-11,** 1130 (1975).

134. L. P. Ivanov, A. S. Logginov, V. V. Randoshkin, and R. V. Telesnin, *JETP Lett.* **23,** 576 (1976); R. V. Telesnin, S. M. Zimacheva, and V. V. Randoshkin, *Sov. Phys. Solid State* **19,** 528 (1977).

134a. L. P. Ivanov, A. S. Logginov, V. V. Randoshkin, and R. V. Telesnin, *Sov. Phys. Solid State* **19,** 1097 (1977).

135. F. H. de Leeuw, *IEEE Trans. Magn.* **MAG-13,** 1172 (1977).

135a. M. Labrune, J. Miltat, and M. Kléman, *J. Appl. Phys.* **49,** 2013 (1978).

135b. L. M. Dedukh, V. I. Nikitenko, A. A. Polyanskii, and L. S. Uspenskaya, *JETP Lett.* **26,** 324 (1978).

136. M. Hirano, M. Kaneko, T. Yoshida, and T. Tsushima, *Japan J. Appl. Phys.* **16,** 661 (1977).

137. T. M. Morris, G. J. Zimmer, and F. B. Humphrey, *J. Appl. Phys.* **47,** 721 (1976).

138. K. Ju, G. J. Zimmer, and F. B. Humphrey, *Appl, Phys. Lett.* **28,** 741 (1976).

139. A. P. Malozemoff and S. Maekawa, *J. Appl. Phys.* **47,** 3321 (1976).

140. D. J. Craik and G. Myers, *Phil. Mag.* **31,** 489 (1975).

141. G. A. Jones and P. Dunk, *Proc. International Conference on Magnetism 1976*, p. 1045. North-Holland, Amsterdam, 1977.

142. T. Ikuta and R. Shimizu, *J. Phys.* **E9,** 721 (1976).

143. P. J. Grundy, D. C. Hothersall, G. A. Jones, B. K. Middleton, and R. S. Tebble, *Phys. Stat. Sol. (a)* **9,** 79 (1972).

144. P. J. Grundy and S. R. Herd, *Phys. Stat. Sol. (a)* **20,** 295 (1973).

145. P. Chaudhari and S. R. Herd, *IBM J. Res. Develop.* **20,** 102 (1976).

146. P. J. Grundy, G. A. Jones, and R. S. Tebble, *AIP Conf. Proc.* **24,** 541 (1975).

147. M. I. Darby and P. J. Grundy, *AIP Conf. Proc.* **24,** 610 (1975).

148. B. Barbara, J. Magnin, and H. Jouve, *Appl. Phys. Lett.* **31,** 133 (1977).

149. J. Engemann and T. Hsu, *Appl. Phys. Lett.* **30,** 125 (1977).

150. E. B. Moore, B. A. Calhoun, and K. Lee, *J. Appl. Phys.* **49,** 1879 (1978).

150a. G. P. Vella-Coleiro, F. B. Hagedorn, S. L. Blank, and L. C. Luther, *J. Appl. Phys.* **50,** 2176 (1979).

150b. B. Keszei and M. Pardavi-Horvath, *IEEE Trans. Magn.* **MAG-14,** 605 (1978).

150c. J. C. Walling, *J. Appl. Phys.* **50,** 2179 (1979).

151. K. J. Sixtus and L. Tonks, *Phys. Rev.* **37,** 930 (1931); **39,** 357 (1932); **42,** 419 (1932); **43,** 70, 931 (1933).

152. J. K. Galt, *Phys. Rev.* **85,** 664 (1952).

153. F. B. Hagedorn and E. M. Gyorgy, *J. Appl. Phys.* **32,** 282S (1961).

154. G. Asti, M. Colombo, M. Giudici, and A. Levialdi, *J. Appl. Phys.* **36,** 3581 (1965); *J. Appl. Phys.* **39,** 2039 (1968).

155. M. A. Wanas, *J. Appl. Phys.* **38,** 1019 (1967).

156. H. Umebayashi and Y. Ishikawa, *J. Phys. Soc. Japan* **20,** 2193 (1965).

157. C. H. Tsang and R. L. White, *AIP Conf. Proc.* **24,** 749 (1975).

158. C. H. Tsang, R. L. White, and R. M. White, *AIP Conf. Proc.* **29,** 552 (1976).

159. G. P. Vella-Coleiro, D. H. Smith, and L. G. Van Uitert, *IEEE Trans. Magn.* **MAG-7,** 745 (1971).

160. G. P. Vella-Coleiro, D. H. Smith, and L. G. Van Uitert, *J. Appl. Phys.* **43,** 2428 (1972).

161. T. Hashimoto, T. Miyoshi, T. Hamasaki, T. Okada, Y. Suemune, and T. Hattanda, *Appl. Phys. Lett.* **25,** 356 (1974).

162. I. Tomáš, *Phys. Stat. Sol. (a)* **30,** 587 (1975).

163. B. A. Calhoun, E. A. Giess, and L. L. Rosier, *Appl. Phys. Lett.* **18,** 287 (1971); L. L. Rosier and B. A. Calhoun, *IEEE Trans. Magn.* **MAG-7,** 747 (1971).

164. A. Ashkin and J. M. Dziedzic, *Appl. Phys. Lett.* **21**, 253 (1972).
165. B. R. Brown, *AIP Conf. Proc.* **29**, 69 (1976).
166. S. Konishi, T. Hsu, and B. R. Brown, *Appl. Phys. Lett.* **30**, 497 (1977).
167. J. Kaczér and R. Gemperle, *Czech. J. Phys.* **B11**, 157 (1961).
168. T. M. Morris and A. P. Malozemoff, *AIP Conf. Proc.* **18**, 242 (1974).
169. G. J. Zimmer, T. M. Morris, and F. B. Humphrey, *IEEE Trans. Magn.* **MAG-10**, 651 (1974).
170. L. Gál, G. J. Zimmer, and F. B. Humphrey, *Phys. Stat. Sol.* (a) **30**, 561 (1975).
171. A. H. Bobeck, I. Danylchuk, J. P. Remeika, L. G. Van Uitert, and E. M. Walters, *Ferrites: Proceedings of the International Conference*, p. 361. University of Tokyo Press, Tokyo, 1971.
172. S. Konishi, K. Mizuno, F. Watanabe, and K. Narita, *AIP Conf. Proc.* **34**, 145 (1976); S. Konishi, K. Mizuno, and K. Narita, *J. Appl. Phys.* **47**, 3759 (1976).
173. J. W. F. Dorleijn and W. F. Druyvesteyn, *Appl. Phys.* **1**, 167 (1973).
174. G. P. Vella-Coleiro, *AIP Conf. Proc.* **24**, 595 (1975).
175. A. P. Malozemoff, *J. Appl. Phys.* **44**, 5080 (1973).
176. J. A. Cape, W. F. Hall, and G. W. Lehman, *Phys. Rev. Lett.* **30**, 801 (1973).
177. J. A. Cape, W. F. Hall, and G. W. Lehman, *J. Appl. Phys.* **45**, 3572 (1974).
178. A. P. Malozemoff and J. C. Slonczewski, *Phys. Rev. Lett.* **29**, 952 (1972).
179. O. Voegeli and B. A. Calhoun, *IEEE Trans. Magn.* **MAG-9**, 617 (1973).
180. T. J. Gallagher and F. B. Humphrey, *Appl. Phys. Lett.* **31**, 235 (1977).
181. T. H. O'Dell, *Phil. Mag.* **27**, 595 (1973).
182. K. R. Papworth, *IEEE Trans. Magn.* **MAG-10**, 638 (1974).
183. M. A. Wanas, *J. Appl. Phys.* **44**, 1831 (1973).
184. M. A. Wanas, *J. Phys.* **D7**, 739 (1974).
185. E. Della Torre and M. Y. Dimyan, *IEEE Trans. Magn.* **MAG-6**, 487 (1970).
186. S. Shiomi, T. Fujii, and S. Uchiyama, *Japan J. Appl. Phys.* **14**, 1911 (1975).
187. T. J. Beaulieu and B. A. Calhoun, *Appl. Phys. Lett.* **28**, 290 (1976).
188. J. A. Copeland, *IEEE Trans. Magn.* **MAG-9**, 660 (1973).
189. G. J. Zimmer, L. Gál, and F. B. Humphrey, *J. Appl. Phys.* **48**, 362 (1977).
190. M. Hirano, M. Kaneko, and T. Tsushima, *IEEE Trans. Magn.* **MAG-13**, 1175 (1977).
191. M. H. Kryder, L. J. Tao, and C. H. Wilts, *IEEE Trans. Magn.* **MAG-13**, 1626 (1977).
192. R. F. Waites, *IEEE Trans. Magn.* **MAG-12**, 694 (1976).
193. D. M. Hannon, *J. Appl. Phys.* **49**, 1847 (1978).
194. R. Wolfe J. C. North, W. A. Johnson, R. R. Spiwak, L. J. Varnerin and R. F. Fischer, *AIP Conf. Proc.* **10**, 339 (1973).
195. J. A. Copeland, J. G. Josenhans, and R. R. Spiwak, *IEEE Trans. Magn.* **MAG-9**, 489 (1973).
196. R. M. Goldstein and J. A. Copeland, *AIP Conf. Proc.* **10**, 383 (1973).
197. R. M. Goldstein, M. Shoji, and J. A. Copeland, *J. Appl. Phys.* **44**, 5090 (1973).
198. P. K. George and J. L. Archer, *AIP Conf. Proc.* **18**, 116 (1974); P. K. George and A. J. Hughes, *IEEE Trans. Magn.* **MAG-12**, 137 (1976).
199. G. S. Almasi and Y. S. Lin, *IEEE Trans. Magn.* **MAG-12**, 160 (1976).
200. S. Fujiwara, K. Yoshimi, and T. Furuoya, *AIP Conf. Proc.* **5**, 165 (1972).
201. T. J. Walsh and S. H. Charap, *AIP Conf. Proc.* **24**, 550 (1975).
201a. M. E. Jones and R. D. Enoch, *IEEE Trans. Magn.* **MAG-10**, 832 (1974).
202. F. C. Rossol and A. A. Thiele, *J. Appl. Phys.* **41**, 1163 (1970).
203. Y. S. Chen, W. J. Richards, and P. I. Bonyhard, *IEEE Trans. Magn.* **MAG-9**, 670 (1973).
204. J. A. Copeland and R. R. Spiwak, *IEEE Trans. Magn.* **MAG-7**, 748 (1971).
205. D. M. Heinz, P. J. Besser, J. M. Owens, J. E. Mee, and G. R. Pullman, *J. Appl. Phys.* **42**, 1243 (1971).

206. N. F. Borrelli, S. L. Chen, J. A. Murphy, and F. P. Fehlner, *AIP Conf. Proc.* **10**, 398 (1973).
207. G. P. Vella-Coleiro and W. J. Tabor, *Appl. Phys. Lett.* **21**, 7 (1972).
208. O. Voegeli, C. A. Jones, and J. A. Brown, *AIP Conf. Proc.* **29**, 71 (1976).
209. G. P. Vella-Coleiro, *AIP Conf. Proc.* **18**, 217 (1974).
210. A. P. Malozemoff and J. C. DeLuca, *Appl. Phys. Lett.* **26**, 719 (1975).
211. G. P. Vella-Coleiro, *Appl. Phys. Lett.* **28**, 744 (1976).
211a. E. I. Il'yashenko, S. N. Matveyev, and N. I. Karmatsky, *J. Appl. Phys.* **49**, 1933 (1978).
212. R. M. Josephs, *Appl. Phys. Lett.* **25**, 244 (1974).
213. F. H. de Leeuw and J. M. Robertson, *AIP Conf. Proc.* **24**, 601 (1975).
214. T. J. Beaulieu and O. Voegeli, *AIP Conf. Proc.* **24**, 627 (1975).
215. C. A. Jones, M. Stroomer, O. Voegeli, and F. J. Friedlander *IEEE Trans. Magn.* **MAG-15**, 926 (1979).
216. L. Berger, *AIP Conf. Proc.* **18**, 918 (1974).
217. L. Berger, *J. Phys. Chem. Solids* **35**, 947 (1974).
218. W. J. Carr, Jr., *J. Appl. Phys.* **45**, 394 (1974).
219. W. J. Carr, Jr., *J. Appl. Phys.* **45**, 3115 (1974).
220. S. H. Charap, *J. Appl. Phys.* **45**, 397 (1974).
221. P. R. Emtage, *J. Appl. Phys.* **45**, 3117 (1974).
221a. J. C. DeLuca, R. J. Gambino, and A. P. Malozemoff, *IEEE Trans. Magn.* **MAG-14**, 500 (1978); J. C. DeLuca and R. J. Gambino, *J. Appl. Phys.* **50**, 2212 (1979).
221b. L. Berger, *J. Appl. Phys.* **49**, 2156 (1978); **50**, 2137 (1979).
222. D. L. Partin, M. Karnezos, L. C. deMenezes, and L. Berger, *J. Appl. Phys.* **45**, 1852 (1974).
223. E. M. Gyorgy and F. B. Hagedorn, *J. Appl. Phys.* **39**, 88 (1968).
224. J. C. Slonczewski, *J. Appl. Phys.* **44**, 1759 (1973); *AIP Conf. Proc.* **5**, 170 (1972).
225. J. C. Slonczewski, *J. Appl. Phys.* **45**, 2705 (1974).
226. F. B. Hagedorn, *J. Appl. Phys.* **45**, 3129 (1974); *AIP Conf. Proc.* **18**, 222 (1974).
227. A. Hubert, *AIP Conf. Proc.* **18**, 178 (1974).
228. A. Rosencwaig, W. J. Tabor, and T. J. Nelson, *Phys. Rev. Lett.* **29**, 946 (1972).
229. P. Dekker and J. C. Slonczewski, *Appl. Phys. Lett.* **29**, 753 (1976).
230. W. J. Tabor, A. H. Bobeck, G. P. Vella-Coleiro, and A. Rosencwaig, *Bell Syst. Tech. J.* **51**, 1427 (1972).
231. A. P. Malozemoff, *Appl. Phys. Lett.* **21**, 149 (1972). (Note: The observation of quantization in the static domain properties has not been confirmed in subsequent studies.)
232. T. Kobayashi, H. Nishida, and Y. Sugita, *J. Phys. Soc. Japan* **34**, 555 (1973).
233. R. F. Lacey, R. B. Clover, L. S. Cutler, and R. F. Waites, *AIP Conf. Proc.* **10**, 488 (1973).
234. H. Nishida, T. Kobayashi, and Y. Sugita, *AIP Conf. Proc.* **10**, 493 (1973).
235. W. J. Tabor, A. H. Bobeck, G. P. Vella-Coleiro, and A. Rosencwaig, *AIP Conf. Proc.* **10**, 442 (1973).
236. F. G. West and D. C. Bullock, *AIP Conf. Proc.* **10**, 483 (1973).
237. R. Suzuki and Y. Sugita, *IEEE Trans Magn.* **MAG-14**, 210 (1978).
238. D. H. Smith and A. A. Thiele, *AIP Conf. Proc.* **18**, 173 (1974).
239. E. Schlömann, *Appl. Phys. Lett.* **21**, 227 (1972).
240. E. Schlömann, *J. Appl. Phys.* **44**, 1837 (1973).
241. E. Schlömann, *J. Appl. Phys.* **44**, 1850 (1973).
242. E. Schlömann, *J. Appl. Phys.* **45**, 369 (1974).
243. W. J. DeBonte, *J. Appl. Phys.* **44**, 1793 (1973).
244. G. R. Henry and B. R. Brown, *AIP Conf. Proc.* **24**, 751 (1975).
245. A. Hubert, *J. Appl. Phys.* **46**, 2276 (1975).
245a. T. Suzuki and M. Takahashi, *Japan J. Appl. Phys.* **17**, 1371 (1978).
246. G. M. Nedlin and R. Kh. Shapiro, *Sov. Phys. Solid State* **17**, 1357 (1975).

247. E. Schlömann, *AIP Conf. Proc.* **10,** 478 (1973).
248. J. C. Slonczewski and Y. S. Lin, unpublished.
249. B. E. Argyle, S. Maekawa, P. Dekker, and J. C. Slonczewski, *AIP Conf. Proc.* **34,** 131 (1976).
250. T. J. Beaulieu, B. R. Brown, B. A. Calhoun, T. Hsu, and A. P. Malozemoff, *AIP Conf. Proc.* **34,** 138 (1976).
250a. V. A. Gurevich and Ya. A. Monosov, *Sov. Phys. Solid State* **18,** 1690 (1976).
251. C. C. Shir and G. R. Henry, *IBM Res. Report* RJ 1896, San Jose, CA (1977).
251a. G. Toulouse and M. Kléman, *J. de Physique* **37,** L149 (1976).
251b. M. Kléman, L. Michel, and G. Toulouse, *J. de Physique* **38,** L195 (1977).
252. E. Feldtkeller, *Z. Angew. Phys.* **19,** 530 (1965).
253. W. Döring, *J. Appl. Phys.* **39,** 1006 (1968).
254. J. Reinhardt, *Int. J. Magnetism* **5,** 263 (1973).
255. J. C. Slonczewski, *AIP Conf. Proc.* **24,** 613 (1975).
256. M. Margulies and J. C. Slonczewski, *J. Appl. Phys.* **49,** 1912 (1978), (abstract only).
257. A. A. Thiele, *Phys. Rev. Lett.* **30,** 230 (1973).
258. A. Hubert, *J. Magnetism and Magnetic Materials* **2,** 25 (1976) (in German).
259. D. C. Bullock, *AIP Conf. Proc.* **18,** 232 (1974).
260. R. Hasegawa, *AIP Conf. Proc.* **24,** 615 (1975).
261. R. M. Josephs and B. F. Stein, *AIP Conf. Proc.* **29,** 65 (1976).
262. T. Obokata, K. Yamaguchi, and K. Asama, *AIP Conf. Proc.* **29,** 74 (1976).
263. M. A. Rozenblat and S. E. Yurchenko, *Intermag Conference, Los Angeles, June 1977* (unpublished).
264. J. Haisma, K. L. L. van Mierloo, W. F. Druyvesteyn, and U. Enz, *Appl. Phys. Lett.* **27,** 459 (1975).
265. R. W. Patterson, *AIP Conf. Proc.* **24,** 608 (1975).
266. V. G. Kleparsky and V. V. Randoshkin, *Sov. Phys. Solid State* **19,** 1900 (1977); V. G. Kleparsky and S. E. Yurchenko, *Sov. Phys. Solid State* **18,** 172 (1976).
266a. V. A. Bokov, V. V. Volkov, T. K. Trofimova, and E. S. Sher, *Sov. Phys. Solid State* **17,** 2338 (1976).
267. R. Wolfe and J. C. North, *Bell Syst. Tech. J.* **51,** 1436 (1972); J. C. North and R. Wolfe, *in* "Ion Implantation in Semiconductors and Other Materials" (B. L. Crowder, ed.), p. 505. Plenum, New York, 1973.
268. R. D. Henry, P. J. Besser, R. G. Warren, and E. C. Whitcomb, *IEEE Trans. Magn.* **MAG-9,** 514 (1973).
269. K. Hoshikawa, K. Minegishi, H. Nakanishi, N. Inoue, K. Takimoto and Y. Suemune, *Japan J. Appl. Phys.* **15,** (1976).
270. K. Hoshikawa, T. Hattanda, and H. Nakanishi, *Japan J. Appl. Phys.* **13,** 2071 (1974).
271. A. B. Smith, M. Kestigian, and W. R. Bekebrede, *AIP Conf. Proc.* **18,** 167 (1974).
272. J. L. Su, E. B. Moore, T. L. Hsu, and B. A. Calhoun, *AIP Conf. Proc.* **29,** 72 (1976).
273. M. H. Kryder, C. H. Bajorek, and R. J. Kobliska, *IEEE Trans. Magn.* **MAG-12,** 346 (1976).
274. J. M. Robertson, M. G. J. van Hout, J. C. Verplanke, and J. C. Brice, *Mat. Res. Bull.* **9,** 555 (1974).
275. R. C. LeCraw and R. Wolfe, *AIP Conf. Proc.* **18,** 188 (1974).
276. R. C. LeCraw, E. M. Gyorgy, and R. Wolfe, *Appl. Phys. Lett.* **24,** 573 (1974).
277. R. Wolfe, J. C. North, and Y. P. Lai, *Appl. Phys. Lett.* **22,** 683 (1973).
278. Y. S. Lin and G. E. Keefe, *Appl. Phys. Lett.* **22,** 603 (1973).
279. M. Takahashi, H. Nishida, T. Kobayashi, and Y. Sugita, *J. Phys. Soc. Japan* **34,** 1416 (1973).

280. M. Takahashi, H. Nishida, T. Kobayashi, and Y. Sugita, *AIP Conf. Proc.* **18,** 172 (1974).
281. J. Engemann, (unpublished).
282. P. F. Tumelty, R. Singh, and M. A. Gilleo, *AIP Conf. Proc.* **29,** 99 (1976).
283. G. R. Henry and J. Gitschier, private communication.
283a. D. M. Hannon and H. L. Hu, *J. Appl. Phys.* **50,** 2198 (1979).
283b. W. Menz, E. B. Moore, and H. L. Hu, *IEEE Trans. Magn.* **MAG-14,** 599 (1978).
284. W. Döring, *Z. Naturforschung* **3a,** 373 (1948).
285. L. R. Walker, unpublished; cf. J. F. Dillon, Jr., *in* "A Treatise on Magnetism" (G. T. Rado and H. Suhl, eds.), Vol. III, pp. 450–453. Academic Press, New York, 1963.
286. G. R. Henry, *J. Appl. Phys.* **42,** 3150 (1971).
287. J. C. Slonczewski, *Int. J. Magnetism* **2,** 85 (1972).
288. E. Schlömann, *AIP Conf. Proc.* **5,** 160 (1972).
289. H. C. Bourne, Jr., and D. S. Bartran, *IEEE Trans. Magn.* **MAG-8,** 741 (1972).
290. W. J. Carr, Jr., *AIP Conf. Proc.* **24,** 747 (1975).
291. T. Ikuta and R. Shimizu, *J. Phys.* **D8,** 322 (1975).
292. H. E. Khodenkov, *Phys. Lett.* **58A,** 135 (1976).
293. A. Aharoni, *IEEE Trans. Magn.* **MAG-14,** 118 (1978).
294. A. Emura, T. Fujii, S. Shiomi, and S. Uchiyama, *IEEE Trans. Magn.* **MAG-13,** 1169 (1977); T. Fujii, T. Shinoda, S. Shiomi and S. Uchiyama, *Jap. J. Appl. Phys.* **17,** 1997 (1978).
294a. V. A. Gurevich, *Sov. Phys. Solid State* **19,** 1696 (1977).
294b. G. M. Nedlin and R. Kh. Shapiro, *Sov. Phys. Solid State* **19,** 1707 (1977).
294c. T. H. O'Dell, *Phys. Stat. Sol.* (*a*) **48,** 59 (1978).
294d. V. M. Eleonskii, N. N. Kirova, and V. M. Petrov, *Sov. Phys. JETP* **41,** 966 (1976).
295. T. Obokata, K. Sokura, and T. Namikata, *IEEE Trans. Magn.* **MAG-9,** 373 (1973).
296. T. T. Chen, J. L. Archer, R. A. Williams, and R. D. Henry, *IEEE Trans. Magn.* **MAG-9,** 385 (1973).
297. J. E. Mee, G. R. Pulliam, D. M. Heinz, J. M. Owens, and P. J. Besser, *Appl. Phys. Lett.* **18,** 60 (1971).
298. R. Hiskes, *AIP Conf. Proc.* **34,** 166 (1976).
299. A. P. Malozemoff, *J. Appl. Phys.* **48,** 795 (1977).
300. W. A. Bonner, J. E. Geusic, D. H. Smith, L. G. Van Uitert, and G. P. Vella-Coleiro, *Mat. Res. Bull.* **8,** 1223 (1973).
301. H. L. Hu and E. A. Giess, *AIP Conf. Proc.* **24,** 605 (1975).
302. D. M. Heinz, R. G. Warren, and M. T. Elliott, *AIP Conf. Proc.* **29,** 101 (1976).
303. W. A. Bonner, J. E. Geusic, D. H. Smith, L. G. Van Uitert, and G. P. Vella-Coleiro, *Mat. Res. Bull.* **8,** 785 (1973).
304. T. Obokata, H. Tominaga, T. Mori, and H. Inoue, *AIP Conf. Proc.* **29,** 103 (1976).
305. W. A. Bonner, J. E. Geusic, D. H. Smith, F. C. Rossol, L. G. Van Uitert and G. P. Vella-Coleiro, *J. Appl. Phys.* **43,** 3226 (1972).
306. R. W. Shaw, J. W. Moody, and R. M. Sandfort, *J. Appl. Phys.* **45,** 2672 (1974).
307. R. S. Sery and H. R. Irons, *AIP Conf. Proc.* **18,** 90 (1974).
308. R. Suzuki and Y. Sugita, *Appl. Phys. Lett.* **25,** 587 (1975).
309. R. Suzuki and Y. Sugita, *J. Phys. Soc. Japan* **41,** 701 (1976).
309a. E. I. Il'yashenko, F. V. Lisovskii, V. I. Scheglov, and S. E. Yurchenko, *Sov. Phys. Solid State* **19,** 522 (1977).
309b. A. Ya. Chervonenkis, A. M. Balbashov, S. G. Pavlova, and A. P. Cherkasov, *Sov. Phys. Solid State* **20,** 851 (1978).
310. J. C. DeLuca and A. P. Malozemoff, *AIP Conf. Proc.* **34,** 151 (1976).
311. R. Hiskes and R. A. Burmeister, *AIP Conf. Proc.* **10,** 304 (1973).

312. D. M. Heinz, P. J. Besser, and J. E. Mee, *AIP Conf. Proc.* **5**, 96 (1972).

313. A. J. Kurtzig, R. C. Le Craw, A. H. Bobeck, E. M. Walters, R. Wolfe, H. J. Levinstein and S. J. Licht, *AIP Conf. Proc.* **5**, 180 (1972).

314. G. P. Vella-Coleiro, F. B. Hagedorn, Y. S. Chen, B. S. Hewitt, S. L. Blank and R. Zappulla, *J. Appl. Phys.* **45**, 939 (1974).

315. D. C. Bullock, J. T. Carlo, D. W. Mueller, and T. L. Brewer, *AIP Conf. Proc.* **24**, 647 (1975).

316. K. Yamaguchi, H. Inoue, and K. Asama, *AIP Conf. Proc.* **34**, 160 (1976).

317. D. Hafner and F. B. Humphrey, *Appl. Phys. Lett.* **30**, 303 (1977).

318. R. I. Potter, V. J. Minkiewicz, K. Lee, and P. A. Albert, *AIP Conf. Proc.* **29**, 76 (1976).

319. M. H. Kryder and H. L. Hu, *AIP Conf. Proc.* **18**, 213 (1974).

320. M. Takahashi, H. Nishida, T. Kobayashi, and Y. Sugita, *J. Phys. Soc. Japan* **35**, 615 (1973).

321. R. M. Sandfort, R. W. Shaw, and J. W. Moody, *AIP Conf. Proc.* **18**, 237 (1974).

322. R. E. Fontana, Jr., and D. C. Bullock, *AIP Conf. Proc.* **34**, 170 (1976).

323. P. W. Shumate, Jr., *IEEE Trans. Magn.* **MAG-7**, 479 (1971).

324. P. W. Shumate, Jr., *J. Appl. Phys.* **42**, 5770 (1971).

325. A. Rosencwaig, *J. Appl. Phys.* **42**, 5773 (1971).

326. J. W. Moody, R. W. Shaw, R. M. Sandfort, and R. L. Stermer, *IEEE Trans. Magn.* **MAG-9**, 377 (1973); J. W. Moody, R. W. Shaw, R. M. Sandfort, and R. L. Stermer, *Mat. Res. Bull.* **9**, 527 (1974).

327. F. H. de Leeuw, R. van den Doel, and J. M. Robertson, *J. Appl. Phys.* **49**, 768 (1978).

328. G. P. Vella-Coleiro, F. B. Hagedorn, Y. S. Chen, and S. L. Blank, *Appl. Phys. Lett.* **22**, 324 (1973).

329. G. T. Rado, R. W. Wright, and W. H. Emerson, *Phys. Rev.* **80**, 273 (1950).

330. G. T. Rado, *Phys. Rev.* **83**, 821 (1951).

331. H. P. J. Wijn, *Physica* **19**, 555 (1953).

332. T. M. Perekalina, A. A. Askochinskii, and D. G. Sannikov, *Soviet Physics JETP* **13**, 303 (1961).

333. N. L. Schryer and L. R. Walker, *J. Appl. Phys.* **45**, 5406 (1974).

334. H. E. Khodenkov, *Fiz. Metal. Metalloved.* **39**, 466 (1975).

335. F. H. de Leeuw, *J. Appl. Phys.* **45**, 3106 (1974).

336. G. J. Zimmer, L. Gal, and F. B. Humphrey, *AIP Conf. Proc.* **29**, 85 (1976).

337. E. Schlömann, *J. Appl. Phys.* **47**, 1142 (1976).

338. F. C. Rossol, *Phys. Rev. Lett.* **24**, 1021 (1970).

339. S. Konishi, T. Kawamoto, and M. Wada, *IEEE Trans. Magn.* **MAG-10**, 642 (1974).

340. T. Ikuta and R. Shimizu, *J. Phys. D: Appl. Phys.* **6**, 633 (1973).

341. T. Ikuta and R. Shimizu, *J. Phys. D: Appl. Phys.* **7**, 726 (1974).

342. T. Ikuta and R. Shimizu, *J. Phys. D: Appl. Phys.* **7**, 2386 (1974).

343. S. Uchiyama, S. Shiomi, and T. Fujii, *AIP Conf. Proc.* **34**, 154 (1976).

344. S. Konishi and T. Miyama, *AIP Conf. Proc.* **24**, 740 (1975).

344a. M. V. Chetkin and A. de la Campa, *JETP Lett.* **27**, 157 (1978); M. V. Chetkin, A. N. Shalygin, and A. de la Campa, *Sov. Phys. Solid State* **19**, 2029 (1977).

345. E. Schlömann, *AIP Conf. Proc.* **18**, 183 (1974).

346. N. Hayashi and K. Abe, *Japan J. Appl. Phys.* **14**, 1705 (1975); **15**, 1683 (1976); **16**, 789 (1977).

347. N. Hayashi, H. Mikami, and K. Abe, *IEEE Trans. Magn.* **MAG-13**, 1345 (1977).

348. C. C. Shir, *J. Appl. Phys.* **49**, 1841, 3413 (1978).

349. J. C. Slonczewski, *J. Magn. Magnetic Materials* **12**, 108 (1979).

350. A. M. Balbashov, P. I. Nabokin, A. Ya. Chervonenkis, and A. P. Cherkasov, *Sov. Phys.*

*Solid State* **19**, 1102 (1977).

350a. V. V. Volkov, V. A. Bokov, G. A. Smolenskii, and S. M. Grigorovich, *Sov. Phys. Solid State* **20**, 525 (1978).

351. A. A. Thiele, *J. Appl. Phys.* **45**, 377 (1974).

352. D. J. Craik, D. H. Cottey, and G. Myers, *J. Phys.* **D8**, 99 (1975).

353. G. P. Vella-Coleiro, A. Rosencwaig, and W. J. Tabor, *Phys. Rev. Lett.* **29**, 949 (1972).

354. J. C. Slonczewski, *Phys. Rev. Lett.* **29**, 1679 (1972).

355. A. A. Thiele, F. B. Hagedorn, and G. P. Vella-Coleiro, *Phys. Rev.* **B8**, 241 (1973).

356. R. W. Patterson, *J. Appl. Phys.* **45**, 5018 (1974).

357. R. W. Patterson, A. I. Braginski, and F. B. Humphrey, *IEEE Trans. Magn.* **MAG-11**, 1094 (1975).

358. L. Gál, G. J. Zimmer, and T. H. O'Dell, *Conference on Magnetism and Magnetic Materials, Minneapolis, Minnesota, November 1977* (unpublished).

359. H. Nishida, T. Kobayashi, and Y. Sugita, *J. Phys. Soc. Japan* **34**, 833 (1973).

360. H. Nishida, T. Kobayashi, and Y. Sugita, *IEEE Trans. Magn.* **MAG-9**, 517 (1973).

361. H. Nishida, T. Kobayashi, and Y. Sugita, *J. Phys. Soc. Japan* **34**, 266 (1973).

361a. S. Konishi, F. Watanabe, N. Shibata, and K. Narita, *Appl. Phys. Lett.* **33**, 471 (1978).

362. J. C. DeLuca, A. P. Malozemoff, J. L. Su, and E. B. Moore, *J. Appl. Phys.* **48**, 1701 (1977).

363. T. Kusuda, S. Honda, and T. Ideshita, *AIP Conf. Proc.* **29**, 84 (1976).

364. S. Honda, Y. Ikeda, S. Hashimoto, and T. Kusuda, *Conference on Magnetism and Magnetic Materials, Minneapolis, Minnesota, November 1977.*

364a. S. Konishi, T. L. Hsu, and B. R. Brown, *IEEE Trans. Magn.* **MAG-15**, 885 (1979).

364b. H. E. Khodenkov, *Fiz. Metal. Metalloved.* **46**, 472 (1978).

365. A. P. Malozemoff, J. C. Slonczewski, and J. C. DeLuca, *AIP Conf. Proc.* **29**, 58 (1976). (Note: Figs. 12 and 13 of this reference are in error and should be replaced by Fig. 18.2 of this review.)

366. B. E. Argyle, J. C. Slonczewski, and A. F. Mayadas, *AIP Conf. Proc.* **5**, 175 (1972).

367. V. G. Kleparsky, N. P. Dymchenko, and S. K. Kukharskaya, *Phys. Stat. Sol. (a)* **33**, K117 (1976).

368. G. J. Zimmer, K. Vural, and F. B. Humphrey, *J. Appl. Phys.* **46**, 4976 (1975).

369. H. Callen, R. M. Josephs, J. A. Seitchik, and B. F. Stein, *Appl. Phys. Lett.* **21**, 366 (1972).

369a. F. H. de Leeuw, *IEEE Trans. Magn.* **MAG-14**, 596 (1978).

370. P. J. Rijnierse and F. H. de Leeuw, *AIP Conf. Proc.* **18**, 199 (1974).

371. T. Ideshita, Y. Fukushima, S. Honda, and T. Kusuda, *IEEE Trans. Magn.* **MAG-13**, 1166 (1977); *J. Appl. Phys.* **49**, 1853 (1978).

372. S. Honda, Y. Fukushima, N. Fukuda, and T. Kusuda, *J. Appl. Phys.* **50**, 1465 (1979); S. Honda, S. Hashimoto, and T. Kusuda, *J. Appl. Phys.* **50**, 2206 (1979).

373. B. E. MacNeal and F. B. Humphrey, *IEEE Trans. Magn.* **MAG-13**, 1348 (1977); *J. Appl. Phys.* **48**, 3869 (1977).

374. R. M. Josephs and B. F. Stein, *AIP Conf. Proc.* **18**, 227 (1974).

375. G. J. Zimmer, T. M. Morris, K. Vural, and F. B. Humphrey, *Appl. Phys. Lett.* **25**, 750 (1975).

376. R. D. Henry, M. T. Elliott, and E. C. Whitcomb, *J. Appl. Phys.* **47**, 3702 (1976).

376a. S. Konishi, F. Watanabe, and K. Narita, *IEEE Trans. Magn.* **MAG-15**, 890 (1979).

377. J. C. Slonczewski and V. Moruzzi, (unpublished).

378. G. R. Henry, T. Hsu, and A. Hubert, *International Conference on Magnetic Bubbles, Eindhoven, The Netherlands, September 13–15, 1976* (unpublished).

379. A. P. Malozemoff and J. C. Slonczewski, *IEEE Trans. Magn.* **MAG-11**, 1094 (1975).

380. A. A. Thiele, *J. Appl. Phys.* **47**, 2759 (1976).

381. J. C. Slonczewski, (unpublished).

382. S. Konishi, T. Hsu, and B. R. Brown, *J. Appl. Phys.* **49**, 1894 (1978).
383. E. I. Il'yashenko, V. G. Kleparskii, and S. E. Yurchenko, *Phys. Stat. Sol.* (*a*) **28**, K153 (1975).
384. K. Ju and F. B. Humphrey, *J. Appl. Phys.* **48**, 4656 (1977); K. Ju, thesis, California Inst. of Tech., Pasadena, CA (1978).
385. J. C. DeLuca, A. P. Malozemoff, and S. Maekawa, *J. Appl. Phys.* **48**, 4672 (1977).
386. H. Nakanishi and C. Uemura, *Japan J. Appl. Phys.* **13**, 1183 (1974).
387. Y. A. Monosov, P. I. Nabokin, and L. V. Nikolaev, *Sov. Phys. JETP* **41**, 913 (1975).
388. R. M. Josephs, *Appl. Phys. Lett.* **25**, 244 (1974).
389. R. M. Josephs and B. F. Stein, *AIP Conf. Proc.* **24**, 598 (1975).
390. R. Suzuki, A. Asano, F. Ishida, Y. Sugita, and M. Takahashi, *J. Appl. Phys.* **49**, 1853 (1978), (abstract only).
391. A. P. Malozemoff and J. C. DeLuca, *J. Appl. Phys.* **49**, 1844 (1978).
392. K. Ju and F. B. Humphrey, *IEEE Trans. Magn.* **MAG-13**, 1190 (1977).
393. S. Maekawa and P. Dekker, *AIP Conf. Proc.* **34**, 148 (1976).
394. H. Nakanishi and C. Uemura, *Japan J. Appl. Phys.* **13**, 191 (1974).
395. J. C. DeLuca and A. P. Malozemoff, (unpublished).
395a. A. V. Markelis, A. V. Antonov, and L. I. Pranyavichyus, *Sov. Phys. Solid State* **20**, 212 (1978).
396. R. Kinoshita, S. Matsuyama, and S. Orihara, *IEEE Trans. Magn.* **MAG-13**, 1342 (1977).
397. T. L. Hsu, B. R. Brown, and M. D. Montgomery, *AIP Conf. Proc.* **29**, 67 (1976).
398. B. R. Brown and T. L. Hsu, *Appl. Phys. Lett.* **29**, 813 (1976).
399. P. K. George, *J. Appl. Phys.* **49**, 1850 (1978).
400. T. Kobayashi, P. J. Besser, K. Ju, and F. B. Humphrey, *IEEE Trans. Magn.* **MAG-12**, 697 (1976).
401. K. Nakao, S. Iida, M. Hirano, and T. Tsushima, *J. Appl. Phys.* **49**, 1909 (1978).
402. B. A. Boxall, *IEEE Trans. Magn.* **MAG-10**, 648 (1974).
403. W. C. Hubbell, *AIP Conf. Proc.* **24**, 552 (1975).
404. B. E. Argyle, J. C. Slonczewski, and O. Voegeli, *IBM J. Res. and Develop.* **20**, 109 (1976).
405. A. P. Malozemoff and J. C. Slonczewski, *AIP Conf. Proc.* **24**, 603 (1975).
406. B. E. Argyle, J. C. Slonczewski, P. Dekker, and S. Maekawa, *J. Magnetism and Magn. Materials* **2**, 357 (1976).
407. B. E. Argyle and P. Dekker, *Intermag Conference, Los Angeles, California* June 6–9, 1977 (unpublished).
407a. S. Iwata, S. Shiomi, S. Uchiyama, and T. Fujii, *Japan J. Appl. Phys.* **17**, 1681 (1978); *J. Appl. Phys.* **50**, 2195 (1979).
407b. K. Ju and F. B. Humphrey, *J. Appl. Phys.* **50**, 2212 (1979).
408. B. E. Argyle, *International Conference on Magnetic Bubbles, Eindhoven, The Netherlands, September 13–15, 1976* (unpublished).
409. V. G. Kleparsky, E. I. Il'yashenko, and S. N. Matveyev, *IEEE Trans. Magn.* **MAG-12**, 700 (1976).
410. H. E. Khodenkov and V. K. Raev, *Phys. Stat. Sol.* (*a*) **28**, K29 (1975).
411. M. H. Kryder, K. Y. Ahn, G. S. Almasi, G. E. Keefe, and J. V. Powers, *IEEE Trans. Magn.* **MAG-10**, 825 (1974).
411a. Z. Aziz, W. Clegg, and R. M. Pickard, *IEEE Trans. Magn.* **MAG-11**, 1133 (1975).
412. W. D. Doyle, W. E. Flannery, and J. A. Coleman, *AIP Conf. Proc.* **18**, 152 (1974).
413. F. B. Hagedorn, S. L. Blank, and R. J. Peirce, *Appl. Phys. Lett.* **26**, 206 (1975).
413a. W. D. Doyle, R. M. Josephs, and A. B. Smith, *IEEE Trans. Magn.* **MAG-14**, 303 (1978).
414. P. W. Shumate and R. J. Peirce, *Appl. Phys. Lett.* **23**, 140 (1973).
415. P. W. Shumate, P. C. Michaelis, and R. J. Peirce, *AIP Conf. Proc.* **18**, 140 (1974).

416. K. Yoshimi and S. Fujiwara, *AIP Conf. Proc.* **18**, 157 (1974).
417. E. H. L. J. Dekker, K. L. L. van Mierloo, and R. de Werdt, *IEEE Trans. Magn.* **MAG-13,** 1261 (1977).
418. R. W. Damon and J. Eshbach, *J. Phys. Chem. Solids* **19**, 308 (1961).
419. I. A. Gilinskii, *Sov. Phys. JETP*, **41**, 511 (1976); *Zh. Eksp. Teor. Fiz.* **68**, 1032, (1975).
420. A. E. Borovik, V. S. Kuleshov, and M. A. Strzhemechnyi, *Sov. Phys. JETP* **41**, 1118 (1975); *Zh. Eksp. Teor. Fiz.* **68**, 2236 (1975).
421. G. M. Nedlin and R. Kh. Shapiro, *Sov. Phys. Solid State* **18**, 985 (1976); *Fiz. Tverd. Tela* (*Leningrad*) **18**, 1696 (1976).
422. J. F. Janak, *Phys. Rev.* **134**, A411 (1964); Thesis, Massachusetts Inst. Technol., Cambridge, Mass. (1964).
423. J. M. Winter, *Phys. Rev.* **124**, 452 (1961).
424. A. A. Thiele, *Phys. Rev.* **B7**, 391 (1973).
425. A. A. Thiele, *Phys. Rev.* **B14**, 3130 (1976).
426. G. R. Henry, *Appl. Phys. Lett.* **29**, 63 (1976).
427. L. Brillouin, "*Wave Propagation in Periodic Structures,*" 2nd edition. Dover, New York, 1953.
428. M. H. H. Höfelt, *J. Appl. Phys.* **44**, 414 (1973).
429. M. M. Sokoloski and T. Tanaka, *J. Appl. Phys.* **45**, 3091 (1974).
430. M. M. Sokoloski and T. Tanaka, *IEEE Trans. Magn.* **MAG-10**, 646 (1974).
431. V. G. Bar'yakhtar, V. V. Gann, and Y. I. Gorobets, *Sov. Phys. Solid State* **18**, 1158 (1976); *Fiz. Tverd. Tela—Leningrad* **18**, 1990 (1976).
431a. V. G. Bar'yakhtar, Y. I. Gorobets, O. V. Il'chishin, and M. V. Petrov, *Sov. Phys. Solid State* **19**, 1658 (1977).
432. V. G. Bar'yakhtar, Y. I. Gorobets, and Y. V. Melikhov, *Sov. Phys. Solid State* **18**, 1162 (1976); *Fiz. Tverd. Tela—Leningrad* **18**, 1996 (1976).
433. V. G. Bar'yakhtar, Y. I. Gorobets, and Y. V. Melikhov, *Sov. Phys. Solid State* **18**, 476. (1976); *Fiz. Twerd. Tela* **18**, 832 (1976).
433a. J. H. Spreen and F. R. Morgenthaler, *J. Appl. Phys.* **49**, 1590 (1978).
434. P. Wigen, *in* "Physics of Magnetic Garnets," *Proc. Int. School of Physics "Enrico Fermi,"* Course LXX (A. Paoletti, ed.), p. 196. North Holland, New York, 1978.
435. J. Kaczér and R. Gemperle, *Physica* **86-88B,** 1313 (1977).
436. J. Kaczér, L. Murtinova, I. Tomáš, and R. Gemperle, *Proc. International Conf. Magnetism, Moscow, Vol. 5, p. 415,* 1974.
437. J. O. Artman and S. H. Charap, *J. Appl. Phys.* **49**, 1587 (1978); *J. Appl. Phys.* **50**, 2024 (1979).
437a. P. S. Limaye, A. A. Parker, D. G. Stroud, and P. E. Wigen, *J. Appl. Phys.* **50**, 2027 (1979).
437b. L. V. Mikhailovskaya and I. V. Bogomaz, *Sov. Phys. Solid State* **19**, 1315 (1977).
438. H. Dötsch, *in* "Magnetic Bubbles," *Proc. of the Winter School on New Magnetic Materials,* (H. Lachowicz, M. Maszkiewicz, and H. Szymczak, eds.), p. 113, 1976.
439. H. Dötsch, H. J. Schmitt, and J. Muller, *Appl. Phys. Lett.* **23**, 639 (1973).
440. H. Dötsch, *AIP Conf. Proc.* **29**, 78 (1976).
441. R. W. Damon, *in* "Magnetism," Vol. I (G. T. Rado and H. Suhl, eds.), p. 551. Academic Press, New York, 1963.
442. H. Dötsch and H. J. Schmitt, *Appl. Phys. Lett.* **24**, 442 (1974).
442a. A. M. Mednikov, S. I. Ol'khovskii, V. G. Red'ko, V. I. Rybak, V. P. Sondaevskii, and G. K. Chirkin, *Sov. Phys. Solid State* **19**, 698 (1977).
443. O. G. Bendik, B. A. Kalinikos, and D. N. Chartorizhskii, *Sov. Phys. Solid State* **16**, 1785 (1975).
444. H. Dötsch, *Phys. Stat. Sol.* (*a*) **39**, 589 (1977).

445. H. Dötsch, *J. Magnetism and Magn. Materials* **4,** 180 (1977).
446. E. Schlömann and J. D. Milne, *IEEE Trans. Magn.* **MAG-10,** 791 (1974).
447. E. Schlömann, *IEEE Trans. Magn.* **MAG-11,** 1051 (1975).
448. A. K. Zvezdin and V. G. Red'ko, *JETP Lett.* **21,** 203 (1975).

# List of Commonly Used Symbols

| | | | |
|---|---|---|---|
| $A$ | exchange stiffness | $H_r$ | radial field, usually stray field of bubble |
| $A$ | wall surface area | | |
| $D$ | Dzialoshinski exchange | $H_{ri}$ | run-in field of stripe |
| $d$ | bubble diameter | $H_{qs}$ | limiting drive field for validity of quasisteady theory |
| $E_L$ | energy per unit length of Bloch line | | |
| $\mathbf{F}$ | force | $h$ | film thickness |
| $\mathbf{F}_d$ | dynamic reaction force | $I$ | winding integer of a closed domain on a specific plane parallel to film plane |
| $\mathbf{F}_L$ | force per unit length on a Bloch line | | |
| $\mathbf{f}_g$ | gyrotropic force density | | |
| $f$ | frequency | $J$ | exchange energy, or total angular momentum quantum number |
| $g$ | Landé g-factor | | |
| $\mathbf{g}$ | gyrovector density | $K$ | uniaxial anisotropy constant |
| GGG | gadolinium gallium garnet | $K_1$ | cubic anisotropy constant |
| $H$ | general field, or effective field driving wall motion, i.e. applied field minus all other effective fields of static origin | $K_p$ | in-plane anisotropy constant (in plane of film) |
| | | $k$ | restoring force constant, or Boltzmann's constant, or wavevector, or an index |
| $H'$ | field gradient | | |
| $H_a$ | applied field, less static bias term | $L$ | length of stripe, or Lagrangian |
| $H_b$ | bias field (static applied field normal to film plane) | $l$ | material length parameter $\sigma_o/4\pi M^2$, or total number of Bloch lines |
| $H_c$ | coercive field of plane wall | $M$ | magnetization |
| $H_{cb}$ | net coercive field of a bubble $4\pi^{-1}H_c$ | $M_n$ | component of magnetization normal to a wall |
| $H_d$ | demagnetizing field | | |
| $H_g$ | gradient drive field on a bubble $r|\nabla H_z|$ | $m$ | wall mass per unit area |
| | | $m_D$ | Döring mass $(2\pi\gamma^2\Delta_o)^{-1}$ |
| $H_k$ | restoring force field | $N_h$ | homotopy number (net number of Bloch points) |
| $H_K$ | uniaxial anisotropy field $2K/M$ | | |
| $H_{max}$ | critical drive field corresponding to a peak velocity | $n_i$ | signature of $i$th Bloch line |
| | | $P$ | pressure |
| $H_p$ | in-plane field (in plane of film) | $Q$ | quality factor $K/2\pi M^2$ |

| | | | |
|---|---|---|---|
| $q$ | wall-displacement coordinate normal to wall plane | $\gamma$ | gyromagnetic ratio |
| | | $\Delta$ | general wall width |
| $r$ | bubble radius | $\Delta_0$ | Bloch-wall width parameter $(A/K)^{1/2}$ |
| $r_2$ | elliptic distortion from circular bubble shape | | |
| | | $\theta$ | polar angle of local magnetization |
| $S$ | winding number of closed domain [mean of $I(z)$] | $\Lambda$ | general Bloch-line width |
| | | $\Lambda_0$ | Bloch-line width parameter $(A/2\pi M^2)^{1/2}$ |
| $s$ | separation between Bloch lines, or contour coordinate | | |
| | | $\mu$ | mobility |
| $T$ | temperature, or pulse length | $\mu_h$ | high-drive differential mobility |
| $\mathbf{T}$ | torque density | $\mu_p$ | rate of increase of maximum wall velocity with in-plane field |
| $t$ | time | | |
| $\mathbf{t}$ | tangent unit-vector of a Bloch line | $\rho$ | usually, skew deflection angle of propagating bubble relative to gradient |
| $V$ | general domain velocity, or volume, or magnetostatic potential | | |
| | | $\sigma$ | general wall energy per unit area |
| $V_n$ | velocity for Bloch line nucleation | $\sigma_0$ | Bloch-wall energy $4 (AK)^{1/2}$ |
| $V_p$ | peak velocity | $\sigma^{\pm}$ | $(1, 2)^{\pm}$ bubble state |
| $V_{po}$ | Bloch-line instability velocity $24\gamma A/hK^{1/2}$ for flat wall | $\phi$ | azimuthal angle of local magnetization |
| $V_s$ | saturation velocity | | |
| $V_{so}$ | high-drive velocity back-extrapolated to zero drive field | $\Phi$ | full spin-rotation angle of a Bloch line |
| | | $\chi$ | susceptibility |
| $V_w$ | Walker's critical velocity $2\pi\gamma\Delta_0 M$, for $Q \rightarrow \infty$ | $\chi^{\pm}$ | $(1, 0)^{\pm}$ bubble state |
| | | $\psi$ | azimuthal angle (relative to film normal) of wall magnetization |
| $W$ | total energy | | |
| $w$ | energy per unit volume, or stripe width | $\psi_H$ | azimuthal angle of applied field |
| | | $\psi_p$ | azimuthal angle of easy in-plane anisotropy axis |
| $X_T$ | domain displacement at end of drive pulse of length $T$ | | |
| | | $\Psi$ | areal integral of wall-magnetization precession angle of a closed domain |
| $X_\infty$ | final domain displacement long after drive pulse | | |
| YIG | yttrium iron garnet | $\omega$ | circular frequency of oscillation |
| $z$ | coordinate usually normal to film plane | $\overset{\cdot}{}$ | time derivative |
| | | $\overline{\phantom{-}}$ | spatial average |
| $\mathbf{z}_0$ | unit vector of magnetization outside closed domain | $< >$ | time average |
| | | $(S, l, p)^\alpha$ | bubble-state specification (p. 91) |
| $\alpha$ | Gilbert viscous damping constant | | |

# Author Index

Numbers in parentheses are reference numbers and indicate that an author's work is referred to although his name is not cited in the text. Numbers in italics show the page on which the complete reference is listed.

## A

Abe, K., 147(346, 347), *302*
Abram, R. A., 15(49), *294*
Aharoni, A., 123(293), *301*
Ahn, K. Y., 41(102), 118(102), 264(411), *296, 304*
Albert, P. A., 131(318), *392*
Almasi, G. S., 29(75), 30(75), 63(75, 199), 263 (199), 264(199, 411), 265(199), *295, 298, 304*
Antonov, A. V., 248(395a), *304*
Archer, J. L., 63(198), 129(296), 130(296), 131 (296), 132(296), 236(296), 263(198), 264 (198), *298, 301*
Argyle, B. E., 29(75b), 30(75b), 41(102), 43 (112, 113), 55(112), 103(249), 104(249), 118(102), 137(112), 199(113, 366), 200 (366), 211(112), 255, 256, 257, 258(249), 259, 260, 261(249, 407), *295, 296, 300, 303, 304*
Artman, J. O., 280(437), *305*
Asama, K., 28(72), 29(79, 80), 30(79, 80), 52 (79), 114(262), 118(79), 131(316), *295, 300, 302*
Asano, A., 240(390), 265(390), *304*
Ashkin, A., 55(164), *297*
Askochinskii, A. A., 137(332), *302*
Asti, G., 52(154), *297*

Avaeva, I. G., 44(118), *296*
Aziz, Z., 265(411a), *304*

## B

Bajorek, C. H., 9(34), 116(273), *294, 300*
Balbashov, A. M., 159(350), 248(309b), 269 (350), 270(350), *301, 303*
Barbara, B., 51(148), *297*
Barns, R. L., 8(32), 130(32), 131(32), 266(32), *294*
Bartels, G., 28(70), 117(70), *295*
Bartran, D. S., 123(289), *301*
Bar'yakhtar, V. G., 279(431, 431a), 280(432, 433), *305*
Beaulieu, T. J., 4(27a), 62(187), 70(187, 214), 103(250), 104, 112(250), *294, 298, 299, 300*
Bekebrede, W. R., 116(271), 129(271), 130 (271), 131(271), *300*
Bendik, O. G., 284(443), *306*
Berger, L., 72, 73(217), 74(221b, 222), *299*
Besser, P. J., 65(205), 116(268), 118(268), 129 (205, 297), 130(268), 131(312), 253(400), *298, 300, 301, 302, 304*
Biolsi, W. A., 8(32), 130(32), 131(32), 266(32), *294*
Blake, T. G. W., 23(59a), *295*
Blank, S. L., 8(32), 15(45), 27(64, 66), 35(84,

**309**

# Subject Index

## A

Amorphous materials
   dynamics, 35, 51, 74, 118, 131, 254, 264–265
   static characteristics, 9, 41, 48–49
Anisotropy, *see* Uniaxial; Cubic; In plane, Tilt
Angular-momentum compensation, 34–37, 142, 197, 266
Array, *see* Stripe or Bubble lattice
Automotion, 94–95, 168, 173, 254–262
   general theory, 258
   type 1 (bias pulses, dc in-plane field), 258–261
   type 2 (in-plane field pulses), 255–256, 261

## B

Ballistic overshoot
   experiment, 67–68, 133, 212–213, 239–245, 268
   theory, 156, 175, 212–213, 233–235
Bias compensation, in gradient propagation, 67, 240, 250
Bias jump effect, 95, 242, 246, 250
Bistable hard bubbles, 97
Bitter pattern, *see* Ferrofluid
Bloch Line, *see also* Mass; Mobility; Skew deflection; Hard bubbles; Dynamic reaction force; Momentum, Vertical or horizontal or curved Bloch line; Coercivity
   annihilation, 98, 113–114, 177, 195, 253
   bunching, 173–174
   experimental observation, 47–49
   general effects on wall dynamics, 149–152
   instability velocity $V_{po}$, 191–193, 197, 208–209, 228, 236, 263–266
   interactions, 89–91
   model of bubble motion, 218–221
   model of wall motion, 189–197
   nucleation, 99, 103, 191, 197, 208–212, 219, 228–229, 239–240, 268
   pile-up or wind-up, 175–177, 194, 228, 229–230, 243
   punch-through, 57, 194–195, 231, 241, 244–245, 267
   signature, 93
   static structure, width and energy, 83–91, 101–102, 208–211
   tangent vector, 87, 93
Bloch-line approximation, 84, 101, 177, 192, 206, 219
Bloch point
   annihilation, 113
   generation, 114–116, 118, 120, 177, 232–233
   hard bubble suppression, 116–118
   mobility, 115
   role in bubble dynamics, 112–113, 164, 184
   state transitions, 120, 177, 232–233
   static structure, 27, 102, 104–115
Bloch wall, 1, 77–83, 123–143
Breakdown, *see* Walker velocity; Peak velocity
Bubble, isolated
   dynamics, 94–95, 149, 166, 168–177, 180–188, 199–206, 218–269
   gradient force, 24, 168–169, 180–187, 226
   statics, 17, 19–23, 96–98
   wall states, statics, 91–98, 112–114, 119–121
Bubble collapse, static, 20–22, 96–98, 116
Bubble collapse, dynamic
   experimental results, 59, 129–131, 167, 185–187, 198–201, 205–206